High Seas Governance

# Publications on Ocean Development

A SERIES OF STUDIES ON THE INTERNATIONAL, LEGAL, INSTITUTIONAL, AND POLICY ASPECTS OF OCEAN DEVELOPMENT



VOLUME 86

The titles published in this series are listed at *brill.com/pood*

# High Seas Governance

*Gaps and Challenges*

*Edited by*

Robert C. Beckman
Millicent McCreath
J. Ashley Roach
Zhen Sun

BRILL
NIJHOFF

LEIDEN | BOSTON

The Library of Congress Cataloging-in-Publication Data is available online at http://catalog.loc.gov

Names: Beckman, Robert C. (Law teacher), editor. | McCreath, Millicent, editor. | Roach, J. Ashley, editor. | Sun, Zhen, editor.
Title: High seas governance : gaps and challenges / Edited by Robert C. Beckman, Millicent McCreath, J. Ashley Roach, Zhen Sun.
Description: Leiden : Brill Nijhoff, 2018. | Series: Publications on ocean development ; 86 | Includes index.
Identifiers: LCCN 2018034973 (print) | LCCN 2018035198 (ebook) | ISBN 9789004373303 (E-book) | ISBN 9789004373310 (hc : alk. paper)
Subjects: LCSH: Law of the sea. | Marine pollution--Law and legislation--International cooperation. | Marine debris--Environmental aspects. | Marine resources conservation--Law and legislation. | Marine resources development--Law and legislation.
Classification: LCC K3590.4 (ebook) | LCC K3590.4 .H54 2018 (print) | DDC 341.4/5--dc23
LC record available at http://lccn.loc.gov/2018034973

Typeface for the Latin, Greek, and Cyrillic scripts: "Brill". See and download: brill.com/brill-typeface.

ISSN 0924-1922
ISBN 978-90-04-37331-0 (hardback)
ISBN 978-90-04-37330-3 (e-book)



This book is printed on acid-free paper and produced in a sustainable manner.

# Contents

# Acknowledgements

The editors would first like to thank all the authors who, despite their busy schedules, have given up a great deal of time and resources to provide their expertise and have been instrumental in making this project successful. We wish to thank Muzi Xu, Research Assistant at the Centre for International Law (CIL), National University of Singapore, for her assistance in checking the references. Appreciation goes to Amber Rose Maggio, Research Fellow at CIL, for her comments during the final stage of the project. Finally, we would like to express our gratitude to Brill for their cooperation and support on this project.

# List of Contributors

*Robert Beckman*

is the Head of the Ocean Law and Policy Programme and the former Director of the Centre for International Law (CIL). He is also an Associate Professor at the NUS Faculty of Law, an Adjunct Senior Fellow at the S. Rajaratnam School of International Studies (RSIS) at Nanyang Technological University, and a member of the National Executive Committee of CSCAP Singapore. Since 2009 he has lectured in the summer programme at the Rhodes Academy of Oceans Law and Policy in Rhodes, Greece, and he is a member of the Governing Board for the Rhodes Academy. He also lectures in the international maritime boundary workshops organised by CIL. He has published widely on ocean law and policy issues.

*Aldo Chircop*

(J.S.D., Dalhousie 1988) is Professor of Law and Canada Research Chair (Tier I) in Maritime Law and Policy at Dalhousie University, based at the Marine & Environmental Law Institute (MELAW, Schulich School of Law). He is also a researcher at the Ocean Frontier Institute, a faculty member at the International Development Studies Programme, and a research fellow at the Centre for the Study of Security and Sustainable Development. In 1988–92 he held directorships of the Mediterranean Institute (University of Malta) and International Ocean Institute, and was a member of Malta's delegation to the Preparatory Commission established by the UN Convention on the Law of the Sea. On joining Dalhousie in 1992, he was Director of the Marine Affairs Program until 2001. In 2003–05 he held the Canadian Chair in Marine Environmental Protection at the IMO's World Maritime University, Sweden. Professor Chircop has published extensively on the law of the sea, maritime law and coastal management fields. His most recent book is *Canadian Maritime Law* (Aldo Chircop, William Moreira, Hugh Kindred and Edgar Gold (eds), Irwin Law 2016), and he is a co-editor of the *Ocean Yearbook* (Martinus Nijhoff). Professor Chircop's research focuses on interstitial issues of the international law of the sea and maritime law, in particular the regulation of polar shipping, use of maritime regulatory tools for marine spatial planning and conservation, and development of Canadian maritime law. Among his several professional affiliations, Professor Chircop is a member of the Nova Scotia bar and Chair of the International Working Group on Polar Shipping of the Comité Maritime International.

*Nicholas Gaskell*

is Professor of Maritime and Commercial Law at the TC Beirne School of Law, University of Queensland and a member of its Marine and Shipping Law Unit (MASLU). Prof Gaskell previously worked for 33 years at the University of Southampton and its Institute of Maritime Law. He is a Barrister and Associate Tenant at Quadrant Chambers (London). For 15 years, he represented the International Union for the Conservation of Nature and Natural Resources (IUCN) at the International Maritime Organization's Legal Committee and the IOPC Funds. He attended and participated in IMO Legal Committee and IMO diplomatic conferences, which resulted in international maritime law conventions, including the Salvage Convention 1989, 1992 Protocols to the CLC and Fund Convention, Hazardous and Noxious Substances Convention 1996, Limitation Convention (LLMC) 1996, Arrest Convention 1999, Bunker Pollution Convention 2001, Athens Convention 2002 and Nairobi Wreck Removal Convention 2007. He is a Titulary member of the Comité Maritime International (CMI) and a member of its International Working Group On 'Unmanned Ships'. Prof. Gaskell has taught maritime law to governments, practitioners and students within the maritime legal and shipping professions both within Australia and internationally. He is a member of the editorial boards of Lloyd's *Maritime and Commercial Law Quarterly*, the *International Journal of Marine and Coastal Law* and *Journal of Maritime Law and Commerce*. Prof. Gaskell has authored/edited seven books, 41 book chapters and over 50 articles on maritime and commercial law, including maritime liabilities (e.g., collisions, salvage, wreck, passenger claims and limitation of liability); environmental claims (e.g., oil, bunker, chemical and nuclear pollution); carriage of goods by sea; and shipping regulation.

*Günther Handl*

is the Eberhard P. Deutsch Professor of Public International Law at Tulane University Law School. He holds law degrees from the University of Graz (Dr. iur.), Cambridge (LL.B.) and Yale (S.J.D.). He is the author of several books and many articles in the field of public international law, international environmental law, law of the sea and nuclear energy law. Professor Handl is the recipient of a number of awards, including the Prix Elisabeth Haub, 1997. He has served in an advisory capacity to governments and international organisations, including the United Nations Environment Programme, the Organisation for Economic Co-operation and Development, the Asian Development Bank and the World Bank. He is Senior Project Consultant on 'Policy and Law for Nuclear Safety and Security' at the National University of Singapore.

*Youna Lyons*
is a bilingual former French litigator with more than 12 years of international practice in Asia and Europe. Passionate about the ocean, Ms. Lyons later studied oceanography and holds a Master's degree in Marine Affairs from the University of Washington. She is a member of the CIL Ocean Law and Policy Programme and currently also pursuing a Ph.D. with the Australian National Centre for Ocean Resources and Security at the University of Wollongong titled 'Defining and Identifying Valuable and Sensitive Marine Areas in the South China Sea and States' Obligations under International Law'. Her dual expertise in law and marine sciences uniquely positions her to integrate these fields for marine policy-making purposes and has led her to spearhead and coordinate several multidisciplinary research projects at CIL including the prospects for re-use of obsolete offshore installations as artificial reefs in Southeast Asia, sensitivity mapping for Singapore in the context of oil spill preparedness and response, and the mapping of shallow features in the South China Sea. Besides her Master's degree in Marine Affairs, she also holds subject degrees in oceanography, an LL.M. and two Bachelor's degrees from La Sorbonne in public and private international law and in procedural law respectively.

*Millicent McCreath*
is a member of the CIL Ocean Law and Policy Programme. She has an LL.M. (Law of the Sea) from the University of Tromsø, Norway, and a LL.B. (Hons)/B.A. (Environmental Studies) from the University of New South Wales in Sydney, Australia. Millicent is admitted as a solicitor in New South Wales and completed her practical legal training at the Environmental Defenders Office, a public interest environmental law centre in Sydney. Prior to joining CIL Millicent was Tipstaff (judicial clerk) at the Land and Environment Court in Sydney, a specialist superior court with environment and planning jurisdiction. Her main research interests are focussed on marine environmental law and international dispute resolution.

*Nilüfer Oral*
is a member of the Law Faculty at Istanbul Bilgi University. She was elected to the United Nations International Law Commission for 2017–21. She is also Distinguished Senior Scholar at the Law of the Sea Institute, University of California Berkeley Law School, and was Scholar-In-Residence at the University of Virginia Law School in 2005. She has lectured on several occasions at the Rhodes Academy of Oceans Law and Policy. Dr. Oral was elected to the International Union for Conservation of Nature (IUCN) Council for Western

Europe (2012–16) and is the IUCN Oceans Focal Point. She serves as Chair of the IUCN Academy on Environmental Law and Co-Chair of the Specialist Group on Oceans, Coasts and Coral Reefs for the IUCN World Commission on Environmental Law. Dr. Oral serves as a legal advisor to the Turkish Foreign Ministry on law of the sea and climate change and has participated in the Turkish delegation at the International Maritime Organization and as a negotiator at the climate change negotiations. She has also been involved as a legal expert in a number of projects with the Turkish Science Council on marine environmental issues, EU and UNDP. Dr. Oral is a member of the Board of Directors of the Jon Van Dyke Institute, Richardson School of Law, University of Hawai'i; the American Society of International Law; and the European Society of International Law. She is the editor of the *International Straits of the World* series (Brill/Nijhoff) and author of numerous other publications.

*J. Ashley Roach*
Captain, JAGC, US Navy (retired) was attorney adviser in the Office of the Legal Adviser, US Department of State, from 1988 until he retired in 2009, where he was responsible for law of the sea matters. Since 2014 he has been Visiting Senior Principal Research Fellow at CIL, where he has written, lectured and taught Arctic and other law of the sea issues in support of CIL's Ocean Law and Policy Programme and CIL's research programme for the Singapore Maritime and Port Authority. He has taught, advised and published extensively on national maritime claims and other law of the sea issues, including piracy and armed robbery at sea, at institutions such as the Rhodes Academy of Oceans Law and Policy. While at the State Department he negotiated and participated in the negotiation of numerous international agreements involving law of the sea issues. He received his LL.M. (highest honors in public international law and comparative law) from the George Washington University School of Law in 1971 and his J.D. from the University of Pennsylvania Law School in 1963.

*Erik Røsæg*
is a Professor at PluriCourts—Centre for the Study of the Legitimate Roles of the Judiciary in the Global Order, and is formerly the Director of the Scandinavian Institute of Maritime Law, University of Oslo. He teaches and writes in the fields of maritime law and third party interests in commercial law. He has been very involved in the negotiations of liability conventions at the IMO. He chairs the Norwegian Maritime Law Commission and has also chaired other committees preparing draft legislation. He has published extensively in national and international journals and books, and is frequently a speaker and consultant in

a number of jurisdictions. He is currently involved in the Oslo Law of the Sea Forum and chairs the Arctic Committee of the University of Oslo.

*Karen N. Scott*
is Professor of Law at the University of Canterbury in New Zealand. Her research focuses on the law of the sea, the polar regions and international environmental law. She has published over 60 journal articles and book chapters in these areas, on issues such as ocean management, environmental governance, climate change and geoengineering. Recent publications include *The Oxford Handbook of the Law of the Sea* (Donald R. Rothwell, Alex G. Oude Elferink, Karen N. Scott and Tim Stephens (eds), OUP 2015). Prof. Scott was the General Editor of the *New Zealand Yearbook of International Law* from 2009 to 2012 and remains a member of the Editorial Board. She was Vice President of the Australian and New Zealand Society of International Law (ANZSIL) from 2011 to 2016 and was Head of the School of Law from 2015 to 2018.

*Zhen Sun*
is a member of the CIL Ocean Law and Policy Programme, in which she focuses on legal issues in the Arctic and the protection of submarine cables. Her other research interests include maritime security, jurisdictional conflicts within the law of the sea, and international dispute resolution. She received her LL.B. from Hainan University Law School, an LL.M. in International Law from the China University of Political Science and Law, a further LL.M. (Distinction) in International Law from the University of Edinburgh, and a Ph.D. from the University of Cambridge, United Kingdom. She is admitted to the bar in the People's Republic of China.

*Robin Warner*
is Professor and Deputy Director at the Australian National Centre for Ocean Resources and Security (ANCORS), University of Wollongong, Australia. She was formerly Assistant Secretary of the International Crime Branch in the Attorney General's Department from 2002 to 2007 and Director of International Law for the Australian Defence Force from 1997 to 1999. Her research interests include law of the sea, oceans governance, marine environmental law, climate law, and maritime regulation and enforcement. She is the author of *Protecting the Oceans Beyond National Jurisdiction: Strengthening the International Law Framework* (Martinus Nijhoff 2009), lead editor (with Simon Marsden) of *Transboundary Environmental Governance: Inland Coastal and Marine Perspectives* (Ashgate 2012), lead editor (with Clive Schofield) of *Climate Change and*

*the Oceans: Gauging the Legal and Policy Currents in the Asia Pacific* (Edward Elgar 2012) and lead editor (with Stuart Kaye) of the *Routledge Handbook of Maritime Regulation and Enforcement* (Routledge 2015). Since becoming a full time academic in 2007, she has published over 70 books, journal articles, book chapters and other publications.

# Abbreviations

| | |
|---|---|
| ABNJ | Areas beyond National Jurisdiction |
| AIS | Automatic Identification System |
| APEI | Areas of Particular Environmental Interest |
| APM | Associated Protective Measure |
| BBNJ | Biodiversity beyond National Jurisdiction |
| BPC | Bunker Pollution Convention |
| BSC | Brussels Supplementary Convention to the Paris Convention on Third Party Liability in the Field of Nuclear Energy |
| CBD | Convention on Biological Diversity |
| CCAMLR | Commission for the Conservation of Antarctic Marine Living Resources |
| CCZ | Clarion-Clipperton Zone |
| CEP | Committee for Environmental Protection |
| CLC | International Convention on Civil Liability for Oil Pollution Damage |
| CMS | Convention on the Conservation of Migratory Species of Wild Animals |
| $CO_2$ | Carbon Dioxide |
| COLREG | Convention on the International Regulations for Preventing Collisions at Sea |
| COP | Conference of the Parties |
| CPPS | Comision Permanente del Pacifico Sur (Permanent Commission for the South Pacific) |
| CRISTAL | Contract Regarding a Supplement to Tanker Liability for Oil Pollution |
| CSC | Convention on Supplementary Compensation for Nuclear Damage |
| EAAFP | East Asia-Australasian Flyway Partnership |
| EBSA | Ecologically or Biologically Significant Marine Area |
| ECA | Emission Control Area |
| ECJ | European Court of Justice |
| EEZ | Exclusive Economic Zone |
| EIA | Environmental Impact Assessment |
| FAO | Food and Agriculture Organization of the United Nations |
| FNPP | Floating Nuclear Power Plant |
| GBR | Great Barrier Reef |
| HNS Convention | International Convention on Liability and Compensation for Damage in Connection with the Carriage of Hazardous and Noxious Substances by Sea |

| | |
|---|---|
| IAEA | International Atomic Energy Agency |
| IBA | Important Bird Area |
| ICC | International Chamber of Commerce |
| ICJ | International Court of Justice |
| ICPC | International Cable Protection Committee |
| ILBI | International Legally Binding Instrument |
| ILC | International Law Commission |
| ILC Articles on State Responsibility | Articles on Responsibility of States for Internationally Wrongful Acts |
| IMMA | Important Marine Mammal Area |
| IMO | International Maritime Organization |
| IMSAS | IMO Member State Audit Scheme |
| IOPC Fund | International Oil Pollution Compensation Fund |
| IOSEA | Indian Ocean and Southeast Asia |
| ISA | International Seabed Authority |
| ITLOS | International Tribunal for the Law of the Sea |
| ITOPF | International Tanker Owners Pollution Federation |
| IUCN | International Union for Conservation of Nature and Natural Resources |
| IUU Fishing | Illegal, Unreported and Unregulated Fishing |
| IWC | International Whaling Commission |
| KBA | Key Biodiversity Area |
| LLMC | Convention on Limitation of Liability for Maritime Claims |
| LTC | Legal and Technical Commission |
| LTE | Low-Tide Elevation |
| MARPOL | The International Convention for the Prevention of Pollution from Ships, 1973/78 |
| MEPC | Marine Environment Protection Committee |
| MGR | Marine Genetic Resources |
| MOSPA | Agreement on Cooperation on Marine Oil Pollution Preparedness and Response in the Arctic |
| MoU | Memoranda of Understanding |
| MPA | Marine Protected Area |
| MSC | Maritime Safety Committee |
| MSR | Marine Scientific Research |
| mt | Metric Tons |
| NAFO | Northwest Atlantic Fisheries Organization |
| NCSR | Sub-Committee on Navigation, Communication and Search and Rescue |
| NEA | Nuclear Energy Agency |
| NM | Nautical Miles |

| | |
|---|---|
| $NO_x$ | Nitrogen Oxides |
| OECD | Organisation for Economic Co-operation and Development |
| OILPOL | International Convention for the Prevention of Pollution of the Sea by Oil |
| OPRC | International Convention on Oil Pollution Preparedness, Response and Cooperation |
| OSPAR Convention | Convention for the Protection of the Marine Environment of the North-East Atlantic |
| OUV | Outstanding Universal Value |
| P & I | International Group of Protection and Indemnity |
| ppm | Parts Per Million |
| PSC | Port State Control |
| PSSA | Particularly Sensitive Sea Area |
| RFMO | Regional Fisheries Management Organization |
| RSA | Regional Seas Arrangement |
| SAR Convention | International Convention on Maritime Search and Rescue |
| SBTTA | Subsidiary Body on Scientific, Technical and Technological Advice |
| SOLAS | International Convention for the Safety of Life at Sea |
| $SO_x$ | Sulphur Oxides |
| SPRFMO | South Pacific Regional Fisheries Management Organization |
| SRFC | Sub-Regional Fisheries Commission |
| SRM | Solar Radiation Management |
| TOVALOP | Tanker Owners Voluntary Agreement Concerning Liability for Oil Pollution |
| UN | United Nations |
| UNCLOS | United Nations Convention on the Law of the Sea |
| UNEP | United Nations Environment Programme |
| UNESCO | United Nations Educational, Scientific and Cultural Organisation |
| UNFCCC | United Nations Framework Convention on Climate Change |
| UNFSA | United Nations Fish Stocks Agreement |
| UNGA | United Nations General Assembly |
| VME | Vulnerable Marine Ecosystem |
| WCPFC | Western and Central Pacific Fisheries Commission |
| WOW | Wings over Wetlands |
| WRC | Wreck Removal Convention |

# Figures and Tables

## Figures



## Tables

CHAPTER 1

# Introduction

*Robert Beckman, Millicent McCreath, J. Ashley Roach and Zhen Sun*

The 1982 United Nations Convention on the Law of the Sea (UNCLOS)[1] is one of the most significant accomplishments of the international community in the 20th century. UNCLOS establishes a comprehensive set of principles and rules that is intended to regulate all uses of the sea and the exploration and exploitation of its resources in all areas of ocean space. UNCLOS is often referred to as 'a constitution for the oceans' and has been almost universally accepted.[2] As indicated in its preamble, UNCLOS is based on the fundamental idea that the problems of the oceans are closely interrelated and must be addressed as a whole. In most respects, UNCLOS has met the test of time and is as relevant today as it was when it entered into force in 1994.

However in the 36 years since its adoption, the significant advances in science and technology, and intensification of human activities in the oceans have raised issues and challenges that the international community was not aware of when UNCLOS was negotiated. The international community has begun to recognise gaps and loopholes in the legal regime established in UNCLOS, particularly the legal regime governing the high seas. These gaps and loopholes are in part due to the fact that the principles and rules governing activities on the high seas in UNCLOS are essentially the same as those set out in the 1958 Convention on the High Seas.[3] The way we use the high seas has changed dramatically since the 1950s and therefore it is not surprising that the rules from that period may no longer be adequate. The development of technology and the advancement of knowledge has enabled the use and

1 United Nations Convention on the Law of the Sea (adopted 10 December 1982, entered into force 16 November 1994) 1833 UNTS 3 (UNCLOS).

2 'A Constitution for the Oceans', Remarks by Tommy T.B. Koh, of Singapore, President of the Third United Nations Conference on the Law of the Sea, <www.un.org/Depts/los/convention_agreements/texts/koh_english.pdf> accessed 22 May 2018. UNCLOS has 167 state parties and is ratified by the European Union as at 3 April 2018, Chronological Lists of Ratifications of, Accessions and Successions to the Convention and the Related Agreements, <www.un.org/Depts/los/reference_files/chronological_lists_of_ratifications.htm#The%20United%20Nations%20Convention%20on%20the%20Law%20of%20the%20Sea> accessed 22 May 2018.

3 Convention on the High Seas (adopted 29 April 1958, entered into force 30 September 1962) 450 UNTS 11.

exploitation of the high seas and its resources in ways that were not possible in the 1950s. With these increased and different uses of the high seas, the long-standing principle of the freedom of the high seas may no longer be sufficient to regulate these uses and to ensure the protection and the preservation of the marine environment.

Steps are being taken to address some of the gaps in the high seas legal regime outlined in UNCLOS. In 2015, the United Nations General Assembly (UNGA) resolved to develop an international legally binding instrument (ILBI) under UNCLOS on the conservation and sustainable use of marine biological diversity of areas beyond national jurisdiction (BBNJ).[4] A Preparatory Committee was formed to make substantive recommendations on the elements of a draft text. The Preparatory Committee completed its four sessions and reported to the UNGA on its progress in 2017.[5] The UNGA resolved to convene an intergovernmental conference in 2018 to consider the recommendations of the Preparatory Committee and to negotiate the text of an ILBI under UNCLOS, with a view to developing the instrument as soon as possible.[6] The negotiations at the intergovernmental conference will address the following topics:

> the conservation and sustainable use of marine biological diversity of areas beyond national jurisdiction, in particular, together and as a whole, marine genetic resources, including questions on the sharing of benefits, measures such as area-based management tools, including marine protected areas, environmental impact assessments and capacity-building and the transfer of marine technology.[7]

Against this background, the Centre for International Law (CIL) at the National University of Singapore convened an international Conference on High Seas

4 United Nations General Assembly (UNGA), 'Development of an International Legally Binding Instrument under the United Nations Convention on the Law of the Sea on the Conservation and Sustainable Use of Marine Biological Diversity of Areas beyond National Jurisdiction', A/RES/69/292, 6 July 2015, <www.un.org/Depts/los/general_assembly/general_assembly_resolutions.htm> accessed 22 May 2018. See Nicholas Gaskell, Chapter 8 'Liability and Compensation Regimes: Pollution of the High Seas.'

5 Preparatory Committee Established by General Assembly Resolution 69/292, <www.un.org/depts/los/biodiversity/prepcom.htm> accessed 22 May 2018.

6 UNGA, 'International Legally Binding Instrument under the United Nations Convention on the Law of the Sea on the Conservation and Sustainable Use of Marine Biological Diversity of Areas beyond National Jurisdiction', A/RES/72/249, 19 January 2018, para. 1, <www.un.org/Depts/los/general_assembly/general_assembly_resolutions.htm> accessed 22 May 2018.

7 Ibid, para. 2.

Governance: Gaps and Challenges in April 2017.[8] The Conference brought together leading scholars and experts on science, law of the sea and maritime law to discuss and explore some of the gaps in the UNCLOS legal regime for the governance of the high seas. At the conclusion of the Conference, selected papers were commissioned for publication in this edited volume.

Before articulating the content of the book, it is worth mentioning one specific topic that did not make its way into the book. There is increasing global concern over the environmental, social and economic impacts of marine debris, of which a large proportion is plastics and microplastics.[9] Plastic debris is a major issue on the high seas where buoyant plastics are carried by ocean currents and concentrated in the ocean gyres, most prominently the North Pacific Gyre, also known as the Great Pacific Garbage Patch.[10] Marine plastics may have a range of negative social, economic and ecological impacts, due to their ingestion by or entanglement with fish, birds and marine mammals and their concentration in the food chain. However the ecological risks of microplastics, and the human health risks (if any) resulting from their consumption by humans, are currently uncertain.[11]

Tackling marine debris is challenging in many aspects. The concentration of marine debris on the high seas presents a legal challenge, as it is not clear which state or actor has the right or obligation under international law to clean up the debris, or what mechanism is best suited to regulate clean-up activities. In 2013 The Ocean Cleanup, a Dutch non-profit organisation, was founded by inventor Boyan Slat with the aim of cleaning up plastic debris in the oceans.[12] The Ocean Cleanup has developed a free-floating barrier system between one

8 Centre for International Law (CIL) International Conference High Seas Governance: Gaps and Challenges, 24–25 April 2017, <https://cil.nus.edu.sg/event/international-conference-on-high-seas-governance-gaps-and-challenges/> accessed 22 May 2018.

9 United Nations Open-ended Informal Consultative Process on Oceans and the Law of the Sea, Seventeenth Meeting, 13–17 June 2016, Panel, 'Marine debris, plastics and microplastics', <www.un.org/depts/los/consultative_process/documents/ICP-17_Panel_for_website.pdf> accessed 22 May 2018.

10 David Santillo, 'Tracking the Rising Tide of Marine Plastic Pollution in Coastal Waters and on the High Seas: What are the Options?' presented at the CIL Conference on High Seas Governance (n 8). Joint Group of Experts on the Scientific Aspects of Marine Environmental Protection (GESAMP), Working Group 40 (WG 40): Sources, Fate and Effects of Plastics and Micro-Plastics in the Marine Environment, <www.gesamp.org/work/groups/40> accessed 22 May 2018.

11 GESAMP Reports and Studies No.93, *Sources, Fate and Effects of Plastics and Micro-Plastics in the Marine Environment (Part 2)*, Annex VII – Summary of Potential Ecological, Social and Economic Impacts of Marine Plastics and Microplastics, Sub-Divided by Size Range, 213–216, <www.gesamp.org/publications> accessed 22 May 2018.

12 The Ocean Cleanup, About, <www.theoceancleanup.com/about/> accessed 22 May 2018.

and two kilometres in length, equipped with an impermeable collecting screen reaching 4 metres below the ocean's surface. Modelling shows that the system should capture a high proportion (> 95%) of plastics greater than 10 mm in size that will then be brought ashore for recycling.[13] Having previously conducted trials, in mid-2018 The Ocean Cleanup is planning to deploy a first working system that demonstrates a functioning barrier and is capable of completing all steps of the process. It will be deployed in the Great Pacific Garbage Patch, located on the high seas.[14] This initiative has raised several legal questions, including the legal status of the drifting array, which state will have jurisdiction and control over the array, and which state will be responsible or liable in the event of an incident causing harm to the environment, property or life.[15]

The legal issues associated with addressing marine debris on the high seas illustrate gaps in the UNCLOS high seas governance regime. Under the law of the sea, it is states that have rights and duties with respect to the uses of the oceans. However, in practice, it is private entities that often conduct activities. The current legal regime has clear rules on the relationship between the flag state and the ship that is flying its flag. However, in many cases the flag state of a ship may not have any connection to the company or person carrying out the high seas activity. Further, the entity in charge of the deployment may not be the entity that developed the concept, making it difficult to determine which entity is 'carrying out the activity'. While the law of the sea provides that the flag state has exclusive jurisdiction on the high seas, is this appropriate when the flag state has little involvement or control over the activity being conducted by a private entity using its ship?

## 1 The Chapters in the Book

This book consists of eight main chapters, followed by a concluding chapter. Discussions in this book focus on using current legal mechanisms to address the identified gaps, as well as the possibility of developing new protocols or agreements under existing legal instruments, including the prospective ILBI.

---

13 'Any Other Business The Ocean Cleanup's Deployment in the North Pacific Submitted by the Netherlands and Vanuatu' IMO Doc MSC 99/21/15 (13 March 2018), para 7.

14 Ibid, paras 8, 10.

15 Bettina Boschen, 'High Seas Installations for the Large-Scale Removal of Plastic Debris on the High Seas – A Practical Application of the Current High Seas Regime', presented at the CIL Conference on High Seas Governance (n 8). Nick Kilvert, 'Great Pacific Garbage Patch Plastic Removal System could become "World's Biggest Piece of Marine Debris," Critics Say', <www.abc.net.au/news/science/2018-05-08/great-pacific-garbage-patch-plastic-ocean-cleanup-boyan-slat/9714246> accessed 22 May 2018.

It is not the intention of this book to duplicate the BBNJ negotiations, nor will it cover all gaps in the high seas governance regime.

Chapter 2 by Nilüfer Oral addresses the legal questions relating to jurisdiction and control over activities by non-state entities on the high seas, pointing out the difficulties that arise when activities take place without the formal authorisation of any state. It first discusses flag state responsibilities with respect to ships flying its flag, including an analysis of the scope and interpretation of the due diligence obligations. It then discusses the application of the obligation of states to conduct an environmental impact assessment for planned activities under their jurisdiction or control that may cause environmental harm. The chapter then evaluates the legal framework for the exercise of other freedoms and activities on the high seas, including marine scientific research activities, the laying of submarine cables and pipelines and the construction of artificial islands and other installations by non-state entities, with a particular focus on their interaction with the obligation to protect and preserve the marine environment. It acknowledges the legal gaps regarding the jurisdiction and control over activities undertaken by non-state entities on the high seas, and argues that the development of the ILBI presents an opportunity for states to clarify and define the obligations of states concerning such activities. Oral states that when these activities are carried out at least in part by ships, it may be the responsibility of the flag state to regulate the activities. She also argues that the state with jurisdiction and control over the private entity should be responsible for regulating the activity.

Chapter 3 by Karen N. Scott focusses on the regulation of geoengineering on the high seas, which is another activity that was not directly addressed by UNCLOS, and analyses to what extent the current legal framework has responded to the gaps and challenges it presents. It first charts the relationship between climate change, geoengineering and the ocean to define geoengineering. It then discusses geoengineering within the framework of UNCLOS, and argues that the principles, concepts and rules designed to protect the marine environment and to regulate marine scientific research can be, and, to a certain extent have been, adapted to apply to geoengineering activities. It then discusses geoengineering within the framework of the conventions regulating marine dumping, particularly the initiative under the 1996 London Protocol to the Convention on the Prevention of Marine Pollution by Dumping of Wastes and Other Matter, 1972[16] to develop a designated regime to regulate ocean fertilisation and potentially other forms of geoengineering. The chapter

16 1996 Protocol to the Convention on the Prevention of Marine Pollution by Dumping of Wastes and Other Matter, 1972 (adopted 7 November 1996, entered into force 24 March 2006) 36 ILM 1 (London Protocol).

concludes with a discussion of how the law of the sea might respond to the broader moral, ethical and environmental issues associated with marine geoengineering and examines two possible options for future regulatory reform: addressing geoengineering within the prospective ILBI, and the adoption of a UNGA resolution on oceans and climate change, which could include principles and guidance relating to geoengineering.

Chapter 4 by Youna Lyons examines the numerous scientific criteria developed for identifying sensitive marine areas in the context of international treaties and other instruments, and considers how they may apply to high seas areas. Lyons examines the different ecological and non-ecological criteria under the various instruments and then identifies their commonalities and differences. Although few of these instruments were developed specifically for establishing sensitive marine areas on the high seas, the fact that most of the criteria are shared across the sets of criteria developed under different instruments shows that they are generally accepted and can serve as a basis for identifying sensitive marine areas in the high seas. However even if such areas can be established, other issues must be addressed, including who initiates the action to establish a sensitive marine area, who approves the establishment of the area, and who monitors activities in these sensitive area. These are difficult issues, some of which are being addressed in the BBNJ negotiations.

In Chapter 5 Aldo Chircop analyses whether the instruments of the International Maritime Organization (IMO) can be used to establish marine protected areas on the high seas in which constraints would be placed on international shipping activities. He first discusses the established and growing legal and management tools adopted by the IMO for the protection of sensitive marine areas and species. In particular he considers (a) special areas designated under the International Convention for the Prevention of Pollution from Ships, 1973/78 (MARPOL),[17] (b) particularly sensitive sea areas (PSSAs) and associated protective measures (APMs) under various instruments,[18] and (c) permanent or seasonal routeing and reporting measures to avoid or minimise negative impacts on particular species.[19] He concludes that there is a sufficient legal

17 International Convention for the Prevention of Pollution from Ships (adopted 2 November 1973, entered into force 2 October 1983) 1340 UNTS 184, as amended by the Protocol Relating to the International Convention for the Prevention of Pollution from Ships of 1973 (adopted 17 February 1978, entered into force 2 October 1983) 1340 UNTS 61, as amended (MARPOL). Annexes I and II entered into force on the same date as the convention.

18 For example special areas under MARPOL Annexes (n 17), and routeing measures under the International Convention for the Safety of Life at Sea (adopted 1 November 1974, entered into force 25 May 1980) 1184 UNTS 2, as amended (SOLAS) Ch V.

19 Ibid Ch V, regs 10 and 11.

and constitutional basis for applying the measures currently available in IMO instruments to establish marine protected areas on the high seas to prevent ship-source pollution and preserve rare and fragile ecosystems, but that some adjustment to the procedures may be necessary. As with other types of protected areas on the high seas, one issue to be addressed is which state or states have the standing to propose the establishment of such areas.

Chapter 6 by Robin Warner discusses the conservation and sustainable use of marine living resources beyond national jurisdiction, including both pelagic and benthic resources, as well as their habitats and the ecosystems of which they form part. It first provides an overview of the current status of marine living resources and marine biological diversity in areas beyond national jurisdiction, and highlights the negative impacts caused by the depletion of living resources on ecosystem dynamics. It then identifies and examines the regulatory and implementation gaps and deficiencies in the regime for the conservation and sustainable use of marine living resources on the high seas, with a particular focus on the work of a number of the regional fisheries management organisations. Finally the chapter discusses the feasibility of using the ILBI to fill the identified gaps in the current legal framework.

Chapter 7 by Günther Handl looks at the international accountability for nuclear pollution on the high seas. Handl first reviews the various scenarios that may cause nuclear pollution of the high seas, and then examines the legal framework that provides redress for high seas nuclear pollution, comprised of the international nuclear instruments and UNCLOS. There is a liability and compensation scheme for pollution of the marine environment from the carriage of nuclear material, but like the scheme for pollution from oil tankers, it covers damage for pollution of the coastal and marine environment of states, and for loss of life, personal injury or property damage, not damage for the marine environment of the high seas. Handl then turns to the law of state responsibility, which reflects the growing expectation that hazardous activities ought to be subject to a regime of strict liability, that damage to the environment per se be compensable, and that the source or controlling state be held accountable financially, at least on a subsidiary basis, for compensating damage. He then concludes that there is a compelling need for an international mechanism or agency capable of vindicating fully the collective interest in preserving the high seas, including its protection against nuclear pollution.

Chapter 8 by Nicholas Gaskell considers some potential gaps relating to liability and compensation for pollution of the high seas, particularly those relating to shipping activities. It first discusses the current legal regime for maritime liability and compensation under UNCLOS and IMO instruments, and identifies the gaps for addressing pollution on the high seas. After demonstrating the need for maritime liability and compensation systems for pollution of the high seas, this chapter discusses the models of compensation schemes as established under the IMO instruments for oil, hazardous and noxious substances, bunkers, and wrecks, and their scope of application. But Gaskell does not limit his analysis to the liability and compensation schemes for pollution from ships. He also explains why there is no similar scheme for pollution from offshore platforms. The chapter then outlines two options to expand the current legal regime: adopting new high seas protocols to the existing IMO instruments to extend the current shipping liability regimes to the high seas, and adopting a new single multilateral maritime environmental liability regime for all sources of pollution on the high seas. This chapter further considers five other compensation models established by industries and other international instruments, such as the conventions dealing with hazardous waste and the regime for deep seabed mining activities. The last section of the chapter explores the feasibility of incorporating the liability and compensation regime for pollution of the high seas into the prospective ILBI. He concludes by stating that the scheme set out in Protocol VI of the Antarctic Treaty may be model for developing a system to establish a scheme to make shipowners liable for pollution of the marine environment of the high seas.

Chapter 9 by Erik Røsæg discusses the marine pollution preparedness, response and cooperation in the Artic high seas from a maritime law perspective. He examines the legal regimes requiring coastal states to establish a system of emergency preparedness and response in the event of a major oil spill off their coast. The chapter first charts the applicable international and regional legal instruments, followed by an overview of the practical aspects of cleaning up and the requirements for pollution preparedness. This chapter then deals with rights and duties in respect of cleaning up, for example whether states have a duty under international law to instigate clean-up operations and whether a state can instigate a clean-up operation in the EEZ of another state. He points out that like the liability and compensation schemes for oil pollution, the conventions for preparedness, response and cooperation are designed to address pollution of the marine and coastal environment of coastal states, not pollution of the high seas. He concludes that there is no strict obligation on states to instigate clean-up operations if there is an oil spill from a tanker or other ship on the high seas.

In Chapter 10 the co-editors outline the gaps in the high seas governance regime identified by the book's contributors, and discuss the reasons for their emergence by reference to the origins of the high seas legal regime. The chapter focuses on identifying possible ways to fill these gaps, and including the potential role of the BBNJ ILBI.

CHAPTER 2

# Jurisdiction and Control over Activities by Non-State Entities on the High Seas

*Nilüfer Oral*

## Abstract

Freedom of navigation on the high seas and the exclusive jurisdiction of the flag state are well-accepted norms of customary international law. This chapter will evaluate the legal framework under the law of the sea for the exercise of jurisdiction and control over the activities of non-state entities on the high seas, with a focus on the protection of the environment for marine scientific research activities, the laying of cables and pipelines, construction of artificial islands and other installations, and activities of vessels in relation to non-state entities. It will also examine the current status of these activities in relation to the negotiation of an international legally binding instrument for the conservation and sustainable use of marine biological diversity in areas beyond national jurisdiction (BBNJ).

## 1 Introduction

Freedom of navigation on the high seas and the exclusive jurisdiction of the flag state are well-accepted norms of customary international law, codified in the 1958 Convention on the High Seas[1] and later in the 1982 United Nations Convention for the Law of the Sea (UNCLOS).[2] According to Article 87 of UNCLOS, activities subject to the freedom of the high seas comprise inter alia freedom of navigation and freedom to lay submarine cables and pipelines (subject to Part VI of UNCLOS), freedom to construct artificial islands and other installations permitted under international law (subject to Part VI of UNCLOS), and freedom of scientific research (subject to Parts VI and XIII of UNCLOS). However, these freedoms are not without qualification, as each is

1 Convention on the High Seas (adopted 29 April 1958, entered into force 30 September 1962) 450 UNTS 11.

2 United Nations Convention on the Law of the Sea (adopted 10 December 1982, entered into force 16 November 1994) 1833 UNTS 3 (UNCLOS).

 DOI:10.1163/9789004373303_003

subject to 'the conditions laid down by [the] Convention and by other rules of international law'.[3] Consequently, understanding the legal scope of the exercise of freedoms of the high seas requires delineating these conditions and other rules of international law and, in particular, with regard to the duty to protect the marine environment as provided for under Part XII of UNCLOS. While the rights and obligations under Article 87 and UNCLOS, in general, apply directly to states, in practice many of the activities under Article 87 are undertaken by private entities. Consequently, before the scope of such freedoms can be determined, the legal relationship between the state and private or non-state entities must be clarified.

This chapter will evaluate the legal framework under the law of the sea for the exercise of jurisdiction and control over the activities of non-state entities on the high seas, with a focus on the protection of the environment for marine scientific research activities, the laying of cables and pipelines, construction of artificial islands and other installations, and activities of vessels in relation to non-state entities. It will also examine the current status of these activities in relation to the negotiation of an international legally binding instrument for the conservation and sustainable use of marine biological diversity in areas beyond national jurisdiction (BBNJ), the subject matter of United Nations General Assembly (UNGA) Resolution 72/249.[4]

## 2 Flag State Responsibilities on the High Seas

### 2.1 *Effective Jurisdiction and Control*

International shipping accounts for 90 per cent of global trade. For this reason, freedom of navigation on the high seas is one of the cornerstones of international law, founded in the historic work of Hugo Grotius.[5] The right of ships to sail the high seas without the interference of other states is an accepted rule of customary international law that has been codified in Article 2 of the

---

3 Ibid art 87.

4 Previously, on 24 December 2017, the General Assembly decided to convene an Intergovernmental Conference, under the auspices of the United Nations, to consider the recommendations of the Preparatory Committee established by resolution 69/292 of 19 June 2015 on the elements and to elaborate the text of an international legally binding instrument under the United Nations Convention on the Law of Sea on the conservation and sustainable use of marine biological diversity of areas beyond national jurisdiction, with a view to developing the instrument as soon as possible.

5 Hugo Grotius, *Mare Liberum* (first published 1609, Hugo Grotius, Mare liberum, 1609–2009: Original Latin Text And English Translation) (Trans. Robert Feensrea, Brill, 2009).

Convention on the High Seas[6] and Article 87 of UNCLOS. Freedom of the high seas under international law is a right that is accorded to states. On the other hand, private non-state actors, who dominate commercial shipping, the bulk of global shipping, exercise the right of freedom of navigation on the high seas through their relationship with the state by flying the flag of the state under whose laws the ship is registered. Once a ship has been duly registered under the laws of the flag state, it is deemed a 'national' of that state, as stated in Article 91 of UNCLOS.

The right of the flag state to exercise freedom of navigation on the high seas is subject to a number of responsibilities under international law. One of these is the requirement that a genuine link must exist between the state and the ship, according to UNCLOS Article 91(1). Second, according to Article 94 the flag state has the obligation to *exercise effective jurisdiction and control* over its ships. Both of these obligations perhaps raise more questions than provide guidance to flag states. What are the requirements for a 'genuine link' and for the 'exercise of effective jurisdiction and control over its ships'? These are critical aspects of understanding the role of the flag state in relation to controlling the activities of non-state actors on the high seas.

In the well-known *Nottebohm (Liechtenstein v Guatemala)*[7] case, the International Court of Justice (ICJ) rejected the adoption of any minimum requirements to recognise the validity of the nationality of a person, leaving the determination of the conditions for the acquisition of nationality to the domestic laws of states. Subsequent judgments by international courts and tribunals have applied the *Nottebohm* rule to shipping, adopting the same deferential approach of leaving it to the flag state to determine the minimum criteria for ships flying its flag. For example, in the *M/V 'Saiga' (No. 2) (Saint Vincent and the Grenadines v Guinea)* case, the International Tribunal for the Law of the Sea (ITLOS) interpreted Article 91 of UNCLOS as granting the flag state the exclusive right to determine the conditions of nationality.[8] However, ITLOS made an important distinction concerning the function of 'genuine link' for the purposes of shipping. In examining both the Convention on the High Seas and UNCLOS, ITLOS explained

> that the purpose of the provisions of the Convention on the need for a genuine link between a ship and its flag State is to secure more effective

6 Convention on the High Seas (n 1).

7 *Nottebohm Case (Liechtenstein v Guatemala)* (Judgment of 6 April 1955) ICJ Reports 1955, 4.

8 The M/V 'SAIGA' (*No* 2) (Saint Vincent and the Grenadines v Guinea) (Judgment of 1 July 1999) ITLOS Reports 1999, 10.

> implementation of the duties of the flag State, and not to establish criteria by reference to which the validity of the registration of ships in a flag State may be challenged by other States.[9]

In the subsequent case of *M/V Virginia G*,[10] ITLOS reaffirmed the interpretation of Article 91 in the *Saiga* (*No. 2*) case and further clarified the meaning of 'genuine link' to mean that 'once a ship is registered, the flag State is required, under Article 94 of the Convention, to *exercise effective jurisdiction and control over that ship* in order to ensure that it operates in accordance with generally accepted international regulations, procedures and practices'.[11] Thus, according to ITLOS, Articles 91 and 94 of UNCLOS are interlinked.

The textual language of Article 94 of UNCLOS prima facie appears to provide a narrow framework for determining what would constitute 'exercise of effective jurisdiction and control' by the flag state over ships flying its flag, highlighting its application in particular to 'administrative, technical and social matters'.[12] This potential narrow reading of Article 94, however, was rejected by ITLOS, sitting en banc, in its advisory opinion in the *Request for an Advisory Opinion Submitted by the Sub-Regional Fisheries Commission* (*SRFC*) (the *SRFC Advisory Opinion*).[13] The request for an advisory opinion, which was submitted by the Sub-Regional Fisheries Commission (SRFC), consisted of four questions. One of the questions requested ITLOS to advise on '... the obligations of the flag State in cases where illegal, unreported and unregulated (IUU) fishing activities are conducted within the Exclusive Economic Zone of third party States'.[14]

---

9 Ibid para 83. See also *The M/V* 'Virginia G' Case (*Panama/Guinea-Bissau*) (Judgment of 14 April 2014) ITLOS Reports 2014, 4. See also, Tullio Scovazzi, 'ITLOS and Jurisdiction over ships' in Henrik Ringbom (ed) *Jurisdiction Over Ships: Post-UNCLOS Developments in the Law of the Sea* (Brill 2015) 382–404.

10 The M/V '*Virginia G*' *Case* (Panama/Guinea-Bissau) (n 9).

11 Ibid para. 113.

12 *Request for An Advisory Opinion Submitted by the Sub-Regional Fisheries Commission* (*SRFC*) (Advisory Opinion of 2 April 2015) ITLOS Reports 2015, 4 (SRFC Advisory Opinion) para 117.

13 Ibid.

14 Ibid. The other three questions are as follows. To what extent shall the flag state be held liable for IUU fishing activities conducted by vessels sailing under its flag? Where a fishing license is issued to a vessel within the framework of an international agreement with the flag state or with an international agency, shall the state or international agency be held liable for the violation of the fisheries legislation of the coastal state by the vessel in question? What are the rights and obligations of the coastal state in ensuring the sustainable management of shared stocks and stocks of common interest, especially the small pelagic species and tuna?

Article 94 makes no mention of fishing activities or to IUU fishing activities undertaken by privately owned and operated fishing vessels. Nevertheless, ITLOS opted for a broad reading of Article 94 and applied it to fishing and IUU fishing activities. Specifically, ITLOS stated that

> the flag State, in fulfilment of its responsibility to exercise effective jurisdiction and control in administrative matters, must adopt the necessary administrative measures to ensure that fishing vessels flying its flag are not involved in activities which will undermine the flag State's responsibilities under the Convention in respect of the conservation and management of marine living resources. If such violations nevertheless occur and are reported by other States, the flag State is obliged to investigate and, if appropriate, take any action necessary to remedy the situation.[15]

ITLOS interpreted the predicate words of '[i]n particular' in Article 94(2) to be only indicative and not exhaustive.[16]

ITLOS not only provided a broad interpretation of Article 94 but also linked the obligation of the flag state to exercise effective jurisdiction and control to other obligations under UNCLOS, such as the conservation and management of marine living resources and protection of the marine environment,[17] which is discussed further on. The linkage between Article 94 and Part XII is further complemented by Article 87, which states clearly that the freedom of the high seas is to be exercised *under the conditions laid down by this Convention and by other rules of international law*. The obligations under Part XII would constitute such 'conditions laid down by' UNCLOS.

In the *SRFC Advisory Opinion* case, ITLOS also noted that certain provisions of UNCLOS, such as Articles 91, 92 and 94 as well as Article 192, are 'general obligations which are to be met by the flag State *in all maritime areas* regulated by the Convention'.[18] In light of this advisory opinion by ITLOS, it is fair to assume that in addition to fishing activities a similar interpretation would apply to other activities of ships. Consequently, the duty of the flag state to exercise effective control and jurisdiction over ships flying its flag under Article 94 is not limited to navigational safety and protection of the marine environment and fishing, but logically would include other shipping-related activities, such as those related to marine scientific research, the laying of

15 Ibid para. 119.
16 Ibid para 117.
17 Ibid para 120.
18 Ibid. (emphasis added).

submarine cables and pipelines, survey activities, and the construction of installations and artificial islands.

One of the long-standing challenges for international shipping concerns identification of the ship owner. Beneficial ownership—which concerns the entity that controls the ship but may not necessarily be the legally registered owner—is also linked to the problem of flags of convenience, where the state of registry does not apply or enforce generally accepted international regulations and standards for shipping. In addition, there is the problem of shipowners who seek to avert compliance with their obligations by registering in two or more ship registers with lax regulatory systems. International law, as codified in Article 92 of UNCLOS, prohibits a ship from flying more than one flag. However, historically the challenge has been in identifying these ships and their owners. Technology is proving to be a key element in addressing the problem of tracing shipowners. In 2007, at the initiative of the European Commission and the French Maritime Administration, Equasis, an information system, was established.[19] The system provides detailed information on the registration of ships and companies, including information on ownership and management. A user can simply enter the name of the ship or company name, and detailed information is provided, including the history of ownership, management and important port state control inspection information.

In addition, there have been important recent developments at the International Maritime Organization (IMO) to address the problem of substandard shipping caused by lax flag state compliance and enforcement of IMO instruments. Notably, as of 1 January 2016, the IMO Member State Audit Scheme (IMSAS) was made mandatory. IMSAS will assess the extent to which the member states give full and complete effect to their obligations and responsibilities under IMO conventions for safety of navigation and pollution prevention.[20] This is an important milestone in implementing the requirements of Article 94 of UNCLOS and the due diligence obligations that are discussed below. For example, IMSAS requires that the flag state establish resources and processes capable of administering a safety and pollution protection programme that has certain minimum features, including the necessary national legislation and use of an audit and inspection programme.[21] IMSAS also provides a non-exhaustive list of necessary measures the flag state must adopt to secure observance of international rules and standards by ships entitled to fly its flag and

19 See Equasis <www.equasis.org/EquasisWeb/public/HomePage> accessed 20 March 2018.

20 IMO Instruments Implementation Code (III Code), IMO Res A.1070(28) (4 December 2013).

21 Ibid pt 2.

by entities and persons under its jurisdiction, so as to ensure compliance with its international obligations.[22]

### 2.2 *Due Diligence Obligations*

The obligation of flag states to exercise effective jurisdiction and control over ships flying their flag also entails the obligation to exercise due diligence. The obligation of the state to exercise due diligence over activities under its control and jurisdiction, however, is broader than the obligation of flag states to exercise effective jurisdiction and control under Article 94, which applies expressly to shipping activities. The state can also exercise control over non-shipping activities, such as marine scientific research, laying of cables and pipelines, bioprospecting, or the construction of offshore structures and installations.

The obligation of states to exercise due diligence has been affirmed by the ICJ and ITLOS in several cases. According to these cases, the obligation of due diligence is

> an obligation which entails not only the adoption of appropriate rules and measures, but also a certain level of vigilance in their enforcement and the exercise of administrative control applicable to public and private operators, such as the monitoring of activities undertaken by such operators, to safeguard the rights of the other party.[23]

It also includes 'an obligation to deploy adequate means, to exercise best possible efforts, to do the utmost to obtain' the required result.[24] The state is required to exercise administrative control over both public and private operators (not simply ships) and to monitor their activities.

The obligation of states to protect and preserve the marine environment includes the obligation to exercise due diligence. The ICJ in the *Pulp Mills on the River Uruguay* case specifically recognised that the 'obligation to protect and preserve the aquatic environment, and in particular to prevent pollution by prescribing appropriate rules and measures' is an obligation of due diligence over all activities under the 'jurisdiction and control' of the parties (Uruguay

---

22 Ibid pt 2, para. 22.

23 *Pulp Mills on the River Uruguay (Argentina v Uruguay)* (Judgment of 20 April 2010) ICJ Reports 2010, 14, para 197; *Responsibilities and Obligations of States Sponsoring Persons and Entities with respect to Activities in the Area* (Request for Advisory Opinion submitted to the Seabed Disputes Chamber) (Advisory Opinion of 1 February 2011) ITLOS Reports 2011, 10 (Seabed Advisory Opinion) para. 115; SRFC Advisory Opinion (n 12) para. 131.

24 Seabed Advisory Opinion ibid para. 110; SRFC Advisory Opinion (n 12) para. 128.

and Argentina).[25] In relation to the protection of the marine environment under UNCLOS specifically, the obligation of due diligence was first applied in the Advisory Opinion of the Seabed Disputes Chamber of ITLOS on *Responsibilities and Obligations of States Sponsoring Persons and Entities with respect to Activities in the Area* (*Seabed Disputes Chamber Advisory Opinion*),[26] and then by the full bench of ITLOS in the *SRFC Advisory Opinion* case on IUU fishing.[27]

In the *Seabed Disputes Chamber Advisory Opinion,* the Seabed Disputes Chamber explained that the obligation 'to ensure' under Article 139(1) and Article 4(4) of Annex III of UNCLOS

> establishes a mechanism through which the rules of the Convention concerning activities in the Area, although being treaty law and thus binding only on the subjects of international law that have accepted them, become effective for sponsored contractors which find their legal basis in domestic law. This mechanism consist in the creation of obligations which States Parties must fulfil by exercising their power over entities of their nationality and under their control.[28]

In this case, the obligation of the sponsoring state 'to ensure' is an obligation of 'conduct' and not to achieve in each and every case the result that the contractor complies with the substantive obligations, but 'it is an obligation to deploy adequate means, to exercise best possible efforts, to do the utmost, to obtain this result'.[29] The Seabed Disputes Chamber further explained that it would not be satisfactory to 'rely on mere application of the principle that the conduct of private persons or entities is not attributable to the State under international law'.[30] The Seabed Disputes Chamber gave as an example Article 194(2) of UNCLOS where states have the obligation to 'take all measures necessary to ensure that activities under their jurisdiction or control are so conducted as not to cause damage by pollution to other States and their environment …'.[31]

In the *SRFC Advisory Opinion* case, ITLOS, sitting en banc, stated that the obligation of due diligence applies to the flag states of fishing vessels and

25 *Pulp Mills* Case (n 23) para. 197.
26 Seabed Advisory Opinion (n 23).
27 Ibid; SRFC Advisory Opinion (n 12).
28 Seabed Advisory Opinion (n 23) para. 108.
29 Ibid para. 110.
30 Ibid para. 112 (citing from ILC Articles on State Responsibility, art 8(1) of the Commentaries).
31 SRFC Advisory Opinion (n 12) para. 113.

fishing activities, including the responsibility to ensure that vessels flying their flags do not engage in IUU fishing activities in the exclusive economic zone of a (SRFC) state.[32] Importantly, while acknowledging the differences between the sponsoring state and contractor, ITLOS nevertheless found that the same mechanism of 'exercising their power over entities of their nationality and under their control', as clarified by the Seabed Disputes Chamber, applied to the relationship between the flag state and fishing vessels flying its flag.[33]

In addition, the Arbitral Tribunal in the *South China Sea Arbitral* Award expressly found that the due diligence obligation of states applied to Article 194(5), where it identified the obligation of states 'to take those measures "necessary to protect and preserve rare or fragile ecosystems as well as the habitat of depleted, threatened or endangered species and other forms of marine life"'.[34] Furthermore, the Arbitral Tribunal stated that the due diligence obligation is triggered when a state becomes aware that its vessels are inter alia 'inflicting significant damage on *rare or fragile ecosystems or the habitat of depleted, threatened or endangered species*'.[35] This means that when the flag state receives information that a vessel under its jurisdiction is engaged in activities in violation of Article 194(5), the state is required to exercise its due diligence obligations and take active enforcement measures against the violator. Presumably, the same trigger would apply to other activities of the vessel that are in contravention of UNCLOS.

These cases demonstrate that in addition to the obligation of the flag state to exercise effective jurisdiction and control over ships flying its flag, other non-flag states also have the obligation to exercise due diligence over the activities of non-state entities of their nationality and under their control. In addition to mining and fishing activities, the obligation of due diligence would also include the responsibility of the state to ensure that, for example, submarine pipelines do not cause pollution of the marine environment after they have been laid. In this case, the state possessing jurisdiction and control and thus having the responsibility to ensure that the pipeline does not pollute the marine environment would be the state of the nationality of the company that owns and operates the pipeline, not the flag state of the vessel that physically

32 Ibid para. 124.

33 Ibid paras 125–26.

34 *The South China Sea Arbitration (The Republic of Philippines v The People's Republic of China)* Award (12 July 2016) PCA 2013–19, para. 908.

35 Ibid para. 961 (emphasis added).

laid the pipeline, unless the pollution was caused by an act or omission during the laying of the pipeline.

Specific activities and freedom of the high seas will be discussed in greater detail further on.

## 3 Impact Assessments

One of the long-standing principles of international law is the obligation of states not to allow activities under their jurisdiction or control to cause harm to the territory of another state. The no-harm rule is codified in Article 194(2) of UNCLOS.[36] Article 194(2) also creates an obligation of due diligence, as reflected in the use of the language 'to ensure' and furthermore applies to 'activities under the jurisdiction or control' of states. However, by contrast the requirement for states to conduct impact assessments in accordance with Article 206 of UNCLOS is a direct responsibility of the state, albeit qualified. Accordingly,

> [w]hen States have reasonable grounds for believing that planned activities under their jurisdiction or control may cause substantial pollution of or significant and harmful changes to the marine environment, they shall, as far as practicable, assess the potential effects of such activities on the marine environment and shall communicate reports of the results of such assessments in the manner provided in article 205.[37]

According to Article 205, states are to communicate reports '... to the competent international organisations, which should make them available to all States'.

The ICJ in the *Pulp Mills* case[38] and the *Construction of the Road Case*[39] and ITLOS in the *Seabed Disputes Chamber Advisory Opinion*[40] recognised as a rule of customary international law the obligation of states to conduct

36 UNCLOS (n 2) art 194(2).

37 Ibid art 206.

38 The ICJ found that states are under a duty to conduct EIAs in cases of transboundary industrial activities that could cause significant harm to a shared resource. See n 42.

39 *Construction of a Road in Costa Rica Along the San Juan River (Nicaragua v Costa Rica)* (Judgment of 16 December 2015) ICJ Reports 2015, 665.

40 Sherry Broder, 'International Governance of Ocean Fertilization and Other Marine Geoengineering Activities' in Carlos Espósito et al (eds) *Ocean Law and Policy: Twenty Years of Development Under the UNCLOS Regime* (Brill 2016).

environmental impact assessments (EIAs) for activities that could cause significant harm to shared resources. ITLOS in the *Seabed Disputes Chamber Advisory Opinion* concluded that the Article 206 obligation to conduct an EIA is a '*direct* obligation under the Convention and a general obligation under customary international law',[41] thus 'an essential part of a comprehensive environmental management system', and also 'a particular application of the obligation on states, enunciated in Article 194(2)'.[42] However, in the *South China Sea* Arbitral Award, the Arbitral Tribunal also recognised that the obligations under Articles 204 and 206 are qualified with 'to endeavor' and 'as far as is practicable', leaving some discretion to the state. However, the duty for the state to communicate reports of the results of assessments under Article 205 is 'absolute'. The obligation for states to conduct EIAs for shared resources would apply equally to areas beyond national jurisdiction (ABNJ). However, this raises several legal and practical questions, which have been the subject of discussion during meetings of the BBNJ Preparatory Committee. For example, which activities should be subject to EIAs? While pipelines transporting oil and gas may be obvious candidates, the inclusion of submarine cables has proven to be more debatable. The International Cable Protection Committee (ICPC) presented its position to the BBNJ Preparatory Committee as to why submarine cables should be exempt from any additional EIA other than what is required under Article 206 of UNCLOS.[43] However, Papua New Guinea, for example, raised the possibility for the laying of cables in ABNJ to be subject to cumulative impact assessments under the international legally binding instrument (ILBI).[44]

Other questions concern who would be responsible for undertaking an EIA: the flag state or the state of the operator? This is particularly pertinent for non-shipping activities such as for activities related to the physical construction of artificial islands on the high seas, offshore energy operations

41 Seabed Advisory Opinion (n 23) para. 145 (emphasis added).

42 Shabtai Rosenne and Alexander Yankov (eds), *United Nations Convention on the Law of the Sea 1982: A Commentary* (Vol IV, Brill Nijhoff 2002) para. 206.6(b) in Myron H Norquist et al (eds) *United Nations Convention on the Law of the Sea 1982: A Commentary* (7 vols, Brill Nijhoff 1985–2011).

43 The International Cable Protection Committee (ICPC), 'Submarine Cables and BBNJ' (The International Cable Protection Committee 2016) <www.un.org/depts/los/biodiversity/prepcom_files/ICC_Submarine_Cables_&_BBNJ_August_2016.pdf> accessed 22 May 2018.

44 Papua New Guinea emphasised cumulative EIAs for fishing and laying submarine cables. See International Institute for Sustainable Development, 'Prepcom 1 Highlights' 2016 25(102) *Earth Negotiations Bulletin* <http://enb.iisd.org/vol25/enb25102e.html> accessed 23 May 2018.

or other activities. Which would be the responsible state under UNCLOS? Furthermore, what should the threshold of harm or impact be in requiring EIAs? Substantial pollution, significant adverse effects, or less? Another important question concerns whether cumulative impacts should be considered. Use of a lower threshold or one that is cumulative could result in having so-called environmentally neutral activities, such as the laying of cables on the seabed, subject to EIAs.

The final recommendations agreed to by consensus by the BBNJ Preparatory Committee for presentation to the UNGA did not answer these questions.[45] The recommendations provided a broad outline of certain procedural steps but without any details as to which activities would be included, what the necessary threshold is, who would be responsible and whether cumulative impacts would be included.

## 4 Freedoms and Activities on the High Seas

### 4.1 *Marine Scientific Research*

The freedom to conduct marine scientific research on the high seas is emphasised in different provisions of UNCLOS. First, UNCLOS recognises that '[a]ll States, irrespective of their geographical location, and competent international organizations have the right, in conformity with the provisions of Part XI, to conduct marine scientific research in the Area'.[46] However, the meaning of marine scientific research is not provided by UNCLOS. There is no global common understanding of what activities would fall under marine scientific research. An opportunity was presented to the ICJ to provide some guidance in the *Whaling in the Antarctic (Australia v Japan: New Zealand intervening)* case.[47] The Court, however, declined the opportunity to provide any definition or criteria to differentiate scientific research versus commercial activities.[48]

45 United Nations General Assembly (UNGA), 'Report of the Preparatory Committee Established by General Assembly Resolution 69/292: Development of an International Legally Binding Instrument under the United Nations Convention on the Law of the Sea on the Conservation and Sustainable Use of Marine Biological Diversity of Areas Beyond National Jurisdiction' UN Doc A/AC.289/2017/PC.4/2 (31 July 2017) <www.un.org/ga/search/view_doc.asp?symbol=A/AC.287/2017/PC.4/2&referer> accessed 22 May 2018.

46 UNCLOS (n 2) art 256.

47 *Whaling in the Antarctic (Australia v Japan: New Zealand intervening)* (Judgment of 31 March 2014) ICJ Reports 2014, 226.

48 Marc Mangel, 'Whales, Science, and Scientific Whaling in the International Court of Justice' (2016) 113(51) *Proceedings of the National Academy of Sciences of the United States of America* 14523.

The issue of defining marine scientific research goes beyond being an academic exercise. Different obligations and conditions may arise depending on whether the activity is scientific or commercial. An example of the fine line between commercial activities and marine scientific research is highlighted by climate change geoengineering activities[49] in the oceans, such as the controversial ocean iron fertilisation venture by a private company in 2007[50] and more recently in Canada to restore salmon stocks.[51] Ocean geoengineering and ocean iron fertilisation in the high seas were unregulated until they were eventually addressed by the Convention on the Prevention of Marine Pollution by Dumping of Wastes and Other Matter 1972 (London Convention) and its 1996 Protocol (the London Protocol);with application on the high seas.[52] In 2008 the parties to the London Convention and London Protocol adopted a resolution to prohibit ocean fertilisation activities other than legitimate scientific research.[53] In 2010 the parties to the London Convention and London Protocol adopted the Assessment Framework for Scientific Research Involving Ocean Fertilization (Assessment Framework) intended for application by states to assess projects that fall within the scope of application of the London Convention and London Protocol, which is 'all marine waters other than the internal waters', thereby including the BBNJ.[54] The Assessment Framework aims to provide 'a tool for assessing proposed activities on a case-by-case basis to determine if the proposed activity constitutes legitimate scientific research that is not contrary to the aims of the London Convention or Protocol'.[55] The Assessment Framework lists a number of criteria for states to assess whether

49 See Chapter 3 by Karen N Scott, 'Mind the Gap: Marine Geoengineering and the Law of the Sea'.

50 Rosemary Rayfuse, Mark G Lawrence and Kristina M Gjerde, 'Ocean Fertilisation and Climate Change: The Need to Regulate Emerging High Seas Uses' (2008) 23 *International Journal of Marine and Coastal Law* 297.

51 Randall S Abate, 'Ocean Iron Fertilization and Indigenous Peoples' Right to Food: Leveraging International and Domestic Law Protections to Enhance Access to Salmon in the Pacific Northwest' (2016) 20 *UCLA Journal of International Law & Foreign Affairs* 45.

52 Convention on the Prevention of Marine Pollution by Dumping Wastes and Other Matter (adopted 29 December 1972, entered into force 30 August 1975) 1046 UNTS 120 (London Convention); 1996 Protocol to the Convention on the Prevention of Marine Pollution by Dumping of Wastes and Other Matter, 1972 (adopted 7 November 1996, entered into force 24 March 2006) 36 ILM 1 (London Protocol); Amended by the 1996 Protocol; Broder (n 40). See Karen N Scott, Chapter 3 'Mind the Gap: Marine Geoengineering and the Law of the Sea'.

53 On the Regulation of Ocean Fertilization, IMO Res LC-LP.1 (2008) (31 October 2008).

54 On the Assessment Framework for Scientific Research Involving Ocean Fertilization, IMO Res LC-LP.2 (2010) (14 October 2010).

55 Ibid para. 1.2.

the proposed project for ocean fertilisation has proper scientific attributes and a detailed procedure for conducting an EIA.[56]

The question of the international legal framework for marine scientific research in ABNJ gained new prominence with the initiation of the BBNJ Preparatory Committee at the United Nations in 2015, with the scope of the proposed ILBI including marine genetic resources (MGRs).[57] Most of the existing activities on the high seas for MGRs are carried out by private companies. The discussions at the BBNJ Preparatory Committee revealed a division between those states that take the position that the regime of freedom of the high seas under Article 87 applies to MGRs and those that argue for application of the regime of the Common Heritage of Mankind under Part XI of UNCLOS for the Area.

While all states have the right according to Article 256 to conduct marine scientific research in the Area, subject to other conditions under UNCLOS, unlike in the case of the high seas, all marine scientific research within the Area 'shall be carried out … for the benefit of [hu]mankind as a whole'.[58] However there are also general obligations applicable to all states, such as the general principles for the conduct of marine scientific research under Article 240, the promotion of international cooperation under Article 242 and the publication and dissemination by other appropriate channels knowledge resulting from marine scientific research under Article 244(1). Nonetheless, marine scientific research conducted for the benefit of humankind as part of the Common Heritage of Mankind brings with it additional conditions, such as those expressly provided for in Article 143(3), than would be otherwise required as part of the high seas freedoms.

Marine genetic research presents a peculiar case where the difference between scientific research and commercial development can be blurred. The Jamaican delegation's submission to the BBNJ Preparatory Committee provides a very good description of the issue in relation to the marine scientific research. It explains the difference between *pure* and *applied* science: the former is open to the public and transparent, but the latter is principally

56 Such consent, however, is to be without prejudice to international law including as reflected in the relevant provisions of UNCLOS; ibid para. 4.2.

57 Development of an International Legally Binding Instrument under UNCLOS on the Conservation and Sustainable Use of Marine Biological Diversity of Areas Beyond National Jurisdiction, IMO Res A/RES/69/292 (19 June 2015).

58 UNCLOS (n 2) art 143(1); Lyle Glowka, 'Genetic Resources, Marine Scientific Resources and the International Seabed Area' (1999) 8 *Review of European Community & International Environmental Law* 56.

oriented towards eventual commercial uses.[59] Because bioprospecting is seen as an activity with commercial ends, the concern is that the samples collected will not be made public but will be applied for private uses. For purposes of MGRS, a single sample is all that may be necessary for future genetic activities that can be conducted in land-based laboratories. A key part of the debate on MGRS concerns benefit-sharing between developed and developing countries. Benefit-sharing for the scientific research would involve non-monetary benefits, such as joint publications of scientific articles or joint expeditions, whereas the benefit-sharing for commercial exploitation of MGRS raises a host of far thornier questions, including that of intellectual property rights.[60] The BBNJ Preparatory Committee has addressed these issues in broad terms in its recommendations.[61] For example, under the sub-heading of 'Principles and approaches guiding benefit-sharing', the recommendations include promoting marine scientific research, research and development, the types of benefits to be shared, and monitoring of the use of marine genetic resources of ABNJ. Notably, the recommendations do not refer to any distinction between pure and applied science for MGRS in ABNJ.

Freedom to conduct marine scientific research is also subject to other parts of UNCLOS that would include the obligation to protect and preserve the marine environment under Article 192, to take measures to prevent, reduce and control pollution from all sources and activities, which would include activities related to the conduct of marine scientific research, and to protect rare and fragile ecosystems and habitats under Article 194.[62] Such rare and fragile ecosystems and habitats can be found, for example, in the hydrothermal vents, cold water coral and methane seeps of the deep sea, which are all sources of valuable MGRS.

These provisions apply equally to all flag states on the high seas. However, UNCLOS applies to states as subjects of international law and not to non-state entities, at least directly. The question arises as to which state has the obligation to ensure that obligations and conditions of UNCLOS are being fulfilled. For research activities that take place directly at sea, it is clear that the flag state

59 Government of Jamaica, 'Submission of the Government of Jamaica on Access and Benefit Sharing Regime for Marine Genetic Resources in Areas Beyond National Jurisdiction' (2016) <www.un.org/depts/los/biodiversity/prepcom_files/rolling_comp/Jamaica-access_and_benefit_sharing.pdf> accessed 22 May 2017.

60 Claudio Chiarolla, 'Intellectual Property Rights and Benefit Sharing from Marine Genetic Resources in Areas Beyond National Jurisdiction: Current Discussions and Regulatory Options' (2014) 4 *Queen Mary Journal of Intellectual Property* 171.

61 Report of the Preparatory Committee (n 45) p. 10.

62 UNCLOS (n 2) art 194.

must fulfil the requirements under Article 94 to exercise effective jurisdiction and control as well as its due diligence obligations to ensure that obligations, such as the protection of the marine environment, are being fulfilled. However, what about other requirements such as the requirement that research is to be for peaceful purposes? Or the requirement to make available by publication and dissemination though appropriate channels inter alia knowledge resulting from marine scientific research under Article 244? Whose obligation is this? What if the research is being carried out by a non-State actor, such as a university?

The *Seabed Disputes Chamber Advisory Opinion* provides some guidance, although it concerns the responsibility of sponsoring states for the commercial (exploitation) activities of contractors in the Area. Nevertheless, it has articulated the meaning of the language 'to ensure' as a general due diligence obligation under different provisions of UNCLOS. Therefore, the flag state would have the responsibility to take the necessary measures to ensure that the entities conducting scientific research under its flag, and thus jurisdiction and control, fulfill these requirements. This can be achieved through adopting the necessary laws and regulations and ensuring their compliance.

## 4.2 *Submarine Cables, Pipelines and Vessels*

### 4.2.1 Exclusive Economic Zone

Articles 87 and 112 of UNCLOS recognise the right of all states to lay submarine cables[63] and pipelines on the bed of the high seas beyond the continental shelf. Furthermore, as part of the reserved freedoms of the high seas provided for in Article 87, the freedom to lay submarine cables and pipelines also extends to the exclusive economic zone under Article 58(1). Additionally, the freedom of laying submarine cables includes uses associated with the operation of submarine cables and pipelines, such as maintenance and repair. In relation to the laying of submarine cables and pipelines, the exclusive economic zone can be considered quasi-high seas, as the jurisdiction of the coastal state to exercise sovereign rights is restricted against interference with these freedoms.[64] However, Article 58(3) provides that states exercising their rights and performing their duties in the exclusive economic zone, such as the laying of submarine

63 For an overview see, Douglas Burnett, Tara Davenport and Robert Beckman, 'Overview of the International Legal Regime Governing Submarine Cables' in Tara Davenport, Robert C Beckman and Douglas R Burnett (eds) *Submarine Cables: The Handbook of Law and Policy* (Brill 2014) 63–90; Tara Davenport, 'Submarine Communications Cables and Law of the Sea: Problems in Law and Practice' (2012) 43 *Ocean Development & International Law* 201.

64 UNCLOS (n 2) art 56(1)(b)(iii).

cables and pipelines, are required to have *due regard* to the rights and duties of the coastal state, and to comply with the laws and regulations adopted by the coastal state in accordance with the provisions of UNCLOS. Specifically, Article 211(5) recognises that the coastal state may adopt laws and regulations for the prevention, reduction and control of pollution from vessels conforming to and giving effect to generally accepted international rules and standards established through the competent international organisation or general diplomatic conference, and enforce these. However, the requirement to comply with the laws of the coastal state regulating vessel source pollution and exercise due diligence would apply to the shipping activities related to the laying of submarine cables and pipelines and fall under the responsibility of the flag state.

### 4.2.2 Continental Shelf

The freedom to lay submarine cables and pipelines under Article 87 is subject to the conditions of Part VI of UNCLOS on the continental shelf. The continental shelf of a coastal state comprises the seabed and subsoil of the submarine areas up to a distance of 200 nautical miles (NM) from the baselines from which the breadth of the territorial sea is measured.[65] In addition, where the margin extends beyond 200 NM, as provided in Article 76(1), the coastal state may establish the outer edges of its continental shelf not exceeding 350 NM from the baselines from which the breadth of the territorial sea is measured, or 100 NM from the 2500 m isobath. Article 79 also recognises the right of all states to lay submarine cables and pipelines on the continental shelf, which cannot be impeded by the coastal state except to take reasonable measures for inter alia, the prevention, reduction and control of pollution from pipelines.[66] Notably, the right of the coastal state to take reasonable measures applies only to submarine pipelines and not submarine cables. Furthermore, it is the only express reference to environmental protection in UNCLOS pertaining to pipelines. No similar provision exists for submarine cables, implicitly recognising that the environmental risks posed by submarine cables differ from those posed by pipelines used for transport of oil and gas. Nonetheless, does this exempt the cable laying–related activities from environmental standards

65 Article 76 (1) provides in full that 'The continental shelf of a coastal State comprises the seabed and subsoil of the submarine areas that extend beyond its territorial sea throughout the natural prolongation of its land territory to the outer edge of the continental margin, or to a distance of 200 nautical miles from the baselines from which the breadth of the territorial sea is measured where the outer edge of the continental margin does not extend up to that distance'. UNCLOS (n 2) art 76.

66 UNCLOS (n 2) art 79(2).

in general and specifically in relation to the high seas where there is no coastal state? The issue is complicated by the dual nature of cable laying operations. First, there is the ship engaged in the physical laying of the cables, and second the company responsible for the operation and maintenance of the cables afterwards. In the first case, the flag state may be responsible for any environmental consequences during the survey process and laying of the cables. However, who is responsible afterwards? Does the state with jurisdiction and control over the cable company have an obligation of due diligence to ensure that the laying, repair and removal of cable does not cause harm to the marine environment?

The cable industry claims that cables are environmentally neutral.[67] Despite this claim, there are some associated risks. The cable laying process is threefold. First a survey must be conducted for determining the routeing of cable. This requires an understanding of the conditions of the seabed in order to find the most efficient and secure route. This begins first with a desk top survey that is followed with a physical route survey undertaken by a cable survey ship.[68] As water depth is traditionally measured by echo sounding,[69] the surveying process involves the use of acoustic sounding methods that can result in 'underwater noise' and disturb marine species, especially marine mammals.[70]

The second step entails the actual cable laying operation, which requires the use of special purpose ships that carry large coils of fiber optic cables to be laid on the seafloor. These ships are staffed by up to 50 crew and can stay at sea for several months.[71] Submarine cables are laid either on the surface of the seabed or buried below. The latter is done to protect the cables. The burial method is used for the laying of cables that extend across the continental shelf to a depth range of about 1,000–1,500 m.[72] The burial method produces impacts that include sediment displacement, physical disturbance and associated impacts, such as damage, displacement and removal of flora and fauna living on the seabed, and disturbance of benthic communities. According to the ICPC, in water depths over 2000 m, cables that are laid directly on the seabed and

67 ICPC, 'Submarine Cables and BBNJ' (n 43), 16, 18, 21–22.

68 See Davenport (n 63) 204–05.

69 Lionel Carter et al, *Submarine Cables and the Oceans: Connecting the World* (UNEP-WCMC Biodiversity Series No 31, ICPC, UNEP, UNEP-WCMC 2009) 21.

70 Ohta Tsutomu and Nishiyama Tomohisa, 'Route Design/Cable Laying Technologies for Optical Submarine Cables' <www.nec.com/en/global/techrep/journal/g10/n01/pdf/100111.pdf> accessed 14 March 2018.

71 Ibid.

72 Carter et al (n 69) 23.

not buried have a minimal impact on the seabed.[73] The cable itself is relatively small in size, ranging from 17 mm to 21 mm, and protected by inert polyethylene material.[74]

A third activity is the repair and maintenance of cables, which have a life span of 15–20 years.[75] Installation, repair works and/or removal phase of submarine cables that involves the retreating and re-laying of cables can also cause physical disturbance and associated impacts, including damage, displacement and removal of flora and fauna living on the seabed, especially if the burial method is used.

In general, cable survey ships and cable laying ships are privately owned. Furthermore, it is now clearly established under the jurisprudence of ITLOS that the flag state has a broad obligation to exercise control and jurisdiction over all ships flying its flag under Article 94 and obligations to exercise due diligence. While the *SRFC Advisory Opinion* applies to fishing activities in the exclusive economic zone of another state, there is no reason this cannot be extended to apply to the activities of survey vessels and cable laying vessels. In which case the exercise of effective jurisdiction and control over ships flying its flag would require the flag state adopt the necessary administrative measures to ensure those vessels are not involved in activities that would undermine its responsibilities under UNCLOS, in particular those under Part XII to protect and preserve the marine environment. Moreover, in relation to the activities of private operators, the flag state has due diligence obligations to adopt the appropriate rules and measures, and enforce them, as well as to monitor the activities undertaken by such operators.[76]

However, the question in relation to submarine cables and pipelines is to determine the responsible party once the cables or pipelines have been laid. The flag state? Does the responsibility of the flag state to exercise effective jurisdiction and control over activities in the high seas continue indefinitely? Or does it end once the submarine cable or pipeline has been successfully laid? Does the due diligence obligation of the flag state cover all impacts, including cumulative, of the submarine cable or pipeline laying activities? What about in relation to the long-term operations of the submarine cable or pipeline? Who is the responsible state? It is clear that the responsibility of the flag state can only be for those activities attributable to its conduct during the voyage

---

73 ICPC, 'Submarine Cables and BBNJ' (n 43) 22.

74 Ibid.

75 Davenport (n 63) 201.

76 *Pulp Mills* Case (n 23) para. 197; Seabed Advisory Opinion (n 23) para. 115.

itself and not for consequences that take place afterwards unrelated to shipping, such as a leaking pipeline.

However, the obligation to exercise effective jurisdiction and control under Article 94 of UNCLOS applies exclusively to the flag state. Nonetheless, with the exception of only a few provisions, Part XII on the protection and preservation of the marine environment applies to states in general. In other words, Article 192 provides a legal basis under UNCLOS to extend the responsibility to protect and preserve the marine environment for activities relating to submarine cables and pipelines to a state other than the flag state, such as the state of incorporation of the survey company or the cable company itself. The *Seabed Disputes Chamber Advisory Opinion* concerns state responsibility for sponsoring the activities inter alia of natural or juridical persons who would be the contractors for exploitation activities in the Area under Part XI. Such sponsoring states are not necessarily the flag states. However, Article 139 expressly lays out the obligation for states to ensure that activities carried out by states parties, state enterprises, or natural or juridical persons that possess the nationality of states parties or are effectively controlled by them or their nationals are to be carried out in conformity with Part XI. There is no comparable provision requiring states to ensure that the laying of cables or that survey activities are conducted in conformity with Part XII. Nonetheless, in light of the jurisprudence recognising a general due diligence obligation of all states to protect the marine environment, such an obligation would equally apply to the companies engaged in survey and cable laying activities and not be limited to the flag state of the ship that is simply engaged in the physical laying of the cables or pipelines. The obligation of states to protect and preserve the marine environment is a continuous obligation and would continue throughout the lifetime of the cable or pipeline.

### 4.3 *Installations, Structures and Artificial Islands*

Under UNCLOS, coastal states have the exclusive right to construct and to authorise and regulate the construction, operation and use of artificial islands, installations and structures in the exclusive economic zone,[77] and the right to exercise exclusive jurisdiction over them in accordance with Article 60, which enumerates other rights and obligations of the coastal state in relation to artificial islands, installations and structures. The same rights and obligations apply mutatis mutandis to artificial islands, installations and structures on the continental shelf (Article 80). Article 87 of UNCLOS recognises the freedom of all states to construct artificial islands, installations and structures. Once again,

77 UNCLOS (n 2) art 56.

these rights are subject to the conditions provided for in UNCLOS and other rules of international law. Furthermore, such freedoms must be exercised with *due regard* for the interests of other states in their exercise of the freedom of the high seas, and also with due regard for the rights under UNCLOS with respect to activities in the Area. However, other than the rights articulated under Article 87, in contrast to the exclusive economic zone and the continental shelf, UNCLOS provides no express guidance on state obligations in relation to establishing, constructing and regulating activities in relation to artificial islands, installations and structures in ABNJ. In addition, there may be questions regarding whether the mutatis mutandis language in Article 80 applies to the outer limits of the continental shelf under Article 76. Logically it should, as other than the technical and procedural requirements to claim a continental shelf beyond the limits of 200 NM under Article 76, there are no differences under UNCLOS between 'inner' and 'outer' continental shelf. This was a point made by the Arbitral Tribunal in the *Barbados v Trinidad and Tobago* Arbitral Award, in which it stated that '… in any event there is in law only a single "continental shelf" rather than an inner continental shelf and a separate extended or outer continental shelf',[78] a view that has been echoed in other judgments.[79]

Nonetheless, certain questions do arise. For example, following the delineation of the outer limits of the continental shelf margin, what becomes of the status of those artificial islands, installations and structures that were constructed by foreign states in what was the high seas and Area before the boundaries of the outer continental shelf were fixed and giving the coastal state exclusive jurisdiction under Article 80? Does the coastal state have the competence to retroactively prohibit the continued existence of any structures or installations affixed to the continental shelf? Assuming the structures remain, which state is responsible for ensuring that the owner or operator of the structure or installation is complying with the laws and regulations of the coastal state? It would have to be the state of the owner or operator similar to the situation of a sponsoring state in the *Seabed Disputes Chamber Advisory Opinion*. Moreover, if the water column above the extended continental shelf

78 *Barbados v Trinidad and Tobago*, Award of the Arbitral Tribunal (11 April 2006) PCA 2004–02, para 213. See also *Bay of Bengal Maritime Boundary Arbitration between Bangladesh and India*, Award (7 July 2014) PCA 2010–16.

79 See Dispute Concerning Delimitation of the Maritime Boundary Between Bangladesh and Myanmar in The Bay of Bengal (*Bangladesh/Myanmar*) (Judgment of 14 March 2012) ITLOS Reports 2012, 4, para. 362 (citing *Barbados v Trinidad and Tobago*); Bay of Bengal Maritime Boundary Arbitration between Bangladesh and India (n 78) para. 77 (citing *Barbados v Trinidad and Tobago*). See also Bjarni Már Magnússon, *The Continental Shelf Beyond 200 Nautical Miles, Delineation, Delimitation and Dispute Settlement* (Brill 2014).

remains high seas, what is the status of floating structures that are not fixed to the continental shelf? Do the same due diligence requirements for protection of the marine environment, including conducting impact assessments, apply in the high seas to the state of the owner or operator?

The *South China Sea Arbitral Award* provides some guidance on the latter question. China had rejected the jurisdiction of the Arbitral Tribunal on grounds that the case entailed sovereignty issues that fell outside the scope of UNCLOS, and also because the subject matter of the case fell within the declaration filed by China excluding maritime delimitation from the compulsory dispute provisions of UNCLOS.[80] However, the Arbitral Tribunal disagreed and concluded that environmental obligations in Part XII apply irrespective of determination of questions of sovereignty or maritime entitlement.[81] In other words, the obligation of states to protect the marine environment applied irrespective of the maritime zone. The Arbitral Tribunal went on to declare a clear pronouncement of the obligation of states to protect the marine environment in offshore construction activities. Consequently, the obligation to protect the marine environment follows the activity and would fall upon all states, including those that exercise control or jurisdiction over the operators of the activities in question.

Specifically, the Arbitral Tribunal found that China's island construction and land reclamation activities, which had resulted in the destruction of coral reefs and harmed endangered species, had violated Article 192, and that dredging activities were in violation of Article 194(1).[82] The Tribunal found that China had failed to exercise its due diligence obligations over Chinese fishing vessels, to prevent harm to endangered species and habitats in accordance with Article 194(5), and that it failed to adopt rules and measures to prevent such acts and maintain a level of vigilance in enforcing them.[83] Consequently, it can be concluded that states must exercise jurisdiction and control over activities related to island construction conducted by non-state entities in ABNJ.

The question is to what extent these obligations would apply to the establishment of mobile offshore installations, such as wind farms, or the construction of artificial islands in their entirely. In the *South China Sea* case China had been engaged in construction activities on naturally formed offshore features, such as corals and reefs. But what if the offshore feature constructed is

80 *The South China Sea Arbitration* (*The Republic of Philippines v The People's Republic of China*) Award on Jurisdiction and Admissibility (29 October 2015).

81 Ibid. para. 408.

82 *The South China Sea Arbitration* (n 34) para. 983.

83 Ibid paras 960–61.

entirely artificial? For example, one of the options for adaptation to climate change and sea-level rise is to build large artificial islands or very large floating structures for airports or buildings. If these are constructed in the high seas by non-state entities, who is the responsible state? Such large projects would entail multiple companies, including multinationals, and tracing the responsible state could prove to be challenging. This challenge is compounded by the lack of an express provision: the equivalent of Article 94(1) for the flag state that would apply to other states. The question of state responsibility over the activities in ABNJ of non-state entities is an issue that needs to be addressed and recognised by the new ILBI under UNCLOS. The responsibility to protect the marine environment of the high seas under Part XII applies to all states and consequently to all activities that are under their jurisdiction and control that would include natural and juridical persons.

An interesting illustration of the challenges and gaps concerning offshore structures and installations is provided by a current project to collect massive amounts of marine litter from the high seas. According to information provided by the First Global Integrated Oceans Assessment, some 275 million metric tonnes (mt) of plastic waste was generated in 192 coastal countries in 2010, with 4.8 to 12.7 million mt entering the ocean.[84] While 80 per cent of marine litter is from land-based sources, it has piled up in remote areas of the oceans, as a result of ocean currents and gyres. Some gyres are estimated to be the size of countries, perhaps the most notorious being the Great Pacific Garbage Patch.

Ocean Cleanup is a marine-litter retrieval project launched by a private organisation from the Netherlands.[85] The project proposes to place a very large floating barrier, some 50 km in length, in the high seas between Hawai'i and California, where a great deal of plastic is accumulating in the Pacific. Ocean Cleanup estimates that there is approximately 140 mt of plastic in the area. By placing a static barrier against the current the plastic will passively accumulate and be collected to be subsequently recycled. What is the nature of the 50 km long barrier? An offshore installation? Structure? In any case, it would clearly fall under the freedom of the high seas under Article 87. The question is what obligations the Netherlands has under UNCLOS. What will the impact be on other uses and activities on the high seas, such as shipping routes and fishing? A 50 km long barrier in the middle of the ocean could very well create

84 Juying Wang et al, 'Marine Debris' in Lorna Inniss, Alan Simcock and the Group of Experts of the Regular Process, *The First Global Integrated Oceans Assessment: World Ocean Assessment I* (United Nations 2016) <www.un.org/depts/los/global_reporting/WOA_RegProcess.htm> accessed 22 May 2018.

85 See The Ocean Cleanup <www.theoceancleanup.com> accessed 22 May 2018.

interferences with other uses of the high seas. It could also cause a hazard to safety of navigation, especially if there are no internationally agreed upon standards or regulations.

Moreover, does the Netherlands have the obligation to exercise effective jurisdiction and control under Article 94, which applies to flag states? Even if not, does the Netherlands have to adopt laws and regulations to ensure that no harm is done to the environment by private operators that are juridical persons under its laws? Is there an obligation to conduct an assessment under Article 206?

## 5 Conclusion

The requirement for flag states to exercise effective control and jurisdiction over the activities of ships flying their flag in the high seas is clearly established under Article 94(1) of UNCLOS and the jurisprudence of ITLOS. However, questions arise concerning activities undertaken by non-state entities in the high seas that do not involve ships flying flags of a state, such as the construction of artificial islands or installations, or the laying of submarine cables and pipelines. UNCLOS remains silent on this issue. For example, in the case of non-state entity activities in the high seas that do not involve ships, there is no guidance as to which state is required to exercise effective jurisdiction and control. The obligations of the flag state presumably cannot continue or be extended to the ongoing activities of offshore structures, submarine pipelines and cables or other commercial activities. Moreover, UNCLOS provides for obligations that are to be undertaken by all states and that implicitly would apply to those that exercise jurisdiction and control over an activity. Article 194(2) requires, in general, states to ensure that all necessary measures are taken for the prevention of transboundary pollution from activities under their jurisdiction or control. Article 206 creates a direct responsibility, in general, that states undertake an impact assessment for planned activities that may cause substantial pollution or significant harm to the marine environment. These obligations are not limited to the flag state.

Guidance on the scope of obligations has been provided for certain activities, such as seabed mining and fishing, by international adjudicatory bodies. However, what is needed is a wider and clearer articulation of the due diligence obligations of the state with jurisdiction and control over activities in the high seas or ABNJ that may pollute or otherwise harm the marine environment or biological diversity. Due diligence cannot be limited to the flag state for activities in the high seas, as this will exclude the role of other states that exercise control or jurisdiction over activities, such as the operation and management of submarine pipelines, construction of artificial islands or installations, marine scientific research, or other offshore activities that are not directly related to shipping activities.

On 24 December 2017 the UNGA adopted a resolution to hold an intergovernmental conference

> to consider the recommendations of the preparatory committee on the elements and to elaborate the text of an international legally binding instrument under the United Nations Convention on the Law of the Sea on the conservation and sustainable use of marine biological diversity of areas beyond national jurisdiction, with a view to developing the instrument as soon as possible.[86]

The development of a new instrument for conservation of biological diversity in the high seas presents an opportunity for states to clarify and define the obligations of all states concerning all activities that take place in the high seas that are under their respective control and jurisdiction.

86 International Legally Binding Instrument under the United Nations Convention on the Law of the Sea on the Conservation and Sustainable Use of Marine Biological Diversity of Areas Beyond National Jurisdiction, UNGA Res 72/249 (24 December 2017).

CHAPTER 3

# Mind the Gap: Marine Geoengineering and the Law of the Sea

*Karen N. Scott*

## Abstract

This chapter examines the legal framework for marine geoengineering, analysing the extent to which the modern law of the sea has responded to the gaps and challenges in the current regulatory framework. It will focus on adapting existing environmental and other principles within the framework of UNCLOS to geoengineering, before examining the recent initiative under the 1996 London Protocol to develop a designated regime to regulate ocean fertilisation. The chapter will conclude with a discussion of how the law of the sea might respond to the broader moral, ethical and environmental issues associated with marine geoengineering, and examines options for future regulatory reform.

## 1 Introduction[1]

Marine geoengineering demonstrates the gaps, promise and limitations of the modern law of the sea. As geoengineering is an activity that lay beyond the imaginations of the original negotiators of the 1982 United Nations Convention on the Law of the Sea (UNCLOS),[2] there are unsurprisingly no provisions of direct application to geoengineering within UNCLOS. However, this does not mean that geoengineering is unregulated today. Principles, concepts and rules designed to protect the marine environment, regulate scientific research and manage the intersection of the rights and duties of all states with maritime interests can be, and, to a certain extent have been, adapted to apply to geoengineering

1 Material for this paper has been drawn and developed from the following publications: Karen N Scott, 'Geoengineering and the Law of the Sea' in Rosemary Rayfuse (ed), *Research Handbook on International Marine Environmental Law* (Edward Elgar Publishing 2015) 451–72; Karen N Scott, 'International Law in the Anthropocene: Responding to the Geoengineering Challenge' (2013) 34 *Michigan Journal of International Law* 309.

2 United Nations Convention on the Law of the Sea (adopted 10 December 1982, entered into force 16 November 1994) 1833 UNTS 3 (UNCLOS).

 DOI:10.1163/9789004373303_004

activities. Nevertheless, there are limits or challenges to the extent to which evolution and adaptation can take place within the existing institutional framework of the law of the sea. Marine geoengineering is more than an activity that takes place within and impacts upon the marine environment. It is deliberately designed to exploit the relationship between the atmosphere and the hydrosphere and is intended to manipulate either the level of carbon dioxide in the atmosphere or global temperatures or both, in order to mitigate the impacts of climate change. Moreover, geoengineering raises very particular ethical and moral questions about how we should respond to climate change, arguably the greatest global challenge in humankind's history. It is open to question as to whether the law of the sea and its institutions provide appropriate global fora for these debates to take place and indeed whether they are able to appropriately connect and interact with other institutions and instruments, such as the 1992 United Nations Framework Convention on Climate Change (UNFCCC),[3] in the management of geoengineering. In fact, geoengineering merely provides a particularly explicit illustration of what might be characterised as the greatest challenge for the law of the sea going forward in the 21st century: responding to interconnected environmental systems through disconnected political and legal institutions and processes.

This chapter will examine the legal framework for marine geoengineering, analysing the extent to which the modern law of the sea has responded to the gaps and challenges in the current regulatory framework. It will focus on adapting existing environmental and other principles within the framework of UNCLOS to geoengineering, before examining the recent initiative under the 1996 London Protocol to the Convention on the Prevention of Marine Pollution by Dumping of Wastes and Other Matter, 1972 (London Protocol)[4] to develop a designated regime to regulate ocean fertilisation and potentially other forms of geoengineering. The chapter will conclude with a discussion of how the law of the sea might respond to the broader moral, ethical and environmental issues associated with marine geoengineering and examines two possible options for future regulatory reform: addressing geoengineering within the future Agreement on the Conservation and Sustainable Use of Marine Biological Diversity of Areas Beyond National Jurisdiction currently under negotiation, and the adoption of a United Nations General Assembly (UNGA) resolution on

3 United Nations Framework Convention on Climate Change (adopted 9 May 1992, entered into force 21 March 1994) 1771 UNTS 107 (UNFCCC).

4 1996 Protocol to the Convention on the Prevention of Marine Pollution by Dumping of Wastes and Other Matter, 1972 (adopted 7 November 1996, entered into force 24 March 2006) 36 ILM 1 (London Protocol).

oceans and climate change, which includes principles and guidance relating to geoengineering.

## 2 Climate Change, Geoengineering and the Oceans

In its Fifth Assessment Report, the Intergovernmental Panel on Climate Change concluded that 'the globally averaged combined land and ocean surface temperature data as calculated by a linear trend, show a warming of 0.85 [0.65 to 1.06] °C, over the period 1880–2012'.[5] In 2015 the World Meteorological Association confirmed atmospheric concentrations of carbon dioxide reached 400 parts per million (ppm) on average, across the year as a whole,[6] an increase from approximately 280 ppm at the onset of the Industrial Revolution.[7] Approximately 50 per cent of fossil fuel emissions emitted from 1750 onwards have been absorbed by the oceans,[8] which contain almost 50 times more carbon dioxide than the atmosphere[9] and which continue to draw down about 30 per cent of all anthropogenic carbon dioxide annually.[10] Anthropogenic climate change has caused the upper ocean to warm since the 1870s,[11] and the

5 Dennis L Hartmann et al, 'Observations: Atmosphere and Surface' in TF Stocker, D Quin, GK Platter et al, *Climate Change 2013: The Physical Science Basis. Contribution of Working Group I to the Fifth Assessment Report of the Intergovernmental Panel on Climate Change* (Cambridge University Press 2013) 161.

6 World Meteorological Organization, Greenhouse Gas Bulletin (No 12, 24 October 2016) <https://ane4bf-datap1.s3-eu-west-1.amazonaws.com/wmocms/s3fs-public/GHG_Bulletin_12_EN_web_JN161640.pdf?aZaKZhdpDfJdmHvtbSvLwbj6zb_PWwdz> accessed 24 May 2018.

7 M R Raupach et al, 'Global and regional drivers of accelerating $CO_2$ emission' (2007) 104 *Proceedings of the National Academy of Sciences* 10288.

8 United Nations Division for Ocean Affairs and the Law of the Sea, 'The First Global Integrated Marine Assessment (World Ocean Assessment I) by the Group of Experts of the Regular Process under the auspices of the United Nations General Assembly and its Regular Process for Global Reporting and Assessment of the State of the Marine Environment, including Socioeconomic Aspects' (2016) <www.un.org/Depts/los/global_reporting/WOA_RegProcess.htm> accessed 24 May 2018 (WOA I) 17.

9 M Rhein et al, 'Observations: Ocean' in TF Stocker et al (ed), Climate Change 2013: The Physical Science Basis. Contribution of Working Group I to the Fifth Assessment Report of the Intergovernmental Panel on Climate Change (Cambridge University Press 2013) (IPCC Fifth Assessment Report) 260.

10 Ibid.

11 John M Lyman et al, 'Robust Warming of the Global Upper Ocean' (2010) 465 *Nature* 334.

increase in ocean temperatures has reduced the coverage of Arctic Ocean ice[12] and contributed to a global mean sea-level rise of 0.19 m between 1901 and 2010.[13] Climate change has also led to changes in salinity levels, water circulation and ocean biochemistry, notably, decreasing the pH level of the oceans:[14] ocean acidity has increased by 30 per cent since the beginning of the Industrial Revolution.[15] The trio of impacts—temperature increase, sea-level rise and ocean acidification—is collectively affecting vulnerable ecosystems and species including corals and other calcifying organisms (through bleaching[16] and interfering with calcification[17]) and causing populations of fish and phytoplankton to vary in their distribution, abundance and location.[18]

Despite the growing consensus on the existence and, increasingly, the extent of the risks of climate change to the hydrosphere, atmosphere and biosphere, states, corporations and individuals remain reluctant to take concerted and meaningful action to reduce greenhouse gas emissions.[19] Global greenhouse gas emissions actually increased by 40 per cent between 1992 and 2008,[20] notwithstanding the agreement of 195 states to stabilise 'greenhouse gas concentrations in the atmosphere at a level that would prevent dangerous anthropogenic interference with the climate system' under Article 2 of the 1992 UNFCCC, and the targets accepted by most developed states under the 1997 Kyoto Protocol to the Framework Convention on Climate Change.[21] It is

12 J Cohen et al, 'Warm Arctic, cold continents: A Common Pattern Related to Arctic sea Ice Melt, Snow Advance and Extreme Winter Weather' (2013) 26 *Oceanography* 150. See also VN Livina and TM Lenton, 'A Recent Tipping Point in the Arctic Sea-ice Cover: Abrupt and Persistent Increase in the Seasonal Cycle since 2007' (2013) 7 *The Cryosphere* 275.

13 IPCC Fifth Assessment Report (n 9) 5.

14 Ibid 3–5.

15 Secretariat of the Convention on Biological Diversity (CBD), *Scientific Synthesis of the Impacts of Ocean Acidification on Marine Biodiversity* (Technical Series No. 46) (Secretariat of the CBD 2009) 9.

16 Barbara E Brown, 'Coral Bleaching: Causes and Consequences' (1997) 16 *Coral Reefs* (Suppl) S129.

17 Scott C Doney et al, 'Ocean Acidification: A Critical Emerging Problem for the Oceans' (2009) 22 *Oceanography* 16, 18.

18 See Séverine Alvain et al, 'Rapid Climatic Driven Shifts of Diatoms at High Latitudes' (2013) 132 *Remote Sensing of Environment* 195; MJ Salinger, 'A Brief Introduction to the Issue of Climate and Marine Fisheries' (2013) 119 *Climatic Change* 23.

19 See Mike Hulme, *Why We Disagree about Climate Change: Understanding Controversy, Inaction and Opportunity* (Cambridge University Press 2009).

20 Ian Allison et al, *The Copenhagen Diagnosis* (The University of New South Wales Climate Change Research Centre 2009), 7.

21 Kyoto Protocol to the United Nations Framework Convention on Climate Change (adopted 11 December 1997, entered into force 18 June 2001) 2303 UNTS 148.

far from certain that the nationally determined contributions agreed by each of the 143 parties to the 2015 Paris Agreement[22] will achieve the objective of the Agreement to limit global temperature increase to 'well below 2 °C above pre-industrial levels' with the aim of limiting the increase to 1.5 °C.[23] Even if this limit is achieved, the impacts of a 2 °C rise are far from understood.[24] It is within this context of uncertainty and inaction that the idea of geoengineering as a solution to, or at the very least, a means of mitigating, climate change has begun to gain legitimacy.

Geoengineering is the deliberate large-scale manipulation of environmental systems for the purpose of mitigating climate change.[25] Although the term 'geoengineering' was first used in 1977,[26] ideas associated with the manipulation of weather and climate go back almost 100 years.[27] It was the work of the Nobel laureate Paul J Crutzen, published in 2006, advocating the injection of sulphur into the stratosphere in order to reflect sunlight and thus cool the planet,[28] which began the modern debate on geoengineering. Geoengineering is typically divided into two categories: technologies and techniques designed to manage solar radiation, and technologies and techniques designed to remove carbon dioxide from the atmosphere. Solar radiation management (SRM) seeks to offset temperature increases caused by climate change through reflecting or deflecting sunlight from the Earth. Proposed technologies and methods include urban albedo enhancement,[29] whitening marine clouds[30]

---

22 The Paris Agreement (adopted 12 December 2015, entered into force 4 November 2016) 55 ILM 743.

23 Ibid art 2(1)(a).

24 Adrian E Raftery et al, 'Less than 2° Warming by 2100 Unlikely' (2017) 7 *Nature Climate Change* 637.

25 David W Keith, 'Geoengineering the Climate: History and Prospect' (2000) 25 *Annual Review of Energy and the Environment* 245, 247.

26 Cesare Marchetti, 'On Geoengineering and the $CO_2$ Problem' (1977) 1 *Climatic Change* 59.

27 See James Rodger Fleming, *Fixing the Sky: The Checkered History of Weather and Climate Control* (Columbia University Press 2010); David W Keith, 'Geoengineering the Climate: History and Prospect' (2000) 25 *Annual Review of Energy and the Environment* 245.

28 P J Crutzen, 'Albedo Enhancement by Stratospheric Sulfur Injections: A Contribution to Resolve a Policy Dilemma?' (2006) 77 *Climatic Change* 211.

29 Hashem Akbari, Surabi Menon and Arthur Rosenfeld, 'Global Cooling: Increasing World-Wide Urban Albedos to Offset $CO_2$' (2009) 94 *Climatic Change* 275, 277; Robert M Hamwey, 'Active Amplification of the Terrestrial Albedo to Mitigate Climate Change: An Exploratory Study' (2007) 12 *Mitigation and Adaptation Strategies for Global Change* 419.

30 John Latham et al, 'Global Temperature Stabilization via Controlled Albedo Enhancement of Low-Level Maritime Clouds' (2008) 366 *Philosophical Transactions of the Royal Society A* 3969; Stephen Salter, Graham Sortino and John Latham, 'Sea-going Hardware for the Cloud Albedo Method of Reversing Global Warming' (2008) 366 *Philosophical Transactions of the Royal Society A* 3989.

or the stratosphere,[31] or even the deflection of solar radiation through the strategic placement of mirrors or 'sunshades' in space.[32] Crucially, SRM does not address the levels of carbon dioxide in the atmosphere, or indeed the oceans, and does not seek to mitigate the impacts of climate change other than temperature rise. By contrast, carbon dioxide removal is focussed on removing carbon dioxide from the atmosphere and seeks to enhance or add to processes that naturally perform this function. Examples include afforestation or reforestation,[33] soil-carbon sequestration,[34] the use of carbon dioxide absorbing algae on building surfaces[35] and even the capture and storage of atmospheric carbon dioxide by artificial 'trees'.[36]

---

31 Crutzen (n 28); Robert E Dickinson, 'Climate Engineering: A Review of Aerosol Approaches to Changing the Global Energy Balance' (1996) 33 *Climatic Change* 279; Yu A Izrael, 'Field Experiment on Studying Solar Radiation Passing through Aerosol Layers' (2009) 34 *Russian Meterology and Hydrology* 265; Philip J Rasch et al, 'An Overview of Geoengineering of Climate Using Stratospheric Sulphate Aerosols' (2008) 366 *Philosophical Transactions of the Royal Society A* 4007; Alan Robock, Luke Oman and Georgiy L Stenchikov, 'Regional Climate Responses to Geoengineering with Tropical and Arctic $SO_2$ Injections' (2008) 113 *Journal of Geophysical Research* D16101; AF Tuck et al, 'On Geoengineering with Sulphate Aerosols in the Tropical Upper Troposphere and Lower Stratosphere' (2008) 90 *Climatic Change* 315.

32 See Roger Angel, 'Feasibility of Cooling the Earth with a Cloud of Small Spacecraft near the Inner Legrange Point (L1)' (2006) 103(46) *Proceedings of the National Academy of Sciences* 17184; James T Early, 'Space-based Solar Shield to Offset Greenhouse Effect' (1989) 42 *Journal of the British Planetary Society* 567; Takanobu Kosugi, 'Role of Sunshades in Space as a Climate Control Option' (2010) 67 *Acta Astronautica* 241; D J Lunt et al, '"Sunshade World": A Fully Coupled CGM Evaluation of the Climatic Impacts of Geoengineering' (2008) 35 *Geophysical Research Letters* L12710; CR McInnes, 'Space-based Geoengineering: Challenges and Requirements' (2010) 224(3) *Proceedings of the Institute of Mechanical Engineers, Part C: Journal of Mechanical Engineering Science* 571; Jerome Pearson, John Oldson and Eugene Levin, 'Earth Rings for Planetary Environment Control' (2006) 58 *Acta Astronautica* 44.

33 See Josep G Canadell and Michael R Raupach, 'Managing Forests for Climate Change Mitigation' (2008) 320 *Science* 1456; Leonard Ornstein, Igor Aleinov and David Rind, 'Irrigated Afforestation of the Sahara and Australian Outback to End Global Warming' (2009) 97 *Climatic Change* 409; Kenneth R Richards and Carrie Stokes, 'A Review of Forest Carbon Sequestration Cost Strategies: A Dozen Years of Research' (2004) 63 *Climatic Change* 1; Brent Sohngen, 'Forestry Carbon Sequestration' in Bjørn Lomborg (ed), *Smart Solutions to Climate Change: Comparing Costs and Benefits* (Cambridge University Press 2010) 114; Massimo Tavoni, Brent Sohngen and Valentina Bosetti, 'Forestry and the Carbon Market Response to Stabilize Climate' (2007) 35 *Energy Policy* 5346.

34 See Raj K Shrestha and Rattan Lal, 'Ecosystem Carbon Budgeting and Soil Carbon Sequestration in Reclaimed Mine Soil' (2006) 32 *Environment International* 781.

35 Eduardo Jacob-Lobes, Carols Henrique Gimenes Scoparo and Telma Teixeira Franco, 'Rates of $CO_2$ Removal by *Aphanothece microscopic Nägeli* in Tubular Photobioreactors' (2008) 47 *Chemical Engineering and Processing* 1365.

36 KS Lackner, 'Capture of Carbon Dioxide from Ambient Air' (2009) 176 *The European Physical Journal, Special Topics* 93.

As the oceans comprise the largest natural sink for carbon dioxide, it is unsurprising that a significant amount of research to date has focussed on expanding this sink and enhancing the natural processes that 'pump' carbon dioxide from the surface to the depths of the ocean, where it is (hoped to be) sequestered for hundreds of years.[37] The ocean carbon cycle is underpinned by the solubility and the biological pumps, and it is the latter that has been of greatest interest to marine geoengineering advocates to date. Certain areas of the oceans, including the Southern Ocean and the Equatorial Pacific, are known to be relatively unproductive in biological terms, owing to the low availability of nutrients such as iron, nitrogen or phosphate.[38] In a now-seminal article published in 1990 John Martin hypothesised that iron dust[39] artificially added to unproductive ocean ecosystems would stimulate biological productivity, creating algal blooms, which would draw down $CO_2$ from the ocean's surface into its depths where it would be sequestered for hundreds if not thousands of years.[40] His essential hypothesis—that plankton biomass responds to iron artificially added, so that surface levels of carbon dioxide are temporarily reduced—has been proven correct.[41] Whether this technique provides an effective response to climate change, however, is far from proven. There have been 13 official ocean fertilisation experiments to date, but results are mixed as to how much and for how long carbon dioxide is sequestered[42] and how large

37 Anand Gnanadesikan and Irina Marinov, 'Export Is Not Enough: Nutrient Cycling and Carbon Sequestration' (2008) 364 *Marine Ecology Progress Series* 289, 289.

38 HJW de Baar and PW Boyd, 'The Role of Iron in Plankton Ecology and Carbon Dioxide Transfer of the Global Oceans' in Roger B Hansen et al (eds) *The Changing Ocean Carbon Cycle: A Midterm Synthesis of the Joint Global Ocean Flux Study* (Cambridge University Press 2000) 61, 107; Robert A Duce and Neil W Tindale, 'Atmospheric Transport of Iron and Its Deposition in the Ocean' (1991) 36 *Limnology & Oceanography* 1715.

39 Iron is not the only nutrient that may be artificially added to stimulate the biological pump. Other options include volcanic ash, phosphate and urea. See Svend Duggen et al, 'Subduction Zone Volcanic Ash Can Fertilize the Surface Ocean and Stimulate Phytoplankton Growth: Evidence from Biogeochemical Experiments and Satellite Data' (2007) 34 *Geophysical Research Letters* L01612; Julia Mayo-Ramsay, 'Environmental, Legal and Social Implications of Ocean Urea Fertilization: Sulu Sea example' (2010) 34 *Marine Policy* 831.

40 John H Martin, 'Glacial-Interglacial $CO_2$ Change: The Iron Hypothesis' (1990) 5 *Paleoceanography* 1. See also Nicolas Cassar et al, 'The Southern Ocean Biological Response to Aeolian Iron Deposition' (2007) 317 *Science* 1067; PW Boyd et al, 'Mesoscale Iron Enrichment Experiments 1993–2005: Synthesis and Future Directions' (2007) 315 *Science* 612.

41 Philip Williamson et al, 'Ocean Fertilization for Geoengineering: A Review of Effectiveness, Environmental Impacts and Emerging Governance' (2012) 90 *Process Safety and Environmental Protection* 475, 477.

42 Ibid. See also Secretariat of the CBD, *Scientific Synthesis of the Impacts of Ocean Acidification on Marine Biodiversity* (n 15); S Blain, 'Effect of Natural Iron Fertilization on Carbon

an area needs to be fertilised in order for the technique to have a meaningful impact on climate change.[43] The risks of ocean fertilisation are also far from understood but may include the introduction of toxic algae,[44] disruption of the ecosystem food chain,[45] oxygen depletion, ocean acidification[46] and the release of gases such as methane and nitrous oxide.[47]

Ocean iron fertilisation is by no means the only technique of marine geoengineering under scientific investigation. Lovelock and Rapley have suggested the placement of vertical pipes in the oceans to pump deep-water nutrients to the surface, in order to create phytoplankton blooms.[48] Others have suggested artificially creating macroalgal forests (comprising kelp and other seaweeds) to store carbon dioxide and for use as a biofuel[49] or even disposing baled crop residues in the deep ocean.[50] Alternative proposals seek to enhance the solubility ocean pump in particular, by increasing ocean alkalinity through the addition of limestone powder or soda ash, thereby reducing ocean acidification and increasing the ocean's capacity to store atmospheric $CO_2$.[51] Finally, at least one SRM proposal has been developed with respect to the oceans—the creation of

---

Sequestration in the Southern Ocean' (2007) 446 *Nature* 1070; P Boyd et al, 'A Mesocale Phytoplankton Bloom in the Polar Southern Ocean Stimulated by Iron Fertilization' (2000) 407 *Nature* 695; K Caldeira and P Duffy, 'The Role of the Southern Ocean in Uptake and Storage of Anthropogenic Carbon Dioxide' (2000) 287 *Science* 620.

43 K Buesseler et al, 'The Effects of Iron Fertilization on Carbon Sequestration in the Southern Ocean' (2004) 304 *Science* 417.

44 Q Schiermeier, 'The Oresmen' (2003) 421 *Nature* 109, 110.

45 A Strong, 'Ocean Fertilization: Time to Move On' (2009) 461 *Nature* 347.

46 Williamson et al (n 41) 480–82. See also H Damon Matthews et al, 'Sensitivity of Ocean Acidification to Geoengineered Climate Stabilization' (2009) 36 *Geophysical Research Letters* L10706.

47 J Furhman and D Capone, 'Possible Biogeochemical Consequences of Ocean Fertilization' (1991) 36 *Limnology and Oceanography* 1951; M Lawrence, 'Side-effects of Ocean Iron Fertilization' (2002) 297 *Science* 1993.

48 See J Lovelock and C Rapley, 'Ocean Pipes Could Help the Earth to Cure Itself' (2007) 449 *Nature* 403; Andrew Yool et al, 'Low Efficiency of Nutrient Translocation for Enhancing Oceanic Uptake of Carbon Dioxide' (2009) 114 *Journal of Geophysical Research* 1.

49 Antione de Ramon N'Yeurt et al, 'Negative carbon via Ocean Afforestation' (2012) 90 *Process Safety and Environmental Protection* 467.

50 S Strand and G Benford, 'Ocean Sequestration of Crop Residue Carbon: Recycling Fossil Fuel Carbon Back to Deep Sediments' (2009) 43 *Environmental Science & Technology* 1001.

51 LDD Harvey, 'Mitigating the Atmospheric $CO_2$ Increase and Ocean Acidification by Adding Limestone Powder to Upwelling Regions' (2008) 113 *Journal of Geophysical Research* C04028; Haroon S Kheshgi, 'Sequestering Atmospheric Carbon Dioxide by Increasing Ocean Alkalinity' (1995) 20 *Energy* 915. See also Jennie C Stephens and David W Keith, 'Assessing Geochemical Carbon Management' (2008) 90 *Climatic Change* 217, 228.

reflective microbubbles in the sea to enhance the oceans' albedo[52]—and other SRM strategies, such as marine cloud brightening, rely on vessels or a form of ocean craft for the release of substances such as salt.[53]

## 3 Geoengineering and UNCLOS

UNCLOS establishes what has been described as a 'constitution for the oceans'[54] and strives to attain a balance between the various interests of multiple actors including states, scientists, entrepreneurs and the international community as a collective, while also seeking to protect global interests and values in marine environmental protection through the development of relatively detailed rules in this area. This essential framework is augmented by numerous global and regional instruments, which develop complex regulatory regimes in respect of particular issues, such as environmental protection or the safety of shipping, and customary or other principles of international law that constrain or guide state activity in order to achieve particular aims, such as environmental protection, participation in decision-making, or holding states and other actors responsible for harm. This modern law of the sea provides the essential regulatory framework for marine geoengineering and the foundation upon which more detailed rules, such as those being developed under the auspices of the London Protocol, are built.

UNCLOS establishes a largely permissive regime for maritime activities with the ultimate aim of balancing the rights of coastal and other states according to location (zone) and activity (function). Within their territorial seas and exclusive economic zones (EEZ), coastal states have exclusive sovereign rights for the purposes of exploring, exploiting and conserving living and non-living natural resources,[55] as well as exclusive jurisdiction over marine scientific research taking place therein.[56] Any geoengineering activities carried out by the coastal state must have due regard to the rights and freedoms of other states, including navigation, and must also be consistent with their obligations under UNCLOS in relation to marine scientific research and environmental

52 Russell Seitz, 'Bright Water: Hydrosols, Water Conservation and Climate Change' (2011) 105 *Climatic Change* 365.

53 Latham et al (n 30); Salter et al (n 30).

54 Tommy TB Koh, 'A Constitution for the Oceans' (Statement by President Koh at the Final Session of the Conference at Montego Bay, 6 and 11 December 1982, reprinted in United Nations, *The Law of the Sea: United Nations Convention on the Law of the Sea* 1983) E.83.V5.

55 UNCLOS (n 2) art 56(1)(a).

56 UNCLOS (n 2) arts 56(b)(ii), 245 and 246.

protection,[57] including their obligation to minimise transboundary pollution or harm.[58]

On the high seas all states are entitled to exercise high seas freedoms, including the freedom of marine scientific research, as codified in Article 87 of UNCLOS. Moreover, it is generally accepted that any activity that is not prohibited and does not compromise the reservation of the high seas for peaceful purposes[59] is classed as a high seas freedom,[60] and consequently, marine geoengineering for any peaceful purpose arguably constitutes a freedom of the high seas. However, high seas freedoms are far from absolute, as users must have due regard to the interests of other states.[61] Geoengineering activities will be limited if they reduce fish stocks, hinder the freedom to fish or involve the placement of objects in the oceans, and interfere with the freedom of navigation. Moreover, states exercising high seas freedoms must also comply with their obligations under UNCLOS with respect to the protection of the environment, the management of activities in the Area and the regulation of scientific research.[62]

### 3.1 *Geoengineering, UNCLOS and the Protection of the Marine Environment*

The protection and preservation of the marine environment is a fundamental aim and objective of UNCLOS,[63] and Article 192 creates a general obligation to this effect in Part XII of the Convention.[64] The sovereign right of coastal states to exploit their natural resources, which arguably include the capacity of the ocean to sequester carbon dioxide, is acknowledged under Article 193 of UNCLOS but is limited by the coastal state's duty to protect and preserve the marine environment, including, under Article 194(5), rare and fragile

57 UNCLOS (n 2) arts 56(2), 192, 193 and 194.

58 UNCLOS (n 2) art 194(2). This environmental principle is now a principle of customary international law as confirmed by the ICJ. See *Legality of the Threat or Use of Nuclear Weapons* (Advisory Opinion of 8 July 1996) ICJ Reports 1996, 226, para 29; *Gabčikovo-Nagymaros Project (Hungary/Slovakia)* (Judgment of 25 September 1997) ICJ Reports 1997, 7, para 140.

59 UNCLOS (n 2) art 88.

60 Article 87 of UNCLOS uses the term 'inter alia' before listing the six high seas freedoms indicating that those six cannot be considered exclusive. See RR Churchill and AV Lowe, *The Law of the Sea* (3rd edn, Manchester University Press 1999) 206.

61 UNCLOS (n 2) art 87(2).

62 Ibid.

63 Ibid Preamble.

64 Article 192 of UNCLOS stipulates that 'States have the obligation to protect and preserve the marine environment'.

ecosystems. In implementing their obligations under Part XII of UNCLOS, states must apply a precautionary approach to environmental protection and explicitly consider the myriad of uncertainties and knowledge gaps associated with the impact of manipulating the ocean carbon cycle. Although omitted as an explicit principle within UNCLOS itself, the precautionary approach has evolved within both treaty law and custom to provide a fundamental component of the decision-making process in the context of activities likely to have a significant detrimental impact on the marine environment, and this has been confirmed by the International Court of Justice (ICJ)[65] and the Seabed Disputes Chamber of the International Tribunal for the Law of the Sea (ITLOS).[66]

The process of applying a precautionary approach to marine geoengineering is likely, and arguably, must be demonstrated through an environmental impact assessment (EIA). The process of an EIA has developed well beyond the basic requirements set out in Article 206 of UNCLOS and has been endorsed as a principle of general international law to be applied in a transboundary and a commons context by the ICJ[67] and the ITLOS Seabed Disputes Chamber respectively.[68] There is nevertheless significant scope for differing interpretations of what constitutes both precaution and an EIA in this context. It is notable that the precautionary approach does not, in the absence of an explicit direction otherwise, reverse the burden of proof under international law.[69] Moreover, geoengineering activities are unusual in that an assessment of the risk of harm is not straightforward given that *not* geoengineering may also lead to a risk of harm.[70] However, of particular relevance to geoengineering in terms of function and location are the requirements relating to both precaution and

65 See for example the decision of the International Court of Justice in the *Pulp Mills on the River Uruguay (Argentina v Uruguay)* ICJ Reports 2010, 14, where the Court concluded that the precautionary approach may be relevant in the interpretation of the statute in dispute between Argentina and Uruguay relying implicitly on Article 31(3)(c) of the 1969 Vienna Convention on the Law of Treaties (para 164).

66 *Responsibilities and Obligations of States Sponsoring Persons and Entities with Respect to Activities in the Area* (Advisory Opinion of 1 February 2011) ITLOS Reports 2011, 10 (*Advisory Opinion No 17*), para 135.

67 *Pulp Mills on the River Uruguay* (n 65) para 204.

68 *Advisory Opinion No 17*, paras 145 and 148.

69 *Pulp Mills on the River Uruguay* (n 65) para 164.

70 For a discussion of geoengineering and the precautionary principle more generally, see Kevin Elliot, 'Geoengineering and the Precautionary Principle' (2010) 24 *International Journal of Applied Philosophy* 237; Elizabeth Tedsen and Gesa Homann, 'Implementing the Precautionary Principle for Climate Engineering' (2013) 7 *Carbon & Climate Law Review* 90.

EIA set out under the 1996 London Protocol[71] and the 1991 Environmental Protocol to the 1959 Antarctic Treaty[72] respectively, and both these instruments take a conservative approach to environmental risk.

To the extent that marine geoengineering can be categorised as 'pollution of the marine environment' as defined by Article 1(4) of UNCLOS, Article 194 of the Convention imposes an obligation on states to take 'all measures to prevent, reduce and control pollution of the marine environment from *any* source'. More specifically—and of particular relevance to geoengineering—Article 196 of UNCLOS requires parties to 'take all measures necessary to prevent, reduce and control pollution of the marine environment resulting from the use of technologies under their jurisdiction or control'. Pollution is defined under UNCLOS as

> the introduction by man, directly or indirectly, of substances or energy into the marine environment, including estuaries, which results or is likely to result in such deleterious effects as harm to living resources and marine life, hazards to human health, hindrance to marine activities, including fishing and other legitimate uses of the sea, impairment of quality for use of sea water and reduction of amenities.[73]

Techniques where matter is 'introduced' into the marine environment such as fertilisation or weathering may be classified as 'pollution' provided the harm threshold is met, but technologies supporting the transport of nutrients through ocean pipes or macroalgal afforestation and marine cloud brightening are unlikely to be so classified.

Materials introduced into the marine environment through 'dumping' are subject to detailed regulation under Article 210 of UNCLOS, which requires coastal and flag states to adopt rules no less effective than 'the global rules and standards' relating to dumping.[74] Dumping is defined under UNCLOS as the 'deliberate disposal of wastes or other matter from vessels, aircraft, platforms

---

71 See The Assessment Framework for Scientific Research Involving Ocean Fertilization (adopted 14 October 2010) Resolution LC-LP.2. The Assessment Framework will be formally incorporated into the Protocol in Annex 5 once the 2013 amendments enter into force.

72 Protocol on Environmental Protection to the Antarctic Treaty (adopted 4 October 1991, entered into force 14 January 1998) 30 ILM 1455. In particular, see Articles 3, 8 and Annex I of the Protocol.

73 UNCLOS (n 2) art 1(4).

74 Ibid art 210(1) and (6).

or other man-made structures at sea',[75] excluding the disposal of waste incidental to shipping or offshore operations[76] and the placement of matter for a purpose other than the mere disposal thereof, provided that such placement is not contrary to the aims of the Convention.[77] The global rules and standards as referred to in UNCLOS currently refer to those established under the 1972 Convention on the Prevention of Marine Pollution by Dumping of Wastes and Other Matter (London Convention)[78] rather than the 1996 Protocol to the Convention on the Prevention of Marine Pollution by Dumping of Wastes and Other Matter (London Protocol), which imposes significantly more stringent standards on dumping and, as will be discussed below, has adopted specific rules relating to ocean fertilisation. On the basis of Article 210 of UNCLOS and the London Convention, it is debatable whether marine geoengineering can be classified as 'dumping'. Although both fertilisation and weathering techniques involve the introduction of matter into the sea, it is not likely that the matter is 'disposed of', given that its introduction is intended to serve a purpose: to create an algal bloom or to increase ocean alkalinity. Moreover, both instruments exclude 'placement of matter for a purpose other than the mere disposal thereof, provided that such placement is not contrary to the aims' of UNCLOS and the London Protocol. The question therefore is whether placement for geoengineering purposes can be considered contrary to the aims of UNCLOS. In order to facilitate the development of a mandate to regulate marine geoengineering under the London Protocol (discussed below), the parties to the London Convention and London Protocol adopted two non-binding resolutions establishing a moratorium on ocean fertilisation for non-scientific purposes[79] and setting out a risk assessment framework for scientific ocean fertilisation.[80] Using these two instruments as an aid to interpretation, non-scientific ocean fertilisation can arguably be considered contrary to the aims and objectives of UNCLOS and thus should be classified as dumping. Whether a non-binding instrument can be used to interpret a binding convention is open to question, and the resolutions do not apply to other geoengineering technologies that do not constitute ocean fertilisation but nevertheless involve introducing matter into the ocean.

Even if ocean fertilisation using iron or other nutrients were classified as 'dumping' for the purposes of Article 210 of UNCLOS, it is not conclusive that

75 Ibid art 1(5)(a).
76 Ibid art 1(5)(b)(i).
77 Ibid art 1(5)(b)(ii).
78 Churchill and Lowe (n 60) 369.
79 The Regulation of Ocean Fertilization (adopted 31 October 2008) IMO Resolution LC-LP.1.
80 The Assessment Framework for Scientific Research Involving Ocean Fertilization (n 71).

the activity is automatically prohibited under the regime. The London Convention, which provides the standards referred to in Article 210, prohibits the dumping of matter listed in Annex I and permits Annex II matter to be dumped subject to a special permit. The matter associated with ocean fertilisation or weathering techniques—such as iron dust, phosphate, urea or limestone powder—are not currently listed in either Annex I or Annex II and, as such, may potentially be dumped subject to a general permit unless it is deemed harmful owing to the quantities in which it is dumped, in which case it is subject to authorisation by a special permit under Annex II of the London Convention.

The 48 state parties to the London Protocol are subject to a less flexible regime than that permitted under the London Convention and UNCLOS, notwithstanding that the amendments designed to regulate geoengineering more generally (discussed below) have yet to enter into force. Under the reverse listing procedure adopted by the London Protocol, only matter listed in Annex I may be dumped subject to a special permit and relatively stringent conditions.[81] Ocean fertilisers and weathering tools are not explicitly listed in Annex I, although provision is made for the issue of permits in respect of 'inert, inorganic, geological material'. Ocean fertilisers and weathering material are by their nature not inert, as their intended purpose is to stimulate a reaction in the ocean environment, such as the creation of a phytoplankton bloom or the alteration of the chemistry of seawater.

### 3.2 *Marine Geoengineering, UNCLOS and the Regulation of Marine Scientific Research*

Geoengineering activities to date largely comprises research rather than actual climate change mitigation.[82] Marine scientific research is categorised as a freedom of the high seas[83] but must nevertheless be carried out in accordance

81 London Protocol (n 4) art 4.1.1.

82 An exception to this generalisation occurred in July 2012 when the Haida Salmon Restoration Corporation deliberately dumped 100 metric tonnes of iron sulphate into the ocean off the coast of western Canada. The scheme, which dumped five times more iron into the ocean than any of the previous 13 iron fertilization experiments carried out to date, was ostensibly designed to both increase local salmon populations and to sequester carbon dioxide. See Jeff Tollefson, 'Ocean-Fertilization Project off Canada Sparks Furore' (2012) 490 *Nature* 458. The scheme was not authorised by the Canadian government. See the Statement by the Delegation of Canada on the Issue of the Ocean Fertilization Incident off the West Coast of Canada in July 2012 reproduced in the *Report of the Thirty-Fourth Consultative Meeting and the Seventh Meeting of the Contracting Parties* (LC 34/15, 23 November 2012) Annex 3.

83 UNCLOS (n 2) arts 87(1)(f), 239.

with Part XIII of UNCLOS. In particular, under Article 240 of UNCLOS, states must comply with principles promoting openness, transparency and collaboration in research. Moreover, research, including research relating to geoengineering, must be for the benefit of humankind where it takes place in the Area[84] and must comply with Part XII of UNCLOS on environmental protection wherever it takes place. Within maritime zones under the jurisdiction of coastal states, as noted above, those coastal states have exclusive jurisdiction of manage marine scientific research in their waters. While coastal states are normally expected to consent to research in their EEZ when that research is likely to benefit all humankind,[85] consent may be legitimately withheld if research involves the introduction of harmful substances, perhaps fertilisers or weathering-based substances, into the marine environment.[86]

## 4 Geoengineering and the 1996 London Protocol to the 1972 London Convention

In October 2013, at the eighth Meeting of the Parties, the states parties to the London Protocol adopted wide-ranging amendments to the Protocol, in order to provide a regulatory basis for the management and control of marine geoengineering.[87] For the purposes of the London Protocol, geoengineering is defined as 'a deliberate intervention in the marine environment to manipulate natural processes, including to counteract anthropogenic climate change and or its impacts, and that has the potential to result in deleterious effects, especially where those effects may be widespread, long lasting or severe'.[88] Notably, this definition is broadly conceived and does not restrict the potential application of the regulatory mandate of the London Protocol to activities that involve the abandonment, disposal or placement of matter in the marine environment. Furthermore, the definition relies on a low threshold of harm, and while it refers to climate change mitigation, this purpose is not an integral part of the definition of geoengineering. Developing a mandate to regulate

84 Ibid art 140. The Area is defined as 'the seabed and ocean floor and subsoil thereof, beyond the limits of national jurisdiction' (UNCLOS (n 2) art 1(1)(1)) and is subject to the regime established under Part XI of UNCLOS and the Part XI Implementation Agreement.

85 UNCLOS (n 2) art 246(3).

86 Ibid art 246(5)(b).

87 The Amendment to the London Protocol to Regulate the Placement of Matter for Ocean Fertilization and other Marine Geoengineering Activities (adopted 18 October 2013) Resolution LP.4(8) (amendment not yet in force) (2013 Amendment to the London Protocol).

88 Ibid art 1(5)*bis*.

geoengineering for any purpose precludes would-be climate manipulators evading the regulatory regime by ostensibly relying on an alternative and ostensibly permitted purpose, such as enhancing fish stocks for food production purposes.

Article 6*bis* of the London Protocol provides the regulatory framework for geoengineering and precludes 'the placement of matter into the sea from vessels, aircraft, platforms, or other man-made structures at sea for marine geo-engineering activities listed in annex 4, unless the listing provides that the activity or the sub-category of an activity may be authorised under a permit'.[89] Thus, notwithstanding the expansive definition of geoengineering set out in Article 1(5)*bis*, the regulatory regime is currently confined to activities involving the 'placement of matter into the sea', closely reflecting the original mandate of the London Protocol. Activities that do not involve the placement of matter into the sea, such as the creation of microbubbles to increase ocean albedo,[90] and activities that simply use the oceans as a location from which to undertake geoengineering, such as marine cloud brightening, are excluded. Similarly, the regime will not regulate matter released incidentally as part of normal vessel operations[91] and matter released from land-based facilities. Article 6*bis* requires parties to adopt administrative or legislative measures for the issuance of permits and stipulates that permits may only be issued where the activity has undergone an assessment that 'has determined that pollution of the marine environment from the proposed activity is, as far as practicable, prevented or reduced to a minimum'.[92]

Consistent with the reverse listing procedure regulating dumping under the London Protocol, Article 6*bis* establishes a presumption that marine geoengineering is generally not permitted subject to limited exceptions agreed upon by the parties and listed in Annex 4. Currently, only ocean fertilisation—defined as 'any activity undertaken by humans with the principal intention of stimulating primary productivity in the oceans'[93] undertaken for legitimate scientific research—is listed as an activity in Annex 4.[94] The definition of ocean fertilisation is sufficiently broad to include objectives such as food

89 Ibid art 6*bis*(1).

90 Seitz (n 52) 366.

91 The notion of using the release of soluble iron from vessels as part of their normal operations has recently been explored. See Akinori Ito, 'Global Modelling Study of Potentially Bioavailable Iron Input from Shipboard Aerosol Sources to the Ocean' (2013) 27 *Global Biogeochemical Cycles* 1.

92 2013 Amendment to the London Protocol (n 87) art 6*bis*(2).

93 Ibid Annex 4.1.1.

94 Ibid Annex 4.1.

production, but fertilisation for 'conventional aquaculture, or mariculture, or the creation of artificial reefs' is excluded.[95] Permitted activities listed in Annex 4 must comply with the Risk Assessment Framework as set out in Annex 5 of the London Protocol. Originally adopted in 2010 in a non-binding resolution,[96] the Risk Assessment Framework was subject to minor amendments through its incorporation into the London Protocol in 2013, in order to expand its application to any marine geoengineering activity within the remit of the Protocol. The Framework requires parties to undertake an initial assessment to determine whether the proposed activity falls within the scope of Annex 4, ie whether it constitutes ocean fertilisation for legitimate scientific research. Assuming a positive response to this question, the initial assessment must be followed by a full environmental assessment, which covers the location of the proposed experiment, the likely environmental impact including known and unknown risks, and measures to mitigate those risks including monitoring and adaptive management. The Framework adopts a highly cautious approach and requires parties to abandon projects in which adverse impacts are predicted.[97]

In responding to address the gap in the law of the sea with respect to the regulation of geoengineering, the parties to the London Protocol have clearly demonstrated its promise. Although a number of small-scale experiments have taken place, including an unauthorised unscientific incident off the coast of Canada in 2012,[98] the London Protocol has by and large created a pre-emptive regulatory regime that endorses and implements a highly precautionary approach to the management of geoengineering activities.

However, this response also demonstrates the limits of the law of the sea to fully address the risks posed by marine geoengineering. First, practically the amendment to the London Protocol has been ratified by only one party (the United Kingdom) as of April 2017 and is thus some way off entering into force.[99] When it does enter into force, it will only bind the parties to the London Protocol, which currently stand at 48.[100] As discussed above, there is no consensus that the London Protocol standards represent the international standards referred to in Article 210 of UNCLOS, and there is no obvious process for the London Protocol to replace UNCLOS in this respect. In any case, the

95 Ibid.

96 The Assessment Framework for Scientific Research Involving Ocean Fertilization (n 71).

97 Ibid.

98 See the text at n 82.

99 Two-thirds of the contracting parties must ratify the amendment for it to enter into force under Article 21(3) of the London Protocol.

100 By contrast, there are 87 states party to the 1972 London Convention.

terms of Article 210 are confined to 'dumping', and permitted geoengineering activities (currently confined to ocean fertilisation for scientific research) are classified as 'placement', which is not 'dumping' for the purpose of Article 1(5) of UNCLOS.[101] Second, despite the broad definition of geoengineering set out in Article 1(5)*bis* of the London Protocol, the regulatory regime is currently and, arguably, sensibly confined to activities involving the placement of matter in the ocean, which is consistent with the overall aims and objectives of the dumping regime. It is contestable as to whether the London Protocol offers a genuine mandate to regulate other forms of marine geoengineering, and geoengineering beyond the oceans clearly falls outside the remit of the regime. Third, the London Protocol and its Meeting of the Parties do not provide an appropriate forum within which to discuss broader ethical and moral issues associated with geoengineering, such as how geoengineering is related to emissions reductions and other mitigation measures and, fundamentally, whether geoengineering of any sort is a desirable response to climate change, taking into consideration the interests of not only current but also future generations.[102] The fact that these issues have not been debated by the parties to the London Protocol demonstrates that this is not an appropriate forum to discuss these fundamentally important questions. However, by developing a regulatory regime—even one that is pre-emptive and precautionary—the London Protocol parties have arguably pre-empted, if not bypassed these issues, and established a perception, if not a presumption that geoengineering is an appropriate response to climate change, albeit one that needs to be tightly controlled.

## 5 Responding to the Challenge: Two Proposals

As demonstrated above, the law of the sea is far from a static body of rules. It is highly adaptable and comprises core principles designed to protect the marine environment and manage the intersection of competing and sometimes conflicting interests of all states. These principles are undoubtedly applicable to marine geoengineering activities. However, by necessity, most of these

101 Where the activity is not classified as 'placement', i.e., where it falls outside of the regime developed under Article 6*bis* of the Protocol, it may be classified as dumping, in which case it is subject to the regime outline in 3.1 of this chapter, above.

102 For a discussion of these ethical issues as they relate to the developing legal regime see Karen N Scott, 'Engineering the "Mis-Anthropocene": International Law, Ethics and Geoengineering' (2015) 29 *Ocean Yearbook* 61–84.

principles are expressed at a relatively general level and are open to varied interpretation by states and institutions. In addressing the shortcomings of the overall regime, the parties to the London Protocol have developed a robust regulatory framework of application to ocean fertilisation as well as are creating a mandate for other forms of marine geoengineering. The application of this regime is, however, constrained by the normal parameters of treaty law and consent and in practice, is far from creating global standards of general application to manage all forms of marine geoengineering. Moreover, there are risks in regulating—and implicitly approving–ocean fertilisation in isolation of geoengineering more generally and outside the broader issue of climate change.

Consequently, I argue in this chapter that the approach to managing geoengineering, which relies on the adaptation of broad general principles and piecemeal regulation, is insufficient. Instead, a truly global response, which integrates geoengineering into the broader climate change regulatory arena, is required. Potential options include creating a stand-alone instrument on geoengineering,[103] addressing geoengineering within the climate change regime through a protocol or other instrument adopted under the auspices of the 1992 UNFCCC,[104] or developing mechanisms to coordinate better interested treaty and other institutions in order to develop a more coherent response to geoengineering.[105] The latter option acknowledges the potential opportunities that arise through collaboration with instruments such as the 1992 Convention on Biological Diversity (CBD),[106] which in 2008 and 2012 adopted resolutions recommending a moratorium on geoengineering activities.[107] I will conclude with a brief discussion of two possible options designed to bridge the current institutional limitations of the law of the sea: addressing geoengineering within the proposed international instrument on the conservation and sustainable

103 See for example Albert C Lin, 'Geoengineering Governance' (2009) 8 *Issues in Legal Scholarship* 1.

104 I explored this option in Scott, 'International Law in the Anthropocene' (n 1).

105 I explored this option in Karen N Scott, 'Transboundary Environmental Governance and Emerging Environmental Threats: Geo-engineering in the Marine Environment' in Simon Marsden and Robin Warner (eds), *Transboundary Environmental Governance in Inland, Coastal and Marine Areas: Asia and Australian Perspectives* (Ashgate Publishing 2012) 223–45.

106 Convention on Biological Diversity (adopted 5 June 1992, entered into force 29 December 1993) 31 ILM 818 (CBD).

107 See Biodiversity and Climate Change (2008) CBD Decision IX/16, C.4; Biodiversity and Climate Change (2010) CBD Decision X/33, para 8(w); Climate Related Geoengineering (2012) CBD Decision XI/20.

use of marine biological diversity beyond national jurisdiction, and addressing geoengineering within a designated UNGA resolution.

### 5.1 *Geoengineering and the Development of an International Legally Binding Instrument on the Conservation and Sustainable Use of Marine Biological Diversity of Areas beyond National Jurisdiction*

In 2015 the UNGA established a Preparatory Committee tasked to develop an internationally legally binding instrument (ILBI) on the conservation and sustainable use of marine biological diversity in areas beyond national jurisdiction (BBNJ).[108] This resolution followed almost a decade of work by the ad hoc working group on areas beyond national jurisdiction, and if adopted as an implementing agreement under UNCLOS, the ILBI may well be the first formal agreement adopted under the auspices of UNCLOS since 1995. The 2015 UNGA resolution determined that the negotiations will address

> the conservation and sustainable use of marine biological diversity of areas beyond national jurisdiction, in particular, together and as a whole, marine genetic resources, including questions on the sharing of benefits, measures such as area-based management tools, including marine protected areas, environmental impact assessments and capacity-building and the transfer of marine technology.[109]

This mandate is undoubtedly broad enough to address marine geoengineering and the relationship between the oceans and the climate more generally. Negotiations are at an early stage, but in the Chair's non-paper dated 28 February 2017 on elements of a draft text of the ILBI,[110] a number of states have already highlighted the link between climate and the oceans as an important element to address in the preamble to the Instrument.[111] In terms of the potential scope of the ILBI, at least one state has emphasised the importance

108 Development of an International Legally Binding Instrument under the United Nations Convention on the Law of the Sea on the Conservation and Sustainable Use of Marine Biological Diversity of Areas Beyond National Jurisdiction (adopted 19 June 2015) UNGA Resolution 69/292.

109 Ibid 2.

110 Chair of BBNJ PrepCom, 'Chair's Non-paper on Elements of a Draft Text of an International Legally-binding Instrument under the United Nations Convention on the Law of the Sea on the Conservation and Sustainable Use of Marine Biological Diversity of Areas Beyond National Jurisdiction' (28 February 2017) <www.un.org/depts/los/biodiversity/prepcom_files/Chair_non_paper.pdf> accessed 24 May 2018.

111 Ibid 4.

of including activities not adequately addressed by existing international conventions in addition to activities that threaten ecosystems and species beyond national jurisdiction.[112] Unsurprisingly, a range of environmental principles including precaution and ecosystem integrity has been endorsed for likely inclusion,[113] and significant emphasis has been placed on international cooperation between institutions with interested mandates.[114] Relatively detailed suggestions relating to the designation of protected areas and the implementation of EIA have been included in the Chair's non-paper,[115] and these would have undoubted implications for proposed marine geoengineering activities located in areas beyond national jurisdiction. However, thus far neither geoengineering nor climate change and the oceans more generally have emerged as a priority issue for the Instrument.

This notwithstanding, I suggest in this chapter that both climate change generally and geoengineering in particular be addressed in the proposed ILBI. It is acknowledged that climate change arguably represents the greatest long-term threat to the oceans, but it is the archetypical issue that cuts across and falls between climate and ocean regimes. This feature has led to the impacts of climate change being inadequately addressed to date, as demonstrated by 1992 UNFCCC's failure to specifically regulate carbon dioxide and ocean acidification and the absence of any real discussion on the implications of a 2°C temperature rise for the oceans. The proposed ILBI provides a genuine opportunity to develop global principles of universal application to address climate change and, specifically, geoengineering as part of the law of the sea.

### 5.2 *Geoengineering and a UNGA Resolution on Oceans, Climate and Geoengineering*

An alternative mechanism is the adoption of a UNGA resolution on oceans, climate change and geoengineering. Climate change and ocean acidification are already an important focus of the annual UNGA resolution on oceans and the law of the sea. For example, in the 2016 resolution,[116] the UNGA highlighted recent research on the impact of climate change on the oceans, commended a number of collaborative initiatives and meetings, and encouraged states to cooperate in relation to climate-related research and mitigation efforts with particular emphasis on the oceans.[117] The issue of climate change and the oceans was identified as a central theme in the 2012 UNGA Resolution The Future We

112 Ibid 12.
113 Ibid 16–19.
114 Ibid 19–21.
115 Ibid 33–77.
116 Oceans and the Law of the Sea (adopted 23 December 2016) UNGA Resolution 71/257.
117 Ibid 185–95.

Want,[118] which also expressed concern about the potential environmental impacts of ocean fertilisation.[119] Ocean acidification is explicitly referred to in Sustainable Development Goal 14, which was adopted by the UNGA in 2015, as part of the 2030 Agenda for Sustainable Development.[120] The central theme of the 2017 UN Secretary General's Report on Oceans and the Law of the Sea is climate and the oceans,[121] and the focus of the 18th United Nations Open-ended Informal Consultative Process on Oceans and the Law of the Sea in 2017, is 'The Effects of Climate Change on the Oceans'.

I suggest that building on this increasingly extensive global work programme, the UNGA should adopt a specific resolution on oceans and the climate and, within that resolution, develop meaningful targets (such as carbon dioxide limits) and initiatives. It would be appropriate within such a resolution for the UNGA to explicitly address marine and other forms of geoengineering and to set out principles and guidance on whether and in what circumstances such technologies should be developed taking into consideration moral and ethical issues and their relationship to emissions reductions and climate change mitigation more generally. These guidelines would build on norms developed in other forums, including the London Protocol and the CBD. UNGA resolutions are non-binding, but they do have the capacity to influence and develop norms and ultimately to change behaviour. An example of the potential influence of UNGA resolutions is illustrated by the adoption of resolutions in the late 1980s and early 1990s[122] calling for a moratorium on large-scale pelagic drift net fishing, which led to the adoption of binding measures prohibiting or restricting drift net fishing in the South Pacific,[123] the Southern Ocean[124] and European waters.[125] It is equally possible that a UNGA resolution on oceans and climate

118 The Future We Want (adopted 27 July 2012) UNGA Resolution 66/288, 165–66.

119 Ibid 167.

120 Transforming Our World: The 2030 Agenda for Sustainable Development (adopted 25 September 2015) UNGA Resolution 70/1.

121 United Nations Secretary General, 'Oceans and the Law of the Sea. Report of the Secretary-General' (6 March 2017) A/72/70.

122 Large-scale Pelagic Driftnet Fishing and its Impact on the Living Marine Resources of the World's Oceans and Seas (adopted 22 December 1989) UNGA Resolution 44/225; Large-scale Pelagic Drift-net Fishing and Its Impact on the Living Marine Resources of the World's Oceans and Seas (adopted 20 December 1991) UNGA Resolution 46/215; Large-scale Pelagic Drift-net Fishing (adopted 24 November 1998) UNGA Resolution 53/33.

123 1991 Convention for the Prohibition of Fishing with Long Driftnets in the South Pacific [Reference].

124 Driftnet Fishing in the Convention Area (1990) CCAMLR Resolution 7/IX.

125 Laying Down Certain Technical Measures for the Conservation of Fishery Resources (29 April 1997) Council Regulation (EC) No 894/79, amended by Council Regulation (EC) 1239/98 (8 June 1998) and Council Regulation (EC) 809/2007 (28 June 2007).

change may galvanise a similar response by both global and regional institutions with appropriate mandates.

## 6 Conclusion

Marine geoengineering provides a textbook example of how the modern law of the sea can evolve through interpretation, practice and treaty amendment to address gaps and new challenges. However, it also demonstrates the natural institutional constraints of the discipline and the challenge of connecting and collaborating with other institutions and regimes in respect of crosscutting and global issues. The regulatory challenges associated with marine geoengineering are in fact representative of the broader issue of the relationship between the atmosphere and the hydrosphere and, more particularly, the impact of climate change on the oceans. Neglected for more than two decades by the climate change regime, the oceans, climate change and ocean acidification have, more recently, become a prominent agenda issue for a number of UN and other institutions. Nevertheless, more needs to be done in order to develop a mandate and a platform for concerted and coherent action. This chapter has identified two options, which are not necessarily alternatives to one another: the inclusion of principles relating to climate change and geoengineering in the ILBI, and the adoption of a UNGA resolution on oceans and the climate, which addresses, among other things, geoengineering. The former option has the benefit of establishing legally binding rules within the regulatory framework of UNCLOS. However, so far, climate change has not been identified as a priority issue for the negotiators, and the focus on developing rules around the status of marine genetic resources means that the road towards a binding instrument may be very long indeed. On the other hand, the latter option—a UNGA resolution—could be adopted with relative speed. Although not binding, UNGA resolutions are authoritative and persuasive, and a resolution on the oceans and climate setting out guidance and principles on geoengineering would add to the growing corpus of rules and other instruments adopted by institutions such as the London Protocol and the CBD.

CHAPTER 4

# Identifying Sensitive Marine Areas in the High Seas: a Review of the Scientific Criteria Adopted under International Law

*Youna Lyons*

## Abstract

This chapter focuses on the parallel and inter-related legal development in different international fora of sets of criteria to identify marine areas of particular sensitivity. It focuses primarily on criteria that were legally adopted by States, whether they are binding or not. These criteria can be grouped into two broad categories: (1) criteria developed to identify areas of particular ecological and cultural sensitivity and promote their protection and conservation; and, (2) criteria developed to identify marine areas that are sensitive to particular activities and provide a basis on which to restrict these activities accordingly. Comparison of sets of ecological and non-ecological criteria demonstrates that they are consistent and mutually reinforcing.

## 1 Introduction*

There is mounting pressure from the international community to better protect the marine environment within and beyond national jurisdiction, especially areas that are considered to be particularly sensitive and exposed to human activities at sea. The international paradigm that currently aims to protect the marine environment and especially marine biodiversity has evolved from early approaches of protection of the marine environment centred on pollution from activities at sea. The shift towards a more holistic understanding of the protection of the marine environment and inclusion of marine biodiversity

* This chapter is based on the larger study by the author on the definition and identification of sensitive marine areas and states' obligations under international law. For more details, see the working paper Youna Lyons, 'Defining and Identifying Sensitive Marine Areas in the South China Sea and States' Obligations under International Law' (Parts 1 and 2) (July 2018) <https://cil.nus.edu.sg/publications/?&keyword=youna%20lyons&search_by=author&pag=1> accessed 28 May 2018.

 | DOI:10.1163/9789004373303_005

can be traced to the 1990s with the landmark of the United Nations Conference on Environment and Development (also known as the Earth Summit) and the signature of the Convention on Biological Diversity (CBD) in 1992. The focus on sensitive marine areas and the need for scientific criteria to identify such sensitive areas have also developed greatly since the early 1990s in distinct international fora with a varying degree of exchange between them.

However legal definitions of sensitive marine areas can be traced back to the late 1960s and early 1970s, when environmental sensitivity motivated adoption of areas to be avoided[1] as ships routeing measures at the International Maritime Organization (IMO) and the designation of Special Areas under the International Convention for the Prevention of Pollution from Ships (MARPOL)[2] to further restrict discharges of oil, noxious liquid substances in bulk, and garbage.[3] The Mediterranean Sea, the Baltic Sea, the Black Sea, the Red Sea and the 'Gulfs area' were designated as Special Areas, where oil and garbage discharges were further limited[4] and/or where greater protection against pollution by noxious liquid substances in bulk was applicable.[5] Particularly sensitive sea areas also feature in Resolution 9 of the 1978 International Conference on Tanker Safety and Pollution Prevention, which is devoted to the protection of these areas against pollution from ships and from dumping of wastes. Pioneering yet another mechanism to better protect the marine environment against ship-source pollution, the IMO adopted in 1991 the first set of comprehensive criteria for the identification of Particularly Sensitive Sea Areas (PSSAs)[6] for pollution sources and environmental impacts other than

---

1 The earliest recorded areas to be avoided by shipping traffic that was motivated by environmental sensitivity of these areas was proposed by France on 23 November 1966 for an area in the Region of the Rochebonne Shelf and adopted on 20 November 1973 (NAV IV/2/Add.5 and A.284(VIII)) and by the USSR for the waters off Cape Terpeniya (NAV IV/2/Add.7 and A.284(VIII)). See G Peet, 'Particularly Sensitive Sea Areas—A Overview of Relevant IMO Documents' (1994) 9(4) *The International Journal of Marine and Coastal Law* 556, 564; G Peet (1994) 'Particularly Sensitive Sea Areas—A documentary History' (1994) 9(4) *The International Journal of Marine and Coastal Law* 569, 571.

2 Protocol of 1978 relating to the International Convention for the Prevention of Pollution from Ships, 1973 (adopted 17 February 1978, entered into force 2 October 1983 for Annexes 1 and 2) 1340 UNTS 61 (MARPOL).

3 Ibid Annexes I, II and V.

4 Ibid Annex I Regulation 10 and Annex V Regulation 5.

5 Ibid Annex II Regulation 1(7).

6 These guidelines also include a section on the designation of Special Area under MARPOL to assist state parties in the use of this mechanism. Guidelines for the Designation of Special Areas and Identification of Particularly Sensitive Sea Areas, IMO Res A.720(17) (6 November 1991) <www.imo.org/blast/blastDataHelper.asp?data_id=22581&filename=A720(17).pdf> accessed 28 May 2018.

those regulated under MARPOL. PSSAs could be established on the basis of any IMO instruments, and could include routeing measures.

Other scientific criteria have also been developed in other international fora since the 1990s for the identification of sensitive sea areas. These scientific criteria can be divided into two categories. The first category concerns criteria that have been developed to identify areas of particular environmental and cultural sensitivity and to promote a greater degree of protection in these areas. They are included in or have derived from instruments the main purpose of which is the conservation of different components of the marine environment. These instruments promote conservation values but provide limited specific guidance on activities that may occur or should be restricted within the sensitive areas that they seek to identify.

Sensitive area criteria, which were developed in the context of the implementation of the following instruments, are described in Section 2 of this chapter:

(i) the 1971 Ramsar Convention on the Protection of Wetlands of International Importance especially as Waterfowl habitat (Ramsar Convention)[7] has led to the development of criteria for the designation of Wetlands of International Importance;

(ii) the 1972 UNESCO Convention concerning the Protection of the World Natural and Cultural Heritage, (UNESCO World Heritage Convention)[8] has led to the development of criteria for the designation of areas of Outstanding Universal Value (OUV);

(iii) the 1979 Bonn Convention on the Conservation of Migratory Species of Wild Animals (Convention on Migratory Species or CMS),[9] seeks to protect critical habitats and migration corridors of migratory species; and

(iv) the 1992 CBD has led to the development of criteria for the identification of Ecologically or Biologically Significant Areas (EBSAs).[10]

---

7 Convention on Wetlands of International Importance especially as Waterfowl Habitat (adopted 2 February 1971, entered into force 21 December 1975) 996 UNTS 245 (Ramsar Convention).

8 Convention Concerning the Protection of the World Cultural and Natural Heritage (adopted 23 November 1972, entered into force 15 December 1975) 1037 UNTS 151 (UNESCO World Heritage Convention).

9 Convention on the Conservation of Migratory Species of Wild Animals (adopted 23 June 1979, entered into force 1 November 1983) 1651 UNTS 333 (Convention on Migratory Species).

10 Convention on Biological Diversity (adopted 5 June 1992, entered into force 29 December 1993) 1760 UNTS 79 (CBD).

Two of the sets of criteria developed under these four instruments are not designed to apply in areas beyond national jurisdiction (ABNJ) (under the Ramsar Convention and the UNESCO World Heritage Convention). However, they are not restricted to coastal areas and apply to marine areas and offshore submerged habitats, such as shallow seamounts and offshore coral reef systems, which can occur within as well as beyond national jurisdiction.[11] Where coastal states have a narrow continental shelf,[12] the part of the exclusive economic zone (EEZ) that is located seaward of the foot of the continental shelf is subject to the same legal regime as the high seas.[13] As such, there should be no reason for criteria to identify ecologically sensitive areas within and beyond national jurisdiction to be different. Criteria developed for areas within national jurisdiction are therefore relevant to ABNJ.

The second category of criteria for the identification of sensitive marine areas concerns criteria that have been developed by international organisations and bodies in charge of the regulation, coordination or supervision of specific uses of the sea. These sets of criteria, which are discussed in the third section of this chapter, are designed to identify areas that are exposed and vulnerable to specific activities and are based on particular conservation values. International regulations and guidelines impose or recommend limitations to these activities. The international organisations and bodies that have developed this second category of criteria include the following:

(i) the IMO, which pioneered this movement;

(ii) the Conference of the Parties (COP) to the 1972 London Convention for the Prevention of Marine Pollution by Dumping of Wastes and Other Matters (London Convention), and its 1996 Protocol for dumping (London Protocol);[14]

---

11 The presence of land is not a condition under any of the criteria. Examples include reefs such as the Tubbataha Reefs and parts of the Great Barrier Reef, which are located in the EEZs of the Philippines and Australia respectively and are both Ramsar sites as well as World Heritage Sites. Comparable reefs can for example be found atop seamounts located in ABNJ.

12 The term 'continental shelf' here refers to the submerged physical prolongation of the land mass rather than the legal definition of the continental shelf in UNCLOS article 76.

13 The EEZ extends 200 NM from the baseline of a coastal state, irrespective of the geomorphology of the seabed. See United Nations Convention on the Law of the Sea (adopted 10 December 1982, entered into force 16 November 1994) 1833 UNTS 3 (UNCLOS) art 57.

14 Convention on the Prevention of Marine Pollution by Dumping Wastes and Other Matter (adopted 29 December 1972, entered into force 30 August 1975) 1046 UNTS 120 (London Convention); 1996 Protocol to the Convention on the Prevention of Marine Pollution by Dumping of Wastes and Other Matter, 1972 (adopted 7 November 1996, entered into force 24 March 2006) (London Protocol).

(iii) the Food and Agriculture Organisation (FAO) for fishing;[15] and
(iv) the International Seabed Authority (ISA) for deep seabed mining.

The 1982 United Nations Convention on the Law of the Sea (UNCLOS)[16] is not included in the above list of scientific criteria adopted under the auspices of specialised instruments and competent international organisations, because no technical criteria have been developed by the parties to UNCLOS. However, in order to comply with their obligation to protect and preserve the marine environment under UNCLOS, states must adopt measures 'necessary to protect and preserve rare or fragile ecosystems as well as the habitat of depleted, threatened or endangered species and other forms of marine life'.[17] Although this provision may not strictly qualify as 'scientific criteria', it includes several of the criteria discussed below.

The fourth section compares the different sets of criteria to identify commonalities and differences and to determine whether a common set of criteria could guide identification of sensitive areas in the high seas.

## 2 Scientific Criteria Adopted to Implement International Instruments for the Purposes of Protection and Conservation

The development of EBSAs under the auspices of the CBD and of criteria for their identification are considered in this chapter first, due to the comprehensive nature of the set of criteria that has been developed and the fact that it can be used as a reference to compare with the other sets of criteria. The other

15 The Food and Agriculture Organization of the United Nations (FAO) is a competent international organisation for the management of fisheries under UNCLOS. UNCLOS (n 13) arts 61(1), 61(2), 64 and 119. See Tullio Treves, 'The Law of the Sea "System" of Institutions' in *Max Planck Yearbook of United Nations Law* (Vol 2, Max Planck Institute for Comparative Public Law and International Law 1998) 329. <www.mpil.de/files/pdf2/mpunyb_treves_2.pdf> accessed 28 May 2018; UN Division for Ocean Affairs and the Law of the Sea (DOALOS), '"Competent or Relevant International Organizations" under the United Nations Convention on the Law of the Sea' (1996) 31 *Law of the Sea Bulletin* 79; JD Kingham and DM McRae, 'Competent International Organizations and the Law of the Sea' (1979) 3(2) *Marine Policy* 106.

16 UNCLOS (n 13).

17 Ibid arts 192 and 194(5). In the South China Sea Arbitration Award, the Tribunal confirmed that the due diligence obligation to protect the marine environment in Article 192 read in the context of Article 194(5) entails the obligation to take necessary measures to give effect to Article 194(5) by preventing (i) the direct harvesting of species that are recognised internationally as being at risk of extinction; and (ii) harm that would negatively impact the habitat of such species. *The South China Sea Arbitration* (*The Republic of Philippines v The People's Republic of China*) Award (12 July 2016) PCA 2013–19.

instruments (and their associated criteria) are presented chronologically by date of adoption of the source instrument, irrespective of the adoption date of the different sets of criteria. The development of each set of criteria was started after its source instrument had been adopted and progressed in an iterative manner that generally involved several amendments.

The same analytical framework is used for each set of criteria and their originating legal instruments:[18]

(i) source instrument and legal background of the adoption of scientific criteria;
(ii) status of adoption of the instrument and extent of the application of the criteria by member states as well as the geographic scope of the instrument(s) and related criteria; and
(iii) scientific criteria.

### 2.1 *EBSAs*

#### 2.1.1 Source Instrument and Legal Background

The CBD is based on the premise that the conservation of biological diversity is a common concern of humankind, given its importance for evolution and for maintaining life-sustaining systems of the biosphere.[19] Its objectives include the conservation of biodiversity and the sustainable use of its components.[20] However, the text of the CBD does not distinguish between terrestrial and marine biodiversity, although the measures it provides for encompass both the terrestrial and marine realms.[21] The CBD does not specify which

18 There is merit in also comparing the listing procedure and status, monitoring, reporting and review process as well as expected management measures for each set of criteria and its related legal instrument. However, the word constraints of this chapter did not allow for this to be included.

19 CBD (n 10) Preamble.

20 Ibid art 1.

21 The CBD in fact appears to have been drafted with terrestrial biodiversity in mind. For instance, the convention focusses on species diversity and areas of high endemism to determine areas where protection would be desirable. D Freestone 'The Conservation of Marine Ecosystems under International Law' in Catherine Redgwell and Michael Bowman (eds) *International Law and the Conservation of Biodiversity* (Kluwer Law International 1996) 91–107. This makes less sense in the context of marine biodiversity than it does for terrestrial biodiversity, because only 20% of known species are marine, and endemism is generally low (except for benthic communities), due to the ability of marine organisms to disperse over large geographic scales (unlike terrestrial animals). See SR Palumbi, 'Marine Speciation on a Small Planet' (1992) 7(4) *Trends in Ecology & Evolution* 114; MV Angel, 'Biodiversity of the Pelagic Ocean' (1993) 7(4) *Conservation Biology* 760. Other important differences that result in different geographic representations of

areas, ecosystems or species should be protected or prioritised and leaves the selection process and management method to each party.[22] Nevertheless, it provides for a clear obligation to identify components of biological diversity (Article 7(a)) and to monitor them (Article 7(b)). Article 7(a) further provides that identification should have regard to the indicative list of categories set down in Annex I under the three headings of ecosystems and habitats, species and communities, and described genomes and genes of social, scientific and economic importance.

In 2002, the Johannesburg Plan of Implementation highlighted the importance of maintaining the productivity and biodiversity of important and vulnerable marine and coastal areas, including in ABNJ.[23] This new emphasis on vulnerable marine ecosystems in ABNJ is also found in the 2003 and 2004 United Nations General Assembly (UNGA) resolutions on oceans and the law of the sea with particular concern placed on the conservation and management of vulnerable marine ecosystems in ABNJ. Both resolutions also affirm the need for states to develop tools for conserving and managing vulnerable marine ecosystems, including the possible establishment of marine protected areas (MPAs).[24] In 2003, the UNGA resolution invited relevant global and regional bodies to urgently investigate how to address this concern on a scientific basis and consistent with international law.[25] In 2004, the UNGA established an Ad Hoc Open-ended Informal Working Group to study issues relating to the conservation and sustainable use of marine biological diversity in ABNJ.[26]

---

biodiversity (and a different approach to geographic scale) include the presence of higher orders of the taxonomic classification being more represented in the ocean: 28 phyla of animals compared to 11 on land.

22 CBD (n 10) art 8.

23 United Nations, 'Johannesburg Plan of Implementation: Report of the World Summit on Sustainable Development' UN Doc A/CONF.199/20 (World Summit on Sustainable Development, Johannesburg, South Africa, 26 August–4 September 2002) para 32 <www.un.org/esa/sustdev/documents/WSSD_POI_PD/English/WSSD_PlanImpl.pdf> accessed 28 May 2018.

24 Oceans and the Law of the Sea, UNGA Res 58/240 (23 December 2003) para 54; Oceans and the Law of the Sea, UNGA Res 59/24 (4 February 2005) para 72.

25 UNGA Resolution 58/240 (n 24) para 52.

26 Ibid para 73. See also Sustainable fisheries, including through the 1995 Agreement for the Implementation of the Provisions of the United Nations Convention on the Law of the Sea of 10 December 1982 relating to the Conservation and Management of Straddling Fish Stocks and Highly Migratory Fish Stocks, and Related Instruments, UNGA Res A/RES/59/25 (17 November 2004) para 66 <https://documents-dds-ny.un.org/doc/UNDOC/GEN/N04/477/70/PDF/N0447770.pdf?OpenElement> accessed 28 May 2018.

These developments were the genesis of the work undertaken in the CBD programme on Marine and Coastal Biodiversity and the cross-cutting issue of Protected Areas, triggering expert meetings on the topic. They eventually led to the adoption by the COP to the CBD of criteria for identifying EBSAs in need of protection, and the scientific guidance for designing representative networks of MPAs.[27]

#### 2.1.2 Adoption and Geographic Scope

A total of 196 states are parties to the CBD,[28] which applies to the components of biological diversity in areas within the limits of national jurisdictions.[29] However, it also applies to processes and activities under the jurisdiction or control of a Contracting Party, whether they are carried out within national jurisdiction or beyond the limits of national jurisdiction.[30]

At each meeting since the adoption of the criteria, the CBD COP reviews the status of the EBSA identification process. Following regional workshops organised by the CBD Executive Secretary to facilitate the description of EBSAs, the Subsidiary Body on Scientific, Technical and Technological Advice (SBTTA) prepares reports based on scientific and technical evaluation of information from the workshop. These reports are to be considered and endorsed in a transparent manner by the COP, with a view to including endorsed reports in the repository and to submit them to all relevant international bodies and organisations such as the UNGA, the FAO, the IMO, regional seas organisations,

27 CBD COP9, 'Decision Adopted by the Conference of the Parties to the Convention. IX/20. Marine and Coastal Biodiversity' UN Doc UNEP/CBD/COP/DEC/IX/20 (9 October 2008) (CBD COP9 Decision IX/20) para 14 and Annex I <https://www.cbd.int/decision/cop/?id=11663> accessed 28 May 2018. At the time of the first expert workshop for the development of scientific criteria for the identification of EBSA, Canada had already developed its own criteria for the implantation of its Ocean Act 'to facilitate provision of a greater-than-usual degree of risk aversion in management of activities in such areas'. (Fisheries and Oceans Canada), 'Identification of Ecologically and Biologically Significant Areas' (Ecosystem Status Report 2004/006, March 2005) <www.dfo-mpo.gc.ca/csas/Csas/status/2004/ESR2004_006_e.pdf> accessed 28 May 2018. They also called these areas 'EBSAs'. The Canadian criteria were discussed at the workshops organised under the auspices of the CBD, but new EBSA criteria were developed for the purpose of the implementation of the CBD. On the history of the development of the EBSA criteria, see DC Dunn et al, 'The Convention on Biological Diversity's Ecologically or Biologically Significant Areas: Origins, Developments and Current Status' (2014) 49 *Marine Policy* 137, 138–40.

28 For status of parties to the CBD, see CBD, 'List of Parties' <www.cbd.int/information/parties.shtml> accessed 28 May 2018.

29 CBD (n 10) art 4(a).

30 Ibid art 4(b).

etc. A repository for storing and sharing scientific and technical information and experience related to the application of the scientific criteria on the identification of EBSAs has been established for that purpose. A purpose-built prototype web-based input tool and database is also fully functional.[31]

While the initial focus of the EBSA criteria was on ABNJ, regional workshops have also included areas of EEZ as proposed or agreed on by the state having jurisdiction over that EEZ.[32] Thirteen regional workshops were organised between 2011 and 2017, and 279 EBSAs were identified by CBD COP13.[33] The workshop in April 2017 concerned the Black Sea and Caspian Sea. As at 30 April 2018, the report has not yet been published on the CBD EBSA website. It is expected to be submitted to CBD COP14 in November 2018. A workshop is also planned for the Baltic Sea in February 2018, at which point most of the world's oceans and seas would be considered to have undergone a 'first pass' of the EBSA process.[34] However, this is not to say that no other EBSA may be added later as 'additional regional or sub-regional workshops may be organised when there is sufficient advancement in the availability of scientific information'.[35]

#### 2.1.3 Scientific Criteria for the Identification of EBSAs

CBD COP9 (2008) adopted seven criteria for the identification of EBSAs (Table 4.1).[36] Five separate criteria for the establishment of networks of MPAs were

31 CBD, 'Ecologically or Biologically Significant Marine Areas', <www.cbd.int/ebsa> accessed 28 May 2018.

32 Coastal states have different rights and jurisdictions in different maritime zones generated from the land over which they have sovereignty as well as in ABNJ. Maritime zones, and respective rights and jurisdiction of states are defined in UNCLOS. While coastal states have sovereignty over activities in their internal waters and in 12 NM Territorial Sea (as defined in UNCLOS), they have sovereign rights over the resources in their EEZ, which extends from the seaward limit of the Territorial Sea up to 200 NM from the baseline from which the breadth of the Territorial Sea has been calculated.

33 CBD COP13, 'Decision Adopted by the Conference of the Parties to the Convention on Biological Diversity. XIII/12. Marine and Coastal Biodiversity: Ecologically or Biologically Significant Marine Areas' UN Doc CBD/COP/DEC/XIII/12 (17 December 2016) <www.cbd.int/doc/decisions/cop-13/cop-13-dec-12-en.pdf> accessed 28 May 2018.

34 DE Johnson et al, 'Reviewing the EBSA Process: Improving on Success' (2018) 88 *Marine Policy* 76.

35 CBD COP11, 'Decision Adopted by the Conference of the Parties to the Convention on Biological Diversity. XI/17. Marine and Coastal Biodiversity: Ecologically or Biologically Significant Marine Areas' UN Doc UNEP/CBD/COP/DEC/XI/17 (5 December 2012) paras 9, 12. <www.cbd.int/doc/decisions/cop-11/cop-11-dec-17-en.pdf> accessed 28 May 2018.

36 CBD COP9 Decision IX/20 (n 27).

TABLE 4.1 EBSA scientific criteria

| EBSA | Criteria | Definition |
|---|---|---|
| 1- | **Uniqueness or rarity** | Area contains (i) unique ('the only one of its kind'), rare (occurs only in few locations) or endemic species, populations or communities; (ii) unique, rare or distinct, habitats or ecosystems; and/or (iii) unique or unusual geomorphological or oceanographic features. |
| 2- | **Special importance for life history of species** | Area that is required for a population to survive and thrive. |
| 3- | **Importance for threatened, endangered or declining species and/or habitats** | Area containing habitat for the survival and recovery of endangered, threatened, declining species or area with significant assemblages of such species. |
| 4- | **Vulnerability, fragility, sensitivity, slow recovery** | Area that contains a relatively high proportion of sensitive habitats, biotopes or species that are functionally fragile (highly susceptible to degradation or depletion by human activity or by natural events) or with slow recovery. |
| 5- | **Biological productivity** | Area containing species, populations or communities with comparatively higher natural biological productivity. |
| 6- | **Biological diversity** | Area contains comparatively higher diversity of ecosystems, habitats, communities, or species, or has higher genetic diversity. |
| 7- | **Naturalness** | Area with a comparatively higher degree of naturalness as a result of the lack of or low level of human-induced disturbance or degradation. |

also adopted, EBSAs being one of them.[37] CBD COP9 also decided to convene an expert workshop to provide scientific and technical guidance on the use and further development of biogeographic classification systems and guidance on the identification of areas located beyond national jurisdiction that meet the scientific criteria.[38] CBD COP10 took note of the guidance document that resulted from the 2009 Ottawa Workshop and invited state parties to take it into account when implementing the criteria.[39]

The initial decision to adopt scientific criteria for the identification of EBSAs was driven by a focus 'on provision of scientific and, as appropriate, technical information and advice relating to marine biological diversity, the application of the ecosystem approach and the precautionary approach'.[40] The intention of the CBD governing bodies is that areas that are identified as EBSAs should also be considered as 'sensitive areas' under other instruments. Further it is intended that maritime and marine uses of these areas as well as pollution from these activities be regulated within the regime applicable to each activity (e.g. shipping through the IMO, fisheries through the FAO and relevant Regional

---

37 The other four are representativity, connectivity, replicated ecological features and an adequate viable site (CBD COP9 Decision IX/20 (n 27 Annex II)).

38 CBD COP9 Decision IX/20 (n 27) para 19.

39 CBD COP10, 'Decision Adopted by the Conference of the Parties to the Convention on Biological Diversity at its Tenth Meeting. X/29. Marine And Coastal Biodiversity' UN Doc UNEP/CBD/COP/DEC/X/29 (29 October 2010) para 40 <www.cbd.int/doc/decisions/cop-10/cop-10-dec-29-en.pdf> accessed 28 May 2018; Expert Workshop, 'Report of the Expert Workshop on Scientific and Technical Guidance on the Use of Biogeographic Classification Systems and Identification of Marine Areas Beyond National Jurisdiction in Need of Protection' (CBD Ottawa Report) UN Doc UNEP/CBD/EW-BCS&IMA/1/2* (22 December 2009) <www.cbd.int/doc/meetings/mar/ebsa-np-01/other/ebsa-np-01-ewbcsima-01-02-en.pdf> accessed 28 May 2018. See also Subsidiary Body on Scientific, Technical and Technological Advice (SBTTA), 'Report of the Expert Workshop on Scientific and Technical Guidance on the Use of Biogeographic Classification Systems and Identification of Marine Areas Beyond National Jurisdiction in Need of Protection' UN Doc UNEP /CBD/SBSTTA/14/INF/4* (11 January 2010) <www.cbd.int/doc/meetings/sbstta/sbstta-14/information/sbstta-14-inf-04-en.pdf> accessed 28 May 2018; SBTTA, 'Training Manual for the Description of Ecologically and Biologically Significant Areas (EBSAs) in the Open-Ocean Waters and Deep-Sea Habitats' UN Doc UNEP/CBD/SBSTTA/16/INF/9. (3 May 2012) <www.cbd.int/doc/meetings/sbstta/sbstta-16/information/sbstta-16-inf-09-en.pdf> accessed 28 May 2018.

40 CBD COP8, 'Decision Adopted by the Conference of the Parties to the Convention on Biological Diversity at its Eighth Meeting'. VIII/24. 'Protected Areas' UN Doc UNEP/CBD/COP/DEC/VIII/24 (15 June 2006) para 42.

Fisheries Management Organizations or Arrangements (RFMO/As), deep sea-bed mining through the International Seabed Authority (ISA, etc.).[41]

### 2.2 *Wetlands of International Importance*

#### 2.2.1 Source Instrument and Legal Background

The objective of the 1971 Ramsar Convention is the wise use of wetlands in particular conservation and management of wetlands of international significance in terms of ecology, botany, zoology, limnology or hydrology.[42] The term 'wetland' includes areas of marine waters that are up to 6 m deep at low tide; deeper sea areas may also be included in the regime if they are riparian and coastal zones adjacent to wetlands and islands or bodies of marine water, especially if they support important waterfowl habitat.[43] Each contracting party commits to 'designate suitable wetlands within its territory for inclusion in a List of Wetlands of International Importance'.[44]

In order to fulfil the objective of designating and listing wetlands of international significance, the COP to the Ramsar Convention has progressively developed sets of criteria to identify such wetlands. The focus of concerns has been progressively enlarged to include rare, vulnerable and endangered species, genetic and ecological diversity as well as critical stages of biological cycles that would be supported by the wetland under consideration.[45] Wetlands supporting other wetland-dependent species were also eventually added.[46]

---

41 CBD COP9 Decision IX/20 (n 27) especially Paragraphs 11, 17 and 18, which adopt the scientific criteria to identify EBSAs and emphasise the importance of coordination with all the relevant international organisations such as the FAO and IMO and invite them to apply the EBSAs' criteria and provide their feedback to assist in the implementation of conservation and management measures. See also Dunn et al (n 27).

42 Ramsar Convention (n 7) arts 2(2), 2(6), 3(1) and 4(1).

43 Ibid arts 1(1) and 2(1).

44 Ibid art 2(1).

45 Ibid arts 14–19.

46 At COP6 and COP9. See 9th Meeting of the Conference of the Parties to the Convention on Wetlands, 'Resolution IX.1 Annex B. Revised Strategic Framework and Guidelines for the Future Development of the List of Wetlands of International Importance' (Resolution IX.1) (Ramsar Guidelines) <www.ramsar.org/document/resolution-ix1-annex-b-revised-strategic-framework-and-guidelines-for-the-future> accessed 28 May 2018. See also Ramsar Convention Secretariat, 'An Introduction to the RAMSAR Convention on Wetlands. International Cooperation on Wetlands' (Ramsar Handbook 1, 5th Edition, 2016) 45. <www.ramsar.org/sites/default/files/documents/library/handbook1_5ed_introductiontoconvention_final_e.pdf> accessed 28 May 2018. This handbook was adopted by Resolution XI.8, annex 2.

### 2.2.2 Adoption and Geographic Scope

While the Ramsar Convention applies only to the marine and coastal areas located within the jurisdiction of a state party and therefore not to marine areas located beyond national jurisdiction, 169 states have adopted it globally as of 14 December 2017. This global acceptance and the comparable criteria and processes developed are the reason for including them in this review. Furthermore, Ramsar sites include shallow and isolated coral reefs which could be found in ABNJ, for example atop shallow seamounts.[47]

As of 21 May 2018, there were 2,308 listed Ramsar sites.[48] They cover a total area of 2.29 mi $km^2$ (equivalent to 229 mi ha), and their size vary greatly from 1 ha to over 6.5 mi ha. One of the five major wetland types is recognised as relevant to this chapter: marine wetlands–coastal wetlands such as lagoons, rocky shores, seagrass beds, atoll and coral reefs.[49] Globally, they represent 41.6% of the number of Ramsar sites and 30.4% of their surface area.

### 2.2.3 Scientific Criteria

The current set of criteria is made up of nine criteria for identifying wetlands of international importance and are divided into two groups (Table 4.2 below).[50] One group (Group A) relates to sites of representative, rare or unique wetland types. Interestingly, rarity and uniqueness are also an EBSA criterion (EBSA Criterion 1), but representativity is not. The other group (Group B) concerns wetland features that are important for the conservation of wetland biological diversity; these features can relate to species and ecological communities, waterbirds, fish or other taxa. This second group of ecological and biological criteria includes quantitative and qualitative elements. For example, a wetland should be considered internationally important if it regularly supports 20,000 or more waterbirds, 1% of the individuals in a population of one species or subspecies of waterbird, or 1% of the individuals in a population of one species or subspecies of wetland-dependent non-avian animal species.[51] This

---

47 Examples include the Tubbataha Reefs Natural Park, a Ramsar site as well as a World Heritage site. UNESCO, 'World Heritage List. Tubbataha Reefs Natural Park' <http://whc.unesco.org/en/list/653/> accessed 28 May 2018.

48 Ramsar, About Ramsar <www.ramsar.org> (accessed 21 May 2018).

49 Ramsar Handbook 1 (n 46), 9.

50 As adopted in COP9 Resolution IX.1 (n 46). See also last revision of the document: Ramsar COP11, 'Resolution XI.8 Annex 2. Strategic Framework and Guidelines for the Future Development of the List of Wetlands of International Importance of the Convention on Wetlands (Ramsar, Iran, 1971)—2012 Revision', <www.ramsar.org/sites/default/files/documents/pdf/cop11/res/cop11-res08-e-anx2.pdf> accessed 28 May 2018. Adopted at COP11 Resolution XI.8 Annex 2.

51 Ramsar COP11 (n 50) 'Appendix D. Criteria for Identifying Wetlands of International Importance' 90 (Ramsar criteria) criteria 5, 6 and 9.

TABLE 4.2 Scientific criteria for the identification of Ramsar Sites

| Criteria group | | | Wetland characteristics |
|---|---|---|---|
| Group A—Sites containing representative, rare or unique wetland types | | 1- | Contains a representative, rare or unique example of a natural or near-natural wetland type found within the appropriate biogeographic region.[a] |
| Group B—Sites of international importance for conserving biological diversity | Species and ecological communities | 2- | Supports vulnerable, endangered or critically endangered species or threatened ecological communities. |
| | | 3- | Supports populations of plant and/or animal species important for maintaining the biological diversity of a particular biogeographic region. |
| | | 4- | Supports plant and/or animal species at a critical stage in their life cycles, or provides refuge during adverse conditions. |
| | Waterbirds | 5- | Regularly supports 20,000 or more waterbirds. |
| | | 6- | Regularly supports 1% of the individuals in a population of one species or subspecies of waterbird. |
| | Fish | 7- | Supports a significant proportion of indigenous fish subspecies, species or families, life-history stages, species interactions and/or populations that are representative of wetland benefits and/or values and thereby contributes to global biological diversity. |
| | | 8- | Is an important source of food for fishes, spawning ground, nursery and/or migration path on which fish stocks, either within the wetland or elsewhere, depend. |
| | Other taxa | 9- | Regularly supports 1% of the individuals in a population of one species or subspecies of wetland-dependent non-avian animal species. |

a In its section devoted to biogeographic regionalisation, Ramsar Guidelines refer to Spalding et al, 'Marine Ecoregions of the World: A Bioregionalization of Coastal and Shelf Areas' (2007) 57(7) *BioScience* 573 <https://academic.oup.com/bioscience/article/57/7/573/238419> accessed 28 May 2018.

quantitative approach differs from the EBSA approach, though quantitative analysis may be necessary in order to inform the qualitative criteria. The Ramsar criteria also include qualitative criteria, such as supporting vulnerable, endangered, or critically endangered species or threatened ecological communities,[52] supporting plant and/or animal species at a critical stage in their life cycles, or providing refuge during adverse conditions.[53] Of these last criteria, the supporting of critical stages in species' life cycles is also an EBSA criterion (EBSA Criterion 2). By contrast, the provision of *refugia* against adverse conditions is not an ESBA criterion.

A wetland needs to meet only one criterion (of either group) to qualify as a wetland of international importance.[54]

### 2.3 *Areas of Outstanding Universal Value* (OUV)

#### 2.3.1 Source Instrument and Legal Background

The 1972 UNESCO World Heritage Convention aims to inventorise, protect, conserve and preserve world cultural and natural heritage. This includes, with respect to natural heritage of the marine realm[55]

> natural features consisting of physical and biological formations or groups of such formations which are of Outstanding Universal Values [OUV] from the aesthetic or scientific point of view; geological and physiographical formations and precisely delineated areas which constitute the habitat of threatened species of animals and plants of [OUV] from the point of view of science or conservation; natural sites or precisely delineated natural areas of [OUV] from the point of view of science, conservation or natural beauty.[56]

Each state party commits to do all it can (to the utmost of its own resources and where appropriate, with any international assistance and co-operation which it may be able to obtain) to identify and delineate such sensitive areas of OUV.[57] A particularity of the World Heritage Convention is that it created an

52 Ibid criterion 2.

53 Ibid criterion 4.

54 Ramsar Convention (n 7) Article 2.2 lists alternative characteristics using the coordinating conjunction 'or'. Similarly, the Ramsar criteria (n 51) are presented as being alternative criteria. See also Ramsar Handbook 1 (n 46).

55 Like the CBD, the World Heritage Convention applies to the terrestrial and marine realms. It also applies to natural heritage as well as cultural heritage. This chapter however focusses primarily on the subset that is devoted to natural heritage in the marine realm.

56 UNESCO World Heritage Convention (n 8) art 2.

57 Ibid art 4.

intergovernmental committee for the protection of world cultural and natural heritage, the World Heritage Committee, which is the main body in charge of the implementation of the convention.[58] Meetings of this committee have discussed the interpretation of what constitutes OUV and made it evolve over time: from a restrictive interpretation that limited areas of OUV to areas that were iconic and whose value was unanimously acknowledged as 'best of the best', to a 'gradual shift towards the representative of the best'.[59] The World Heritage Convention is not intended to ensure the protection of all properties of great interest, importance or value, but only for a select list of the most outstanding of these from an international viewpoint.

### 2.3.2 Status of Adoption and Geographic Scope

As at 31 January 2017, 193 states have adopted the World Heritage Convention. The UNESCO World Heritage List includes 1073 sites globally, 49 of which are marine sites listed by 37 state parties.[60] The first site, the Galapagos Islands, was listed in 1978. Listed marine sites include a variety of marine systems and sizes. The larger sites are the Phoenix Islands Protected Area in Kiribati, Papahanaumokuakea in Hawaii, the Great Barrier of Reef of Australia and the Galapagos Islands in Ecuador.[61] They extend beyond the territorial sea and include coastal as well as underwater seamounts, volcanoes and other deep-sea habitats.

The World Heritage Convention, as well as the powers vested in the bodies it created, is limited to areas within national jurisdiction despite the preamble of the convention focussing on the protection of world heritage with no geographic limits. Different legal mechanisms can be envisaged to enlarge the World Heritage Committee's mandate to encompass World Heritage sites

58 The World Heritage Committee is composed of 15 state representatives elected by state parties to the Convention. UNESCO World Heritage Convention (n 8) art 8.

59 World Heritage Committee, 'Item 9 of the Provisional Agenda: Evaluation of Outstanding Universal Value' UNESCO Doc WHC-06/30.COM/9 (23 June 2006) para 6 <http://whc.unesco.org/archive/2006/whc06-30com-09e.pdf> accessed 28 May 2018.

60 Their surface areas are 408,250 km², 362,075 km², 348,700 km² and 140,665 km². Data compiled from the database of the UNESCO, 'World Heritage List' <http://whc.unesco.org/en/interactive-map/> accessed 28 May 2018. See also UNESCO World Heritage Marine Programme, '2017 Annual Newsletter' <http://whc.unesco.org/archive/websites/unesco_marine_2017_newsletter/?page=1> accessed 28 May 2018; World Heritage Committee, 'Decisions Adopted during the 41st session of the World Heritage Committee (Krakow, July 2017)' UNESCO Doc WHC/17/41.COM/18 (12 July) <http://whc.unesco.org/archive/2017/whc17-41com-18-en.pdf> accessed 28 May 2018.

61 Ibid.

beyond national jurisdiction, but they all require an international agreement.[62] This initiative, which is also supported by the International Union for Conservation of Nature (IUCN),[63] has been brought to the intergovernmental discussions on biodiversity beyond national jurisdiction (the Working Group on BBNJ) launched by UNGA in 2004.[64] UNESCO proposes five sites in ABNJ for consideration: two each in the Pacific and Atlantic Oceans, and one in the Indian Ocean.[65]

### 2.3.3 Scientific Criteria

The criteria developed by the World Heritage Committee over time for the inscription of sites of OUV on the World Heritage List are reflected in the Operational Guidelines for the Implementation of the World Heritage Convention, which are periodically revised to reflect the decisions of the committee.[66] There are ten criteria, not all of which are relevant to the protection of sensitive marine environments due to their focus on types of cultural heritage that are not found at sea. In Table 4.3 below, relevant criteria are highlighted in two shades of grey. Criteria 7 to 10 are generally recognised as relevant for the

---

62 UNESCO, *World Heritage in the High Seas: An Idea Whose Time Has Come* (World Heritage Reports 44, UNESCO 2016) (World Heritage Reports 44) <https://whc.unesco.org/en/series/44> accessed 28 May 2018.

63 Ameer Abdulla et al, Marine Natural Heritage and the World Heritage List Interpretation of World Heritage Criteria in Marine Systems, Analysis of Biogeographic Representation of Sites, and a Roadmap For Addressing Gaps (IUCN 2013).

64 First to the Ad Hoc Open-ended Informal Working Group established in 2004 by UNGA Resolution 59/24 to study issues relating to the conservation and sustainable use of marine biological diversity beyond areas of national jurisdiction. Second to Preparatory Committee (PrepCom) established by UNGA Resolution 69/292 (6 July 2015) with the mandate to make substantive recommendations to the General Assembly on the elements of a draft text of an international legally binding instrument (ILBI) under UNCLOS in BBNJ. UNGA Resolution 72/249 decided to convene an intergovernmental conference, under the auspices of the United Nations, to consider the recommendations of the preparatory committee on the elements and to elaborate the text of an ILBL under the UNCLOS on the conservation and sustainable use of marine biological diversity of ABNJ. UN DOALOS, 'Oceans and the Law of the Sea in the General Assembly of the United Nations. General Assembly Resolutions and Decisions' <www.un.org/depts/los/general_assembly/general_assembly_resolutions.htm> accessed 28 May 2018.

65 See 'The Costa Rica Thermal Dome and the White Shark café in the Pacific Ocean, the Sargasso Sea and The Lost City Hydrothermal Field in the Atlantic Ocean, and the Atlantis Bank in the Indian Ocean' in World Heritage Reports 44 (n 62) 31–41.

66 Mandate of the World Heritage Committee to define the criteria in provided in. UNESCO World Heritage Convention (n 8) art 11(5). UNESCO, 'The Operational Guidelines for the Implementation of the World Heritage Convention' UNESCO Doc WHC.17/01 (12 July 2017) <http://whc.unesco.org/en/guidelines/> accessed 28 May 2018.

selection of World Heritage marine sites and used in the selection process of marine sites included in the World Heritage List.[67]

To be deemed of OUV, an area must also have an adequate protection and management system to ensure its safeguarding. In addition, criteria of authenticity and integrity are required for cultural heritage. While authenticity refers to cultural heritage and is therefore not relevant to this discussion, integrity is relevant.[68] Integrity is a measure of the wholeness and intactness of the natural and/or cultural heritage and its attributes. Examining the conditions of integrity therefore requires assessing the extent to which the site

(i) includes all the elements necessary to express its OUV;
(ii) is of adequate size to ensure the complete representation of the feature and processes that convey the property's significance; and
(iii) suffers from adverse effects of development and/or neglect.

This means that with respect to marine natural heritage, biophysical processes and landform features should be relatively intact and in general the distinctive character of a site should be maintained so that the criteria can continue to be met.[69]

### 2.4 *Critical Habitats and Ecological Networks for Migratory Species Listed in the Convention on Migratory Species and Related Instruments*

#### 2.4.1 Source Instruments and Legal Background

The 1979 Convention on Migratory Species (CMS) highlights the conservation of migratory species of wild animals in general (terrestrial, avian and marine) whilst focussing on two categories of migratory species. First those that are listed in Appendix I to CMS on the basis that they are endangered,[70] meaning that they are 'facing a very high risk of extinction in the wild in the near future'.[71] Second those that are listed in Appendix II to CMS on the basis that

67 See UNESCO World Heritage Marine Programme, 'Marine World Heritage' <http://whc.unesco.org/uploads/activities/documents/activity-13-137.pdf> accessed 28 May 2018.

68 UNESCO, 'The Operational Guidelines for the Implementation of the World Heritage Convention' (n 66) paras 87–95 (Decision 20 COM IX.13).

69 Ibid.

70 Convention on Migratory Species (n 9) art III.

71 CMS COP12, 'Guidelines for Assessing Listing Proposals to Appendices I and II of the Convention' (Manila, October 2017) UN Doc UNEP/CMS/Resolution 11.33 (Rev.COP12) <http://cms.int/en/document/guidelines-assessing-listing-proposals-appendices-i-and-ii-convention-0> accessed 28 May 2018.

TABLE 4.3 World Heritage sites criteria of Outstanding Universal Beauty

| Criteria | Characteristics of outstanding universal beauty for inclusion in the World Heritage list |
|---|---|
| 1- | Represent a masterpiece of human creative genius [*with authenticity and integrity*]. |
| 2- | Exhibit an important interchange of human values, over a span of time or within a cultural area of the world, on developments in architecture or technology, monumental arts, town-planning or landscape design [*with authenticity and integrity*]. |
| 3- | Bear a unique or at least exceptional testimony to a cultural tradition or to a civilization which is living or which has disappeared [*with authenticity and integrity*]. |
| 4- | Outstanding example of a type of building, architectural or technological ensemble or landscape which illustrates (a) significant stage(s) in human history [*with authenticity and integrity*]. |
| 5- | Outstanding example of a traditional human settlement, land-use, or sea-use which is representative of a culture (or cultures), or human interaction with the environment especially when it has become vulnerable under the impact of irreversible change [*with authenticity and integrity*]. |
| 6- | Directly or tangibly associated with events or living traditions, with ideas, or with beliefs, with artistic and literary works of outstanding universal significance [*with authenticity and integrity*]. |
| 7- | Superlative natural phenomena or areas of exceptional natural beauty and aesthetic importance [*with integrity*]. |
| 8- | Outstanding examples representing major stages of earth's history, including the record of life, significant on-going geological processes in the development of landforms, or significant geomorphic or physiographic features [*with integrity*]. |
| 9- | Outstanding examples representing significant on-going ecological and biological processes in the evolution and development of terrestrial, fresh water, coastal and marine ecosystems and communities of plants and animals [*with integrity*]. |
| 10- | Most important and significant natural habitats for in-situ conservation of biological diversity, including those containing threatened species of outstanding universal value from the point of view of science or conservation [*with integrity*]. |
| and | Adequate protection and management system to ensure safeguarding. |

they have an unfavourable conservation status[72] that require or would benefit from international cooperation for their conservation. The CMS provides for obligations for Range States, which include any state (in relation to a particular migratory species) 'that exercises jurisdiction over any part of the range of that migratory species, or a State, flag vessels of which are engaged outside national jurisdictional limits in taking that migratory species'.[73] Should activities be occurring beyond national jurisdiction, states and the COP of this convention would seek cooperation from relevant international authorities (e.g. the relevant RFMO/As in the context of large migratory species that may be caught as by-catch, such as protected sharks under the 2010 Memorandum of Understanding on the Conservation of Migratory Sharks).[74]

With respect to Appendix I species, the obligations of Range States include conservation and, where feasible and appropriate, restoration of the habitat of these species that are of importance in preventing extinction.[75] These include critical habitats in the range of each species. With respect to Appendix II-species, Range States shall endeavour to conclude agreements to restore the migratory species concerned to a favourable conservation status.[76] Such agreements may for instance include provisions for the maintenance of a network of suitable migratory routes and sound management of the taking of these species. Overall the convention seeks to identify threatened and endangered migratory species as well as the nature of the threats they face, their migration routes, and ecological characteristics, in order to set a framework for protection. The listing of new species is adopted at the COPs by a two-thirds majority of parties present and voting,[77] following proposal by a state party and taking into account the advice of the CMS Scientific Council.[78]

72 Convention on Migratory Species (n 9) art IV(1).

73 Ibid art I(1)(h).

74 See CMS COP11, 'Proceedings of the 11th Meeting of the Conference of the Parties' (Quito, Ecuador, 4–9 November 2014) <www.cms.int/sites/default/files/publication/cms_cop11_proceedings_e.pdf> accessed 28 May 2018; Memorandum of Understanding on the Conservation of Migratory Sharks (Costa Rica, February 2016) <www.cms.int/sites/default/files/instrument/Sharks_MOU_Text_annexes.pdf> accessed 28 May 2018.

75 Convention on Migratory Species (n 9) art III(4)(a).

76 Ibid art V(1).

77 Ibid art XI(4).

78 CMS COP11, 'Guidelines for Assessing Listing Proposals to Appendices I and II of the Convention' (Quito, 4–9 November 2014) UN Doc UNEP/CMS/Resolution 11.33 <www.cms.int/sites/default/files/document/Res_11_33_Guidelines_Assessing_Listing_Proposals_E_0.pdf> accessed 28 May 2018.

As at December 2017, 647 species have been listed in CMS Appendix I and/or Appendix II.[79]

Species-specific memoranda of understanding (MOU), agreements, action plans and initiatives aim to set out the framework and invite Range States to adopt measures that will control activities that may have adverse impacts on the species concerned, while they are within their jurisdiction (e.g. limit fishing or shipping within migration corridors). Globally, these include 7 agreements (regional or taxa specific agreements),[80] 19 taxa-specific MOUs[81] and 4 special species initiatives.[82] Of those, the 4 agreements and 7 MOUs concern coastal and marine species, including seabirds.[83]

In the context of the implementation of CBD Aichi Target 11,[84] CMS COP10 (2011), COP11 (2014) and COP12 (2017) engaged in the development of ecological networks to address the needs of migratory species. Target 10 of the CMS

79 See database at CMS, 'Species' <www.cms.int/en/species> accessed 28 May 2018.

80 ACAP (for Albatrosses and Petrels), ACCOBAMS (for cetaceans of the Black Seas, Mediterranean and contiguous Atlantic area), AEWA (for African-Eurasian migratory waterbirds), ASCOBANS (for small cetaceans in the Baltic, Northeast Atlantic, Irish and North Seas), EUROBATS (for European bats), Gorilla Agreement (for gorillas and their habitats) and Wadden Sea Seals. CMS, 'Agreements' <www.cms.int/en/cms-instruments/agreements> accessed 28 May 2018.

81 See CMS, 'Memoranda of Understanding' <www.cms.int/en/cms-instruments/mou> accessed 28 May 2018.

82 See CMS, 'Special Species Initiatives' <www.cms.int/en/cms-intruments/special-initiatives> accessed 28 May 2018.

83 Agreements on the Conservation of Albatrosses and Petrels, on the Conservation of Cetaceans of the Black Seas, Mediterranean and Contiguous Atlantic Area, on the Conservation of Small Cetaceans in the Baltic, North East Atlantic, Irish and North Seas, and on the Conservation of Seals in the Wadden Sea. MoUs Concerning Conservation Measures for Marine Turtles of the Atlantic Coast of Africa, concerning Conservation and Management of Dugong (*Dugon dugon*) and their Habitats Throughout their Range, on the Conservation and Management of Marine Turtles and their Habitats of the Indian Ocean and Southeast Asia, concerning Conservation Measures for the Eastern Atlantic Populations of the Mediterranean Monk Seal (*Monachus monachus*), for the Conservation of Cetaceans and their Habitats in the Pacific Islands Region, on the Conservation of Migratory Sharks and concerning the Conservation of the Manatee and Small Cetaceans of Western Africa and Macaronesia.

84 Aichi Target 11 provides that 'By 2020 … 10 percent of coastal and marine areas, especially areas of particular importance for biodiversity and ecosystem services, are conserved through effectively and equitably managed, ecologically representative and well-connected systems of protected areas and other effective area-based conservation measures, and integrated into the wider landscape and seascape'. CBD COP10, 'Decision Adopted by the Conference of the Parties to the Convention on Biological Diversity at its Tenth Meeting. X/2. The Strategic Plan for Biodiversity 2011–2020 and the Aichi Biodiversity Targets'

Strategic Plan for Migratory Species 2015–2023 (adopted at COP11) also specifically provides that 'all critical habitats and sites for migratory species [are to be] identified and included in area-based conservation measures'.[85] This initiative is based on the recognition that 'habitat destruction and fragmentation are among the primary threats to migratory species and that the identification and conservation of habitats of appropriate quality, extent, distribution and connectivity are thus of paramount importance'.[86]

However, as the designation of protected areas across the entire migration range of migratory species is not practically feasible, CMS COP10 adopted a resolution for the identification of ecological networks for the conservation and management of migratory species that focusses on critical habitats and sites.[87] CMS COP10, COP11 and COP12 recognised that 'the identification and conservation of habitats of appropriate quality, extent, distribution and connectivity' and of all 'sites that perform a critical role … such as core areas, corridors, restoration areas and buffer zones' are critical for the conservation of migratory species and may be linked by 'a concept of ecological networks'.[88] These critical habitats and sites vary from one taxonomic group to another and possibly even species to species. In the context of the flyway approach, these networks would for instance include breeding grounds, stop-over sites, non-breeding areas, and feeding and nesting places, all of which can be included in the flyway.[89]

Recommendations for further advancing the design and implementation of ecological networks to address the needs of migratory species include guiding principles for this purpose. They also highlight that they are work in progress,

UN Doc UNEP/CBD/COP/DEC/X/2 (29 October 2010) <www.cbd.int/doc/decisions/cop-10/cop-10-dec-02-en.pdf> accessed 28 May 2018.

85 CMS COP11, 'Strategic Plan For Migratory Species 2015–2023' (Quito, 4–9 November 2014) UN Doc UNEP/CMS/Resolution 11.2. <www.cms.int/sites/default/files/document/Res_11_02_Strategic_Plan_for_MS_2015_2023_E_0.pdf> accessed 28 May 2018.

86 Ibid.

87 CMS COP10, 'The Role of Ecological Networks in the Conservation of Migratory Species' (Bergen, 20–25 November 2011) UN Doc UNEP/CMS/Resolution 10.3. <www.cms.int/sites/default/files/document/10_03_eco_networks_e_0_0.pdf> accessed 28 May 2018.

88 Ibid. CMS COP12, 'The Role of Ecological Networks in the Conservation of Migratory Species' (Manila, October 2017) UN Doc UNEP/CMS/Resolution 12.7. <http://cms.int/sites/default/files/document/cms_cop12_res.12.7_e.pdf> accessed 28 May 2018. See also Resolution 9.9 (Rev.COP12) where parties, the Scientific Council and the CMS Secretariat are urged to identify priority 'species and habitats in the marine sphere requiring intervention by the CMS'. CMS COP12, 'Marine Migratory Species' (Manila, October 2017) UN Doc UNEP/CMS/Resolution 9.9 (Rev.COP12) <http://cms.int/sites/default/files/document/cms_cop12_res.9.9%28rev.cop12%29_e.pdf> accessed 28 May 2018.

89 Ibid. In the context of migratory birds, identification of critical habitats follows flyways, which are flight paths used by migratory birds.

as further research is needed on animal distributions, movement patterns, and gap analyses of existing networks, all of which will influence the design of optimum ecological networks.[90] More specific criteria to identify critical habitats and sites for migratory species are being developed for some taxa such as birds, marine mammals in general and marine turtles.

### 2.4.2 Adoption of the Convention on Migratory Species and Other Relevant Instruments of the CMS Family of Instruments

#### 2.4.2.1 *Adoption of the Convention on Migratory Species*

As of 1 December 2017, the CMS has 126 state parties globally with an uneven regional representation. Whereas most states from Europe, Middle East, Africa, South America and Oceania are party, North and Central America, Southeast Asia and East Asia are hardly represented.

Few of the CMS family of instruments have developed criteria to identify critical habitats. Criteria that involve large geographic areas as well as Range States from developing and developed countries have been selected for this comparative analysis, which investigates commonalities between sets of criteria in an attempt to identify consistent and mutually reinforcing criteria.

#### 2.4.2.2 *Migratory Marine Mammals and Important Marine Mammal Areas (IMMAs)*

IMMAs are defined as 'discrete portions of habitat, important to marine mammal species that have the potential to be delineated and managed for conservation'.[91] Following the development of the IMMAs criteria by the IUCN Joint SSC/WCPA Marine Mammal Protected Areas Task force (the IMMA Task Force), the Aquatic Mammals Working Group of the CMS Scientific Council recommended that these criteria be enforced by CMS COP12 and integrated directly into CMS Ecological Networks Work Programme.[92] However, the

---

90 UN Doc UNEP/CMS/Resolution 12.7 (n 88) Annex.

91 See definition of the Marine Mammal Protected Areas Task Force at Marine Mammal Protected Areas Task Force, 'Important Marine Mammal Areas' <www.marinemammalhabitat.org/activities/immas/> accessed 28 May 2018; See also Hoyt E and Notarbartolo di Sciara G (eds) 'Report of the Workshop for the Development of Important Marine Mammal Area (IMMA) Criteria' (IUCN Marine Mammal Protected Areas Task Force and International Committee on Marine Mammal Protected Areas, Marseille, France, 22 October 2013) <www.marinemammalhabitat.org/download/report-of-the-workshop-for-the-development-of-important-marine-mammal-area-imma-criteria/> accessed 28 May 2018.

92 1st Meeting of the Sessional Committee of the CMS Scientific Council, 'Important Marine Mammal Areas (IMMAs)' (Bonn, Germany, 18–21 April 2016) UN Doc UNEP/CMS/ScC-SC1/Doc.10.4.2.1 <www.cms.int/sites/default/files/document/cms_scc-sc1_doc-10-4-2-1_immas_e.pdf> accessed 28 May 2018.

wording of CMS COP12 is more cautious, as it 'acknowledges' the importance of these criteria and identification processes described in the IMMA Guidance Document. It also 'requests' parties to identify 'specific areas where the identification of IMMAs could be particularly beneficial'.[93] This resolution also encourages ongoing liaison with the International Whaling Commission on using IMMAs to mitigate threats, such as ship strikes, by jointly approaching IMO and presenting selected IMMAs.[94] Unless they are objected to by state parties at future COPs, these IMMA criteria may therefore become de facto criteria for the implementation of CMS instruments that relate to marine mammals. The IMMA Task Force is identifying IMMAs through regional workshops planned from 2016 to 2021.

#### 2.4.2.3 *Marine Turtles in the Indian Ocean and Southeast Asia: Development of a Network of Sites of Importance*

Marine turtles have been listed on Appendix I and II of the Convention on Migratory Species since 1985 at the latest.[95] Parties to the MOU on the Conservation and Management of Marine Turtles in the Indian Ocean and Southeast Asia (IOSEA Marine Turtle MOU) have adopted the IOSEA Network of Sites of Importance for Marine Turtles based on evaluation criteria that were acknowledged by CMS COP11 and 12, in their resolutions on ecological networks.[96] Adopted in 2001, the IOSEA Marine Turtle MOU has 36 state parties. Only nine Range States are not party, which makes it a successful CMS MOU.[97] The objective of the MOU is 'to protect, conserve, replenish and recover marine turtles and their habitats'.[98]

---

93 CMS COP12, 'Important Marine Mammal Areas (IMMAS)' (Manila, October 2017) UN Doc UNEP/CMS/Resolution 12.13. <http://cms.int/sites/default/files/document/cms_cop12_res.12.13_immas_e.pdf> accessed 28 May 2018.

94 Ibid.

95 Only the loggerhead, hawksbill and Ridley turtles were listed on Appendix I in 1985. All the others were listed on Appendix I and II in 1979. CMS, 'Species' (n 79).

96 CMS COP11, 'Advancing Ecological Networks to Address the Needs of Migratory Species' (Quito, 4–9 November 2014) UN Doc UNEP/CMS/Resolution 11.25 in CMS COP11, 'Proceedings of the 11th Meeting of the Conference of the Parties' (n 74). UN Doc UNEP/CMS/Resolution 12.7 (n 88).

97 CMS, 'IOSEA Marine Turtles' <www.cms.int/en/legalinstrument/iosea-marine-turtles> accessed 28 May 2018.

98 Memorandum of Understanding on the Conservation and Management of Marine Turtles and their Habitats of the Indian Ocean and Southeast Asia (Manila, 23 June 2001) <www.cms.int/sites/default/files/instrument/mou_cmp_2009_e.pdf> accessed 28 May 2018.

### 2.4.2.4 *Sharks and Rays*

The protection of migratory sharks and rays is the subject of CMS COP11 Resolution 11.20, in which states are requested, among other measures, 'to identify and conserve critical habitats and life stages, and migration routes' and are encouraged 'to minimize the impact of fishing in migration corridors and other habitats deemed critical to the recovery and sustainability of shark and rays populations'.[99] Other relevant instruments of international law are also referred to (such as the UNGA Resolutions on fisheries, the FAO's International Plan of Action for Sharks, and Convention on International Trade in Endangered Species of Wild Fauna and Flora (CITES)), and states are urged to implement them.

CMS states parties are also invited to sign the CMS MOU on the Conservation of Migratory of Sharks (Shark MOU),[100] which applies to listed shark species in the CMS including rays.[101] As of January 2018, the MOU has been signed by 47 states. However, 106 Range States have yet to sign.[102] The Conservation Plan of the Shark MOU includes the compilation of data including 'shark habitats', 'aggregations', 'the seasonal spatial migration patterns and routes of sharks', and the identification and prioritisation '(with a view to developing conservation measures) [of] critical shark habitats including critical migration routes'.[103] The parties have met twice since 2010. Protection of critical habitats and corridors for critical life stages is mentioned in the report of the second meeting (2016); however there is no mention of more specific criteria to identify that, among the habitats, are 'critical enough' and should therefore benefit from a greater degree of protection.

### 2.4.2.5 *Marine Important Bird Areas (IBAs) and Flyways*

With respect to migratory bird flyways, no set of general and scientific criteria has yet been established by CMS instruments. The Flyways Working Group of the CMS Scientific Council released three global reviews, which include an

---

99 CMS COP Recommendation 8.16 called for the development of a global migratory shark conservation instrument. CMS COP8, 'Migratory Sharks' UN Doc UNEP/CMS/ Recommendation 8.16 (2005) <www.cms.int/sites/default/files/document/CP8Rec_8_16_Migratory_Sharks_E_0.pdf> accessed 28 May 2018; CMS COP11, 'Conservation of Migratory Sharks and Rays' (Quito, 4–9 November 2014) UN Doc UNEP/CMS/Resolution 11.20 <www.cms.int/sites/default/files/document/Res_11_20_Conservation_Sharks_and_Rays_E.pdf> accessed 28 May 2018.

100 UN Doc UNEP/CMS/Resolution 11.20 (n 99) para 12.

101 Memorandum of Understanding on the Conservation of Migratory Sharks (n 74).

102 CMS Sharks MOU, 'Signatories and Range States' <www.cms.int/sharks/en/signatories-range-states> accessed 28 May 2018.

103 Ibid.

analysis of existing mechanisms and discussion and recommendations for the identification of critical habitats. These will be taken into account in the CMS Programme of Work on Migratory Birds and Flyways 2014–2023, which includes the identification and designation of critical sites.[104] Of note, the reviews endorse and discuss criteria developed under the Ramsar Convention and three specialised initiatives:[105] the identification of Important Bird and Biodiversity Areas (IBAs) and Marine IBAs by the NGO BirdLife International,[106] and the critical sites network developed under the Wings over Wetlands (WOW) UNEP-GEF African-Eurasian Flyways Project, and the East Asia-Australasian Flyway Partnership (EAAFP).[107] Importantly, the East Asian-Australasian Flyway Partnership, which has been agreed outside the CMS Family of Instruments, covers numerous CMS Appendix I and II-listed bird species. Furthermore, CMS COP9 welcomed its establishment and recognised that the partnership fulfils the conditions of a non-binding agreement under the CMS.[108] This partnership works on the basis of voluntary submissions of sites by participating states.

---

104 CMS COP12, 'Programme of Work on Migratory Birds and Flyways (2014–2023)' (Manila, October 2017) UN Doc UNEP/CMS/Resolution 12.11, /Annex1. <www.cms.int/sites/default/files/document/cms_cop12_res.12.11_annex1_pow-2014-2023_e.pdf> accessed 28 May 2018.

105 Colin A Galbraith et al, *A Review of Migratory Bird Flyways and Priorities for Management* (CMS Technical Series Publication No 27, UNEP/CMS Secretariat 2014) 70 <http://cms.int/sites/default/files/publication/CMS_Flyways_Reviews_Web.pdf> accessed 28 May 2018.

106 BirdLife International, 'Marine IBA E-Atlas' <https://maps.birdlife.org/marineIBAs/default.html> accessed 28 May 2018. It is based on standardised techniques for identifying priority sites for the conservation of seabirds at sea. BirdLife International, 'Marine Important Bird Areas Toolkit: Standardised Techniques for Identifying Priority Sites for the Conservation of Seabirds at Sea' (Version 1.2. February 2011) <http://datazone.birdlife.org/userfiles/file/Marine/Marinetoolkitnew.pdf> accessed 28 May 2018. This refers to Ramsar Criteria A4, which aims to identify congregations of birds, as well as to other initiatives. CMS Scientific Council Flyways Working Group Reviews, 'Review 2. Review of Current Knowledge of Bird Flyways, Principal Knowledge Gaps and Conservation Priorities' UN Doc UNEP/CMS/ScC17/Inf.4.2b (11 April 2011) 130.

107 Critical site identification method of WOW. See WINGS over Wetlands UNEP/GEF African-Eurasian Flyways Project, The Critical Site Network: Conservation of Internationally Important Sites for Waterbirds in the African-Eurasian Waterbirds Agreement Area (Wetlands International and BirdLife International 2011). <www.eaaflyway.net/documents/resources/Critical_Site_Network.pdf> accessed 28 May 2018; East Asian-Australasian Flyway Partnership, 'Flyaway Site Network' <http://eaaflyway.net/about-us/the-flyway/flyway-site-network/> accessed 28 May 2018.

108 Convention on Migratory Species (n 9) art IV(4). CMS COP9, 'Priorities for CMS Agreements' UN Doc UNEP/CMS/Resolution 9.2 (2008) Annex VIII. <www.cms.int/sites/default/files/document/Res_9_02_CMS_Agreement_Priotirties_En.pdf> accessed 28 May 2018.

The information form to be submitted in the nomination of a site includes scientific criteria adapted from the Ramsar Convention, with respect to wetlands of international importance, and IBA with respect to congregations, albeit with lower thresholds.

### 2.4.3 Scientific Criteria

While no generic scientific criteria have been developed by the CMS COP to streamline the identification process of critical and essential habitats for listed migratory species, the importance for state parties to proceed with such identification is highlighted in most COP reports. The last two COPs even adopted a resolution for further advancing the design and implementation of ecological networks to address the needs of migratory species with the following guidance:

(i) defining network objectives;
(ii) ensuring that networks have a sufficiently holistic scope;
(iii) ensuring the functional benefits of connectivity;
(iv) considering other design factors such as particular migratory patterns and use of critical sites;
(v) assessing risks;
(vi) further developing knowledge and engagement, and
(vii) ensuring the development of an implementation regime, including an 'adaptive management' approach.[109]

Such a network should address the needs of migratory species throughout their life cycles and migratory ranges.[110] However, the CMS COP welcomes numerous network creations, even though they may rely on different criteria, provided that they are 'scientifically robust criteria'. A good example is the different flyways that rely on similar but different criteria. Two other examples of sets criteria adopted or acknowledged by CMS instruments to identify critical habitats are the IMMA criteria and the criteria for the IOSEA Marine Turtle Site Network.

Despite the fact that the IMMAs criteria (Table 4.4) have been developed by the IUCN outside the mandate of an international instrument for the protection of marine mammals,[111] they are in effect endorsed by CMS COP12, which requires parties to apply them. Their authority also results from the fact that

109 UN Doc UNEP/CMS/Resolution 12.7 (n 88).
110 Ibid para 8.
111 Marine Mammal Protected Areas Task Force, 'IMMA Selection Criteria' <www.marinemammalhabitat.org/activities/immas/imma-criteria/> accessed 28 May 2018.

TABLE 4.4 Scientific criteria to identify International Marine Mammals Areas

| No | Criteria | | Sub-criteria | Description |
|---|---|---|---|---|
| 1 | A | **Species or population vulnerability** | – | Areas containing habitat important for the survival and recovery of threatened and declining species |
| 2.1 | B(i) | **Distribution and abundance** | **Small and resident populations** | Areas supporting at least one resident population, containing an important proportion of that species or population, that are occupied consistently |
| 2.2 | B(ii) | | **Aggregations** | Areas with underlying qualities that support important concentrations of a species or population |
| 3.1 | C(i) | **Key life cycle activities** | **Reproductive areas** | Areas that are important for a species or population to mate, give birth, and/or care for young until weaning |
| 3.2 | C(ii) | | **Feeding areas** | Areas and conditions that provide an important nutritional base on which a species or population depends |
| 3.3 | C(iii) | | **Migration routes** | Areas used for important migration or other movements, often connecting distinct life-cycle areas or the different parts of the year-round range of a non-migratory population |
| 4.1 | D(i) | **Special attributes** | **Distinctiveness** | Areas that sustain populations with important genetic, behavioural or ecologically distinctive characteristics |
| 4.2 | D(ii) | | **Diversity** | Areas containing habitat that supports an important diversity of marine mammal species |

they are designed by the IUCN, are based on a robust and iterative scientific methodology, and are being tested regionally and improved to reflect the lessons learned. They are also the only existing set of criteria developed for marine mammals globally and are likely to remain so.

By contrast, the criteria developed for the identification of critical sites in the IOSEA Marine Turtle Site Network and adopted by the signatory states (Table 4.5) are not limited to species' biological and ecological criteria. They also include governance, socio-economic and political criteria and network-wide ecological criteria. Given the relative value given to different criteria, this methodology results in non-ecological criteria being brought to an equal footing with ecological criteria as they add up respectively, to 72 and 89 points maximum. A total value of 75 points minimum is required for a site to qualify.[112]

There seems to be an overlap between the identification of sensitive marine areas under the CMS and some EBSA identification criteria, which would apply to critical habitats of threatened or endangered migratory species. This observation is based on the language used to describe the following EBSA criteria:

(i) EBSA Criterion 2 on areas that have a special importance for life history for species, i.e. areas that are required for a population to survive and thrive;
(ii) EBSA Criterion 3 on the importance of an area for threatened, endangered or declining species and/or habitats;
(iii) EBSA Criterion 4 on vulnerability, fragility, sensitivity or slow recovery; and
(iv) EBSA Criterion 6 on biological diversity.[113]

A similar overlap exists between the identification of sensitive marine areas under the CMS and the identification of Ramsar sites that are critical for migratory birds, which is why Ramsar criteria have been used in the development of flyway sites. The same can be said for other endangered migratory species that rely on coastal wetlands for survival.[114]

---

112 IOSEA Marine Turtle MoU, 'Criteria for the Evaluation for the Evaluation of Nominations to the Network of Sites of Importance' (9 July 2012) <www.cbd.int/doc/meetings/mar/ebsa-sio-01/other/ebsa-sio-01-iosea-01-en.pdf> accessed 28 May 2018.

113 Consistent with the analysis in the Global Ocean Biodiversity Initiative (GOBI)'s Report to the CMS: CY Kot et al (2014) 'A Review of Marine Migratory Species and the information Used to Describe Ecologically or Biologically Significant Areas (EBSAs). Final Report' in UN Doc UNEP/CMS/COP11/Inf23. (21 October 2014) <www.cms.int/sites/default/files/document/COP11_Inf_23_EBSA_final_report.pdf> accessed 28 May 2018.

114 CMS COP12 adopted a resolution 12.25 promoting conservation of critical intertidal and other coastal habitats for migratory species and highlights the importance of the CBD

TABLE 4.5 Criteria for inclusion in the IOSEA Marine Turtle Site Network

| No | Category | | Criteria | Description |
|---|---|---|---|---|
| 1.1 | EB1a | Ecological and biological criteria (EB) | Abundance at nesting sites | Number of marine turtles constituting a management unit, the size of which is considered to be of regional importance |
| 1.2 | EB1b | | Abundance at foraging sites | Number of marine turtles (of any species) foraging at a site, which is considered to be of regional importance |
| 1.3 | EB2 | | Species or management unit richness | Number of species or marine turtle management units (if known) regularly using a site's nesting habitat or foraging habitat |
| 1.4 | EB3 | | Presence of rare marine turtle species | Presence of a marine turtle species that is considered rare in the IOSEA region |
| 1.5 | EB4 | | Resistance and resilience | Site containing habitat of importance to marine turtles that is likely to be relatively resistant and/or resilient to disturbance |
| 2.1 | G1 | Governance criteria (G) | Legal framework | Provides adequate protection of the site and of the life stage(s) of the marine turtle population found at the site |
| 2.2 | G2 | | Conservation actions | Conservation interventions have been undertaken to mitigate known threats to marine turtles identified at the site |
| 2.3 | G3 | | Collaborative management, surveillance and enforcement | Participatory work with local stakeholders to strengthen local stewardship of marine turtles, and/or to provide for adequate surveillance and enforcement of prevailing regulations |
| 2.4 | G4 | | Research and monitoring | Site is currently used to monitor marine turtle abundance or other critical parameters and/or has marine turtle surveys with standardised data that span > 15 years for the site and/or survey data are used to estimate trends in the size of management units |
| 2.5 | G5 | | Sustainable human and financial resources | Availability of long-term resources (human and financial) to enable effective governance activities, including monitoring, management interventions, |

| No | | Category | Criteria | Description |
|---|---|---|---|---|
| | | | | surveillance and enforcement, and performance evaluation. |
| 3.1 | S1 | Socio-economic and political criteria (S) | Cultural importance | Site contains prehistoric, historic, and/or contemporary resources, or embodies non-consumptive traditional beliefs/practices of cultural, religious and/or spiritual significance, in relation to marine turtles |
| 3.2 | S2 | | Compatible activities | Activities occurring within the vicinity of the site that are compatible with the conservation of marine turtles and their habitats |
| 3.3 | S3 | | Educational values | Existence of actual, or future opportunities for, educational and outreach activities, by virtue of the site's location and other inherent characteristics |
| 3.4 | S4 | | Existing recognition | Length of existing protected status or other national, regional or international recognition for the site's value to marine turtles |
| 3.5 | S5 | | National significance | Significance of the site in a national context, relative to other sites |
| 3.6 | S6 | | Perceived ancillary benefits as a consequence of site inclusion in the network | e.g. for other biodiversity/local communities associated with the site, or other related conservation initiatives |
| 4.1 | N1 | Network-wide Ecological criteria | Representativeness and replication | Site contributes to the network's: (i) adequate representation of the full range of habitat diversity required for the maintenance of marine turtle management units and species of the IOSEA region (representativeness), and/or (ii) inclusion of multiple sites containing identical habitat types (replication) |
| 4.2 | N2 | | Ecological connectivity | Site contributes to protecting functional links among areas of marine turtle habitat |
| 4.3 | N3 | | Area | Area of a site or combined area of functionally-linked sites contributes to protecting the area of marine turtle habitat needed |

## 3 Identification of Sensitive Areas for the Purpose of Limiting Pollution and Adverse Impact from Maritime Activities

The main international instruments that seek to regulate activities at sea globally and protect and preserve the marine environment, including sensitive marine areas, are presented below. They are grouped according to categories of activities at sea that they regulate.[115]

### 3.1 *Shipping Regulations Adopted under the Purview of the IMO*

As the United Nations specialised agency responsible for the promotion of safe, secure, environmentally sound, efficient and sustainable shipping, the IMO has overseen the development of several operational mechanisms to provide increased protection from shipping in areas that are considered to be particularly sensitive. The first area-based mechanism was developed in MARPOL, which defines certain sea areas as 'special areas' in which authorised discharges are further limited. The second mechanism is the PSSA and its Associated Protective Measures (APMs). Direct routeing measures, vessel traffic services and ship reporting systems are not included here, because no specific criteria have been developed for their application.

#### 3.1.1 Special Areas under MARPOL

##### 3.1.1.1 *Source Instrument and Legal Background*

MARPOL is the main IMO instrument for the prevention of pollution from ships. It regulates both operational discharges and accidental pollution through six technical annexes, each devoted to a specific type of pollution: oil (I), noxious liquid substances in bulk (II), harmful substances in packaged form (III), sewage (IV), garbage (V); and air pollution (VI).

In Annexes I, II, IV and V, MARPOL defines certain sea areas as 'Special Areas', where 'for recognised technical reasons in relation to [their] oceanographic

---

and the Ramsar Convention. CMS COP12, 'Promoting Conservation of Critical Intertidal and Other Coastal Habitats for Migratory Species' (Manila, October 2017) UN Doc UNEP/CMS/Resolution 12.25 <www.cms.int/sites/default/files/document/cms_cop12_res.12.25_conservation-intertidal-coastal-habitats_e.pdf> accessed 28 May 2018.

115 Not all maritime activities, whether within or beyond national jurisdiction, are regulated internationally. Within national jurisdiction, offshore oil and gas activities and other types of mineral extractions are activities for which states have failed to fulfil their obligation to agree on global rules and standards. UNCLOS (n 13) art 208. See A Roach, 'International Standards for Offshore Drilling' in MH Nordquist et al (eds), *The Regulation of Continental Shelf Development: Rethinking International Standards* (Brill Nijhoff 2013) 105–52.

and ecological conditions and to the particular character of [their] traffic', the adoption of a higher level of protection (than in other areas of the sea) is necessary.[116] These 'Special Areas under MARPOL' have increased restrictions or complete prohibitions on the discharge of oil, noxious liquid substances, sewage and garbage. 'Emission Control Areas' have also been designated with more stringent restrictions on specific atmospheric emissions from ships (sulphur emissions and nitrogen oxides) in Annex VI.[117]

As of 19 January 2017, there are 20 Special Areas under MARPOL Annexes I, II, IV and V and 4 Emission Control Areas under Annex VI. Some areas are protected under several annexes; these include the Antarctic area, the Mediterranean Sea, the Baltic Sea, the Black Sea and the 'Gulfs' area.[118]

#### 3.1.1.2 *Status of Adoption (Table 4.6)*[119]

#### 3.1.1.3 *Scientific Criteria*

Several guidelines have been adopted by the Marine Environment Protection Committee (MEPC) for the designation of Special Areas. The first guidelines

116 MARPOL (n 2) Annex I Regulation 1(10) and Annex V Regulation 1(3). A similar definition is included in MARPOL Annex II Regulation 1(7): the 'particular character of the traffic' is changed to the 'peculiar transportation traffic'. The initial draft of MARPOL Annex IV did not include specific provisions for Special Areas; these were added by amendment to the Annex by Resolution MEPC 200(62) adopted on 15 July 2011. It defines 'Special Area' with the same language as Annexes I and II. AMENDMENTS TO THE ANNEX Of the Protocol of 1978 Relating to the International Convention for the Prevention of Pollution from Ships, 1973, IMO Res MEPC.200(62) (15 July 2011) <www.imo.org/en/KnowledgeCentre/IndexofIMOResolutions/Marine-Environment-Protection-Committee-(MEPC)/Documents/MEPC.200(62).pdf> accessed 28 May 2018. See Aldo Chircop, Chapter 5 'The Use of IMO Instruments for Marine Conservation on the High Seas'.

117 MARPOL (n 2) Annex VI Regulations for the Prevention of Air. Pollution from Ships, included in Protocol of 1997 to amend International Convention for the Prevention of Pollution from Ships of 2 November 1973, as modified by the Protocol of 17 February 1978 (adopted 26 September 1997. entered into force 19 May 2005). In October 2008 (MEPC 58), a revised Annex VI, which significantly tightened emissions limits, was adopted. It entered into force on 1 July 2010. IMO, 'Prevention of Air Pollution from Ships' <www.imo.org/en/OurWork/Environment/PollutionPrevention/AirPollution/Pages/Air-Pollution.aspx> accessed 28 May 2018.

118 IMO, 'Special Areas under MARPOL' <www.imo.org/en/OurWork/Environment/SpecialAreasUnderMARPOL/Pages/Default.aspx> accessed 28 May 2018.

119 Status as of 28 May 2018 as provided in the IMO, 'Comprehensive Information on the Status of Multilateral Conventions and Instruments in respect of which the IMO or its Secretary-General Performs Depositary or Other Functions' <www.imo.org/en/About/Conventions/StatusOfConventions/Documents/Status%20-%202018.pdf> accessed 28 May 2018.

TABLE 4.6 Adoption of MARPOL Annexes

| MARPOL annexes | Number of state parties |
|---|---|
| I and II | 156 |
| III | 148 |
| IV | 142 |
| V | 153 |
| VI | 91 |

were adopted in 2001 and concerned both the designation of Special Areas and the identification of PSSAs.[120] The latest guidelines (which are currently applicable) are devoted solely to Special Areas under MARPOL: the 2013 Guidelines for the Designation of Special Areas under MARPOL (the Special Area Guidelines).[121] They set out in detail the conditions to be met for an area to qualify as a Special Area under MARPOL Annexes I, II, IV and V.[122]

The conditions for the designation of a Special Area are grouped into three cumulative categories of oceanographic conditions, ecological conditions and vessel traffic characteristics as set out in Table 4.7.[123] While the three categories of conditions need to be met, not all ecological criteria provided for (to determine whether the ecological conditions are met) need to be fulfilled. Meeting one of them (for instance the spawning ground of a commercially important species (criterion 2.3) or large aggregation of an endangered species (criterion 2.1)) would be sufficient, provided that the oceanographic conditions

120 Guidelines for the Designation of Special Areas and the Identification of Particularly Sensitive Areas, IMO Assembly Res A.720(17) (6 November 1991.) (1991 Guidelines). They were revoked and replaced by 2001 Guidelines for the Designation of Special Areas under MARPOL 73/78 and Guidelines for the Identification and designation of PSSAs, IMO Assembly Res A.927(22) (29 November 2001). These were revised in 2005 by IMO Assembly Resolution A.982(24) (removing parts relating to PSSAs), before being again revoked and replaced by the 2013 Guidelines, which are currently applicable. 2005 Revised Guidelines for the Identification and Designation of Particularly Sensitive Sea Areas, IMO Assembly Res A.982(24) (1 December 2005) (2005 Revised PSSA Guidelines) http://www.imo.org/blast/blastDataHelper.asp?data_id=25322&filename=A982(24).pdf accessed 28 May 2018.

121 2013 Guidelines for the Designation of Special Areas under MARPOL, IMO Assembly Res A.1087(28) (4 December 2013).

122 There is no record of guidelines for the designation of Emission Control Areas other than Appendix III to MARPOL (n 2) Annex VI 'Criteria and Procedures for Designation of Emission Control Areas' as amended. See IMO, 'Report of the Marine Environment Protection Committee on its Fifty-Eighth Session' IMO Doc MEPC 58/23/Add.1 (17 October 2008).

123 Based on the Special Area Guidelines (n 121).

TABLE 4.7 Conditions for the designation of Special Areas under MARPOL

| | Conditions | Criteria | Description |
|---|---|---|---|
| 1 | **Oceanographic conditions** | – | May cause concentration or retention of harmful substances in the water or sediments of the areas e.g. convergence zones or gyres, low flushing, extreme ice state, etc. |
| 2.1 | **Ecological conditions** | **Depleted, threatened or endangered marine species** | – |
| 2.2 | | **High natural productivity** | – |
| 2.3 | | **Spawning, breeding and nursery areas for important marine species** | Includes migratory marine species and migratory birds |
| 2.4 | | **Rare or fragile ecosystems** | E.g. corals reefs, mangrove, seagrass beds and wetlands |
| | | **Critical habitats for marine resources** | Includes fish stocks and areas of critical importance for the support of large marine ecosystems |
| 3 | **Vessel traffic characteristics** | – | Discharge would be unacceptable in the light of existing conditions |

and the vessel traffic characteristics demonstrate the threat posed by the discharge(s).

A Special Area may encompass the maritime zones of several states or even an entire enclosed or semi-enclosed sea area (including high seas where applicable),[124] such as the Baltic Sea, the North Sea and the Mediterranean Sea.

124 Special Area Guidelines (n 121) para 2.2.

### 3.1.2 PSSAS

#### 3.1.2.1 *Legal Background*

A PSSA is an area 'which need[s] special protection through action by IMO because of [its] significance for recognized ecological or socio-economic or scientific reasons and which may be vulnerable to damage by maritime activities'.[125] This includes therefore ship-source pollution as well as environmental hazards associated with shipping such as physical damage to marine habitats or organisms.[126] The PSSA concept has been developed by the IMO to attract attention to such sensitive areas and provide a basis on which protective measures may be taken through IMO measures.[127] Originally, as it is clearly stated in the first guidelines adopted in 1991, PSSAs were framed as a means to adopt complementary measures in MPAs,[128] be they national measures or measures adopted in the context of global or regional seas agreements.[129] The 1991 PSSA and Special Area Guidelines refer specifically to MPAs designated in the context of the World Heritage Convention, Ramsar Convention and the CMS.[130] This reference explains why the criteria for the designation of PSSAs include criteria developed under these three instruments to identify sensitive areas.[131]

---

125 Definition from the MEPC 36th session in 1986 as reiterated in the Preface of the 1991 Guidelines (n 120). See Aldo Chircop, Chapter 5 'The Use of IMO Instruments for Marine Conservation on the High Seas'.

126 1991 Guidelines (n 120) para 1.4.8.

127 G Peet (n 1).

128 The PSSA mechanism would be particularly useful to impose restrictions to shipping in sensitive areas that lie beyond the coastal state competence to impose unilateral protective measures on foreign vessels; examples include sensitive areas in EEZs or in International Straits. See Gjerde and Freestone's recount of the early meetings on the legal implications of PSSA. K Gjerde and D Freestone, 'Particularly Sensitive Sea Areas—An Important Environmental Concept at a Turning Point' (1994) 9(4) *The International Journal of Marine and Coastal Law* 431, 433. Nevertheless many PSSAs have been designated within 12 NM of the coast where getting the IMO approval may also be necessary if the protective measure can interfere with the right of innocent passage of foreign vessels.

129 1991 Guidelines (n 120) Ch 1.

130 1991 Guidelines (n 120) para 1.3.2 to 1.3.5. Unlike the following guidelines, the 1991 Guidelines include very useful information on the historical development of the PSSA concept as well as examples of situations it may be used in and specific measures that may be taken.

131 See Sections 2.2 to 2.4 of this chapter. The criteria for the identification of sensitive areas under the CBD have been included in revised guidelines, after entry into force of this convention.

As of 31 December 2017, 15 PSSAs have been designated: 7 are located in the Indo-Pacific, 5 in Europe and the Mediterranean seas, and 3 in the Atlantic Ocean and Caribbean seas.[132]

#### 3.1.2.2 *Conditions and Scientific Criteria*

The PSSA mechanism and procedure are described in guidelines developed by the MEPC, which have been reviewed several times to increase the rigour of the process and to respond to criticisms.[133] The applicable guidelines are the 2005 Revised Guidelines for the Identification and Designation of Particularly Sensitive Sea Areas, as amended in 2015 (the PSSA Guidelines).[134] The PSSA Guidelines set out three cumulative conditions for the designation of PSSAs:

(i) One of the alternative criteria for the identification of a PSSA must be met, i.e. 1 of 11 ecological criteria, 1 of 3 social, cultural and economic criteria, or 1 of 3 scientific and educational criteria (Table 4.8).[135]

(ii) The area demonstrates vulnerability of the area to impact from international shipping. This involves consideration of the several factors, such as vessel traffic characteristics (type of maritime activity, vessel type, volume, etc.), natural factors, evidence of impact from shipping or history of incidents, and analysis of actual or expected benefits from past or prospective measures.[136]

(iii) The area has one or several appropriate APMs that the IMO has competence to approve.[137]

---

132 IMO, 'Particularly Sensitive Sea Areas' <www.imo.org/en/OurWork/Environment/PSSAs/Pages/Default.aspx> accessed 28 May 2018.

133 See for instance KM Gjerde (2001) 'Protecting Particularly Sensitive Sea Areas from Shipping: A Review of IMO's new PSSA Guidelines'(Proceedings of the 12th Biennal Coastal Zone Conference, Cleveland -, USA, 15–19 July 2001) <http://webapp1.dlib.indiana.edu/virtual_disk_library/index.cgi/4916229/FID3790/pdf_files/gjerdek.pdf> accessed 28 May 2018; J Roberts, *Marine Environment Protection and Biodiversity Conservation—The Application and Future Development of the IMO's Particularly Sensitive Sea Area Concept* (Springer 2007) 88–96.

134 MEPC Resolutions MEPC 982(24) and MEPC 267(68) (15 May 2015). See also n 120.

135 Based on the 2005 Revised PSSA Guidelines (n 120 and 134).

136 2005 Revised PSSA Guidelines Section 5 (n 120 and 134).

137 2005 Revised PSSA Guidelines Sections 6 and 7.5 to 7.10 (n 120 and 134).

TABLE 4.8 Criteria for the identification of an area as a potential PSSA

| No | Category | Criteria | Description of corresponding area's characteristics |
|---|---|---|---|
| 1.1 | Ecological | Uniqueness or rarity | 'The only one of its kind'; rare means that it occurs in a few locations or has been seriously depleted across its range. Can be habitats of rare, threatened or endangered species that only occur in one area, or nursery, feeding, breeding or spawning areas |
| 1.2 | | Critical habitat | Essential for the survival, function, or recovery of fish stocks or rare or endangered species, or for the support of large marine ecosystems |
| 1.3 | | Dependency | Ecological processes are highly dependent on biotically structured systems (e.g. coral reefs, mangrove forest, seagrass beds). Includes migratory routes of marine species and birds. |
| 1.4 | | Representativeness | Outstanding and illustrative example of specific biodiversity, ecosystems, ecological or physiographic processes or community or habitat types, etc. |
| 1.5 | | Diversity | May have an exceptional variety of species or genetic diversity or highly varied ecosystems, habitats or communities |
| 1.6 | | Productivity | Particularly high rate of natural biological production; e.g. relatively high biomass in oceanic fronts, upwelling areas and some gyres |
| 1.7 | | Spawning or Breeding grounds | Critical spawning, breeding or nursery area for marine species that may spend the rest of their life cycle elsewhere or for migratory routes for marine species or birds |
| 1.8 | | Naturalness | Relative lack of human-induced disturbance or degradation |
| 1.9 | | Integrity | A biologically functioning unit, and an effective, self-sustaining ecological entity |

| No | Category | Criteria | Description of corresponding area's characteristics |
|---|---|---|---|
| 1.10 | | Fragility | Highly susceptible to degradation by natural events or human activities. Some biotic communities may have low tolerance to changes in environmental conditions or exist close to the limits of their tolerance. Existing stress can justify need for special protection from further stress. |
| 1.11 | | Bio-geographic importance | Contains rare biogeographic qualities or representative of a biogeographic 'type(s)' or contains unique or unusual biological, chemical, physical or geological features |
| 2.1 | Social, cultural and economic | Social or economic dependency | Environmental quality and use of living marine resources are of particular social or economic importance, including fishing, recreation, tourism, people livelihoods, etc. |
| 2.2 | | Human dependency | Particularly important for traditional subsistence or food production or the protection of the cultural resources of the local populations |
| 2.3 | | Cultural heritage | Particularly important due to the presence of significant historical and archaeological sites |
| 3.1 | Scientific and educational | Research | Has high scientific interest |
| 3.2 | | Baseline for monitoring studies | Provides suitable baseline conditions: no substantial perturbations or steady state such that considered to be in a natural or near-natural condition |
| 3.3 | | Education | Offers exceptional opportunity to demonstrate particular natural phenomenon |

#### 3.1.3 Sensitive Area Mapping to Support Oil Spill Response and Preparedness under the 1990 Convention on Oil Pollution Preparedness, Response and Co-operation[138] (OPRC Convention)

The OPRC Convention seeks to ensure that states take all the necessary measures to prepare for and respond to an oil spill incident. It fleshes out the legal duties necessary to ensure adequate preparedness, contingency planning and response to a spill. The measures include on-board oil pollution emergency plans as well as for all areas surrounding oil handling facilities, availability of oil combatting equipment, oil pollution reporting procedures, a national system in place to respond, and a cooperation mechanism. The OPRC Convention has been adopted by 112 states globally.[139]

Several sets of guidelines have been adopted by the IMO in order to support the implementation of the OPRC Convention. Important in the context of the protection of sensitive marine and coastal areas is the guidance on sensitivity mapping for oil spill response, first published in 1996 and subsequently updated by MEPC 63 (The Sensitivity Mapping Guidance).[140] This guidance document provides a methodology to map sensitive marine habitats including marine biodiversity, in order to mitigate the impacts of oil pollution incidents, according to a three-tier approach based on the expected severity of the spill. The approach is different from the approach in sets of criteria previously reviewed. It seeks to map

(i) shoreline types including the substrate and ecosystem type (an exposed rocky shore is less sensitive than a tidal flat or a mangrove area);
(ii) biodiversity-sensitive elements including sensitive areas and species and sub-tidal habitats;
(iii) sensitive socio-economic features;
(iv) oil spill response operational and logistical features; and
(v) potential sources of accidental pollution.[141] (Table 4.9)

138 1990 Convention on Oil Pollution Preparedness, Response and Co-operation (adopted 30 November 1990, entered into force 13 May 1990) 1891 UNTS 77 (OPRC).

139 IMO, 'Comprehensive Information on the Status of Multilateral Conventions and Instruments' (n 119).

140 IMO Doc MEPC 63/23 adopts the IPIECA-IMO-OGP Guidance submitted as MEPC 62/8, IMO, 'Implementation of the OPRC Convention and the OPRC-HNS Protocol and Relevant Conference Resolutions. Guidance on Sensitivity Mapping for Oil Spill Response' (5 April 2011). IMO, 'Report of the Marine Environment Protection Committee on its Sixty-Third Session' IMO Doc MEPC 63/23 (14 March 2012); IPIECA, IMO and OGP, 'Sensitivity Mapping for Oil Spill Response' (OGP Report Number 477, 2012) <www.oilspillresponseproject.org/wp-content/uploads/2016/02/GPG-Sensitivity-Mapping.pdf> accessed 28 May 2018.

141 While shoreline types are not relevant to high seas, the other components are.

TABLE 4.9 Oil Sensitivity Mapping under OPRC Convention

| No | Components | Sub-components | Description |
|---|---|---|---|
| 1- | **Shoreline types** | **Grain, size, slope**<br>**Exposure to wave**<br>**Biological productivity and sensitivity** | Mapping shoreline type and its general environmental sensitivity. Includes exposed rocky shore, wave-cut platforms in bedrock, mud or clay, sand, gravel, mud beaches/flats, sheltered shores, tidal-flats, marshes, mangroves, etc. |
| 2- | **Biodiversity-sensitive elements** | **Sensitive areas and species** | Includes endangered sensitive species, coastal and marine |
| | | **Sub-tidal habitats** | Coral reefs, sea grass beds and kelp beds |
| 3- | **Socio-economic features** | **Living resources**<br>**Non-living resources**<br>**Managed Areas** | Aquaculture, subsistence and commercial fishing or fishing villages<br><br>Water intakes, tourism and recreation areas, industrial activities, infrastructures, cultural sites, etc.Include features that may be affected |

Following the identification and mapping of the different types of shores, sensitive ecosystems and natural ecosystems as well as socio-economic features and their sensitivity to a potential oil spill must be ranked based on their expected recovery time (on a scale of one to ten). This sensitivity includes both intrinsic sensitivity and exposure to oil (depending on oil type and conditions).

## 3.2 *Vulnerable Marine Ecosystems in the Context of Fisheries*

### 3.2.1 Legal Background

UNCLOS was negotiated in the 1970s with the intention to regulate states' rights and obligations with respect to all the uses of the oceans.[142] UNCLOS

142 UNCLOS preamble states that the Convention establishes a legal order for the seas and oceans. See Resolution Adopted by the General Assembly. 47/65 Law of the Sea, UNGA Res 47/65 (11 December 1992) <www.un.org/en/ga/search/view_doc.asp?symbol=A/

provisions on fisheries concern the conservation and the exploitation of living resources. The overriding objective for exploitation is that of 'maximum sustainable yield' through the determination of allowable catch based on best scientific evidence, in order to avoid over-exploitation.[143] While exploited living resources are the primary focus of these provisions, effects on species associated with or dependent upon harvested species must also be taken into consideration 'with a view to maintaining or restoring' their populations 'above levels at which their reproduction may become seriously threatened'.[144]

The 1995 United Nations Fish Stocks Agreement[145] further implements the obligation in UNCLOS to protect and preserve the marine environment in the context of straddling and highly migratory fish stocks. Specifically mentioned in the preamble to the Fish Stocks Agreement is 'the need to avoid adverse impacts on the marine environment, preserve biodiversity [and] maintain the integrity of marine ecosystems'.[146] With these objectives in mind, the Fish Stocks Agreement also explicitly requires parties to adopt a precautionary approach to ensure the conservation of target and non-target species and to protect habitats of special concern.[147]

As of 31 December 2017, 168 states are party to UNCLOS and 87 states are party to the Fish Stocks Agreement.

The 1995 Code of Conduct for Responsible Fisheries adopted by the FAO Conference[148] further recommends that states

RES/47/65> accessed 28 May 2018. See also UN DOALOS 'Third United Nations Conference on the Law of the Sea' <www.un.org/depts/los/convention_agreements/convention_overview_convention.htm> accessed 28 May 2018. The comprehensiveness of UNCLOS and its pre-eminence in international law for all matters pertaining to the oceans are further discussed in Section 4 of this chapter.

143 UNCLOS (n 13) art 61(2) and (3).

144 Ibid art 61(4).

145 United Nations Agreement for the Implementation of the Provisions of the United Nations Convention on the Law of the Sea of 10 December 1982 relating to the Conservation and Management of Straddling Fish Stocks and Highly Migratory Fish Stocks (adopted 4 August 1995, entered into force 11 December 2001) 2167 UNTS 88 (UN Fish Stocks Agreement).

146 Ibid preamble.

147 Ibid arts 5(c) and (g) and 6(3)(d). See Robin Warner, Chapter 6 'Conservation and Management of Marine Living Resources beyond National Jurisdiction: Filling the Gaps'.

148 The FAO Conference is composed of representatives of member states to the 1945 Constitution of the Food and Agriculture Organization of the United Nations. The FAO Conference has 194 member states. FAO, 'FAO Conference (CC-701)' <www.fao.org/unfao/govbodies/gsbhome/conference/en/> accessed 28 May 2018.

(i) protect and rehabilitate 'all critical fisheries habitats in marine … ecosystems', making 'particular effort … to protect such habitats from destruction, degradation, pollution and other significant impacts resulting from human activities that threaten the health and viability of the fishery resources'; and

(ii) take measures to ensure the use of selective and environmentally safe fishing gear and practices (including non-destructive fishing gear and closed seasons and areas) to minimise, inter alia, negative impacts on associated or dependent species and habitats and conserve the biodiversity of aquatic habitats.[149]

In the context of the depletion of deep-sea fish stocks in the high seas, the FAO has also developed the concept of Vulnerable Marine Ecosystems (VMEs) in response to repeated calls by the UNGA. Gjerde traces the origin of these UNGA resolutions and subsequent development of VMEs to a 2001 expert workshop organised by two deep-sea scientists to explore ways 'to protect vulnerable species, habitats and long-term research sites from expanding human uses in the deep and open ocean beyond national boundaries'.[150]

As the lead competent international organisation for fisheries, the FAO developed a new deep-sea high seas programme. The first outcome was the 2008 FAO International Guidelines for the Management of Deep-Sea Fisheries in the High Seas (FAO Deep-Sea Guidelines),[151] which aim to assist states and RFMO/As (RFMO or Arrangements) in sustainably managing deep-sea fisheries, especially with respect to the conservation and management of VMEs, defining VMEs and providing criteria for their identification.[152] A web-based platform

149 FAO Code of Conduct for Responsible Fisheries arts 6(8), 7(2)(2), 7(6)(9), 7(6)(10), 8(4)(7) and 12(10).

150 Gjerde also highlights the importance of prior scientific assessments by the International Council for Exploration of the Seas (ICES), which highlighted the issues faced by deep sea habitats damaged by fishing practices in the late 1990s. Kristina M Gjerde, 'Perspectives on a Developing Regime for Marine Biodiversity Conservation and Sustainable Use beyond National Jurisdiction' in Harry N Scheiber, Nilufer Oral and Moon-Sang Kwon (eds) *Ocean Law Debates: The 50-Year Legacy and Emerging Issues for the Years Ahead* (Brill Nijhoff 2018) 358, 360.

151 FAO, 'Report of the FAO Workshop for the Development of a Global Database for Vulnerable Marine Ecosystems' (Rome, 7–9 December 2011, FAO Fisheries and Agriculture Report No 1018, 2013) <www.fao.org/docrep/017/i3109e/i3109e.pdf> accessed 28 May 2018.

152 FAO, 'International Guidelines for the Management of Deep-Sea Fisheries in the High Seas' (FAO 2009) (FAO Deep-Sea Guidelines) para 6 <www.fao.org/in-action/vulnerable-marine-ecosystems/background/deep-sea-guidelines/en/> accessed 28 May 2018.

on VMEs was subsequently developed to support RFMO/As with the identification and management of VMEs and to provide a global outlook.[153] The global VME database provides an inventory of fisheries measures adopted in ABNJ to prevent significant adverse impact of bottom fisheries in VMEs.[154]

### 3.2.2 Scientific Criteria

According to the FAO Deep Sea Guidelines, the vulnerability of a potential VME should be assessed on the basis of 'the likelihood that a population, community or habitat will experience substantial alteration from short-term or chronic disturbance, and the likelihood that it would recover and in what time frame'. VME features may be physically or functionally fragile. The most vulnerable ecosystems are those that are both easily disturbed and slow to recover or may never recover. The guidelines also indicate that the vulnerability should be assessed for specific threats.[155]

Also important is the fact that these guidelines have been developed for fisheries that occur in ABNJ and have the following characteristics:

(i) they include species that can only sustain low exploitation rates, and

(ii) the fishing gear is likely to contact the sea floor during normal operations.[156]

The management and conservation steps include the identification of VMEs and assessment of significant adverse impacts; a non-exhaustive list of five alternative criteria is set out (Table 4.10).[157] Examples of potentially vulnerable species groups, communities and habitats are certain cold water corals, sponge-dominated communities, communities composed of dense emergent fauna where large sessile protozoans and invertebrates form an important structural component of habitat, and seep and vent communities comprised of invertebrate and microbial species found nowhere else. Examples of features that

153 FAO, 'VME Database' <www.fao.org/in-action/vulnerable-marine-ecosystems/en/> accessed 28 May 2018.

154 Also of note are the 2011 FAO Technical Guidelines for Responsible Fisheries. They provide information and guidance on the use of MPAs in the context of fisheries and include VMEs. FAO, 'Technical Guidelines for Responsible Fisheries' (No 4, Marine Protected Areas and Fisheries, 2011) <www.fao.org/docrep/015/i2090e/i2090e.pdf> accessed 28 May 2018.

155 FAO Deep-Sea Guidelines (n 152) paras 14 and 15.

156 Ibid para 8.

157 For a discussion of possible improvement of the identification and management process, see JA Ardron, 'A Systematic Approach Towards the Identification and Protection of Vulnerable Marine Ecosystems' (2014) 49 *Marine Policy* 146. See also PJ Auster, 'Definition and detection of Vulnerable Marine Ecosystems on the High Seas: Problems with the "Move-On" Rule' (2011) 68 *ICES Journal of Marine Science* 254.

TABLE 4.10 Criteria for the identification of VMEs

| No | Criteria | Description | Examples |
|---|---|---|---|
| 1- | **Uniqueness or rarity** | Rare area or ecosystem or contains rare species | Habitats containing endemic species or threatened or endangered species or nurseries, discrete feeding, breeding, spawning areas, etc. |
| 2- | **Functional significance of the habitat** | Discrete areas or habitats necessary for the survival, function, spawning, recovery, particular life-history stages or of rare threatened or endangered species | |
| 3- | **Fragility** | Highly susceptible to degradation by anthropogenic activities | |
| 4- | **Life-history traits of component species that make recovery difficult** | Populations or assemblages of species with either slow growth rate, late maturity, long or unpredictable recruitment or long-lived | |
| 5- | **Structural complexity** | Complex physical structures created by significant concentrations of biotic and abiotic features; ecological processes highly dependent on these structures | Often systems with high diversity that is dependent on the structuring organisms |

potentially support these species groups or communities include submerged edges and slopes, summits and flanks of seamounts, canyons and trenches.[158]

Noteworthy is the fact that none of these criteria refer expressly to biological diversity, except as a common (but not necessary) trait of criterion 5, structural complexity; the EBSA criteria of biological productivity and naturalness are not

158 FAO Deep-Sea Guidelines (n 152) Annex.

included either. With respect to biodiversity, Ardron et al indicate that 'most areas designated as VMEs have been assessed based on their density as well as biodiversity' and that 'expert reviews by the CBD and FAO have concluded that similar data can be used for both VME and EBSA criteria'.[159] However, it is possible that an area that theoretically qualifies as an EBSA under the CBD may not qualify as a VME, such that adverse impacts from fishing on such area would not fall within the scope of VMEs management framework.[160] The FAO VME database shows some VMEs that are closed to fishing under the management of different RFMO/As, but it also shows large areas of oceans that are not covered, including large parts in the Atlantic, Indian and Pacific Oceans.[161]

### 3.3 *Sensitive Marine Areas in the Context of Deep Seabed Mining in the Area: Areas of Particular Environmental Interest in the Clarion Clipperton-Zone*

#### 3.3.1 Legal Background

Rights to the Area and its resources are 'vested in mankind as a whole, on whose behalf the Authority shall act'.[162] The ISA is charged with the responsibility of organising, regulating and controlling exploration and exploitation of seabed minerals in the Area.[163] Its mandate includes the adoption of necessary measures to ensure effective protection of the marine environment from harmful effects that may arise from deep seabed mining.[164] In this capacity, the ISA has been developing a Mining Code, which includes provisions for the protection of the marine environment, one of the components of which is marine biodiversity.

---

159 Ardron (n 157).

160 David Johnson, Maria Adelaide Ferreira and Ellen Kenchington, 'ATLAS Deliverable 7.2 Report on Priorities for an Expert Assessment of North Atlantic MPAs, EBSAs and VMEs in ABNJ' (27 September 2017) <https://zenodo.org/record/997986#.WtmgJq17HMU> accessed 28 May 2018. This paper contributes to the EU H2020 ATLAS project. The Atlas-EU Project is a Trans-Atlantic assessment and deep-water ecosystem-based spatial management plan for Europe, <www.eu-atlas.org/about-atlas/atlas-work-package-descriptions> accessed 28 May 2018.

161 For a review of the status of closures, see G Wright et al, 'Advancing Marine Biodiversity Protection through Regional Fisheries management: A Review of Bottom Fisheries Closures in Areas Beyond National Jurisdiction' (2015) 61 *Marine Policy* 134.

162 UNCLOS (n 13) art 137(2).

163 UNCLOS (n 13) arts 136, 137 and 153, as amended by the 1994 Agreement Relating to the Implementation of Part XI of the United Nations Convention on the Law of the Sea of 10 December 1982 (adopted 28 July 1994, entered into force 28 July 1996) 1836 UNTS 3.

164 UNCLOS (n 13) arts 145 and 162, as amended by the 1994 Agreement Relating to the Implementation of Part XI of the United Nations Convention on the Law of the Sea of 10 December 1982 (n 163).

In 2012, the ISA approved the first regional-scale environmental management plan for the deep seabed of the Clarion-Clipperton Zone (CCZ), which includes the designation of Areas of Particular Environmental Interest (APEIs), as it had been proposed by the ISA Legal and Technical Commission.[165] The ISA Council's decision decided that the environmental management plan would be implemented over a three-year period and in a flexible manner, so that it may be improved as more scientific, technical and environmental baselines and resource assessment data are made available. The UNGA resolutions on Oceans and the Law of the Sea invite the ISA 'to consider developing and approving environmental management plans in other international seabed area zones, in particular where there are currently exploration contracts'.[166] However, the scientific criteria used to identify APEIs in the CCZ do not appear to have been formalised or been the subject of new discussions in the context of the CCZ.

### 3.3.2 Scientific Criteria

One of the six guiding principles of the environmental plan for the CCZ is the 'conservation and sustainable use of biodiversity' on the basis that 'all States have a duty to conserve and sustainably use marine biodiversity'.[167]

The design of the network of nine APEIs in the CCZ has been guided by best environmental practices and the use of spatial management tools. The following scientific indicators have been taken into consideration:

(i) spatial variation and representativity, to ensure the protection of areas thought to be representative of the full range of habitats and based on nine bio-geographic areas;[168]

(ii) of a size to cover large areas of self-sustaining populations and a broad range of habitat variability, biodiversity and ecosystem structure and

165 ISA Council 'Decision of the Council relating to an environmental management plan for the Clarion-Clipperton Zone', ISA Doc ISBA/18/C/22 (26 July 2012) <www.isa.org.jm/sites/default/files/files/documents/isba-18c-22_0.pdf> accessed 28 May 2018. See also Michael Lodge, 'Seabed Mining: International Seabed Authority Environmental Management Plan for the Clarion-Clipperton Zone. A Partnership approach' (2014) 49 *Marine Policy* 66. On the development of the Environmental Management Plan, see also Aline L Jaeckel, *The International Seabed Authority and the Precautionary Principle: Balancing Deep Seabed Mineral Mining and Marine Environmental Protection* (Brill Nijhoff 2017).

166 UNGA, 'Resolution Adopted by the General Assembly on 9 December 2013. 68/70. Oceans and the Law of the Sea' UN Doc UNGA A/RES/68/70 (27 February 2014) <www.un.org/en/development/desa/population/migration/generalassembly/docs/globalcompact/A_RES_68_70.pdf> accessed 28 May 2018.

167 The other five are (i) common heritage of mankind, (ii) precautionary approach, (iii) protection and preservation of the marine environment, (iv) prior EIA and (v) transparency.

168 Lodge (n 165) 71.

function within the management area (200 km by 200 km surrounded by a buffer of 100 km);

(iii) scientific design that relies on generally accepted and widely applied principles for the design of MPA networks and includes the protection of 30% to 50% of the total management area (this indicator uses geological, oceanographic and biological proxy data[169] based on previous ISA workshops and reports as well as scientific peer-reviewed literature and experience;); and

(iv) flexibility to allow the location and size of the areas to be modified based on improved information.[170]

This approach appears to be holistic and partly qualitative. However, the extent to which species abundance and distribution has been taken into account is not specified. The CCZ management plan itself mentions that areas of special significance valued according to the following criteria have not been incorporated in the scientific design but may be included in the future, as more information becomes available:

(i) their uniqueness;[171]

(ii) biological diversity;

(iii) biological productivity; and

(iv) areas of special importance to the life histories of non-fish species.

## 3.4 *Dump-Site Selection under the London Dumping Convention and Protocol*

### 3.4.1 Legal Background

A fourth forum for the regulation of activities at sea where provisions have been adopted to protect sensitive areas is the COP to the London Convention and the London Protocol. Article IV of the London Convention requires that a permit be issued prior to the disposal at sea of any authorised waste following 'careful consideration of all the factors set forth in Annex III, including prior studies of the characteristics of the dumping site'. These factors include taking into account other areas such as 'amenity areas, spawning, nursery and fishing areas and exploitable resources' as well as the method of deposit, dispersal,

169 On the use of geological, oceanographic and biological proxy data in the identification of the APEIs, see D Johnson and MA Ferreira, 'ISA Areas of Particular Environmental Interest in the Clarion-Clipperton Fracture Zone' (2015) 30 *International Journal of Marine and Coastal Law* 559, 561–62.

170 ISBA Legal and Technical Commission 17th Session, 'Environmental Management Plan for the Clarion-Clipperton Zone' ISA Doc ISBA/17/LTC/7 (13 July 2011).

171 Ibid. This is at odd with the vision statement of the CCZ management plan to preserve 'representative and unique marine habitats and species'.

water and bottom characteristics and possible effects on marine life and other uses of the sea. The London Protocol also requires a similar procedure for the dump-site selection.[172] Generic Waste Assessment Guidelines have been adopted to assist state parties with the implementation of these provisions.[173]

### 3.4.2 Adoption

As of 20 March 2018, 87 states have adopted the London Convention and 50 states have adopted the London Protocol, with a total number of 99 states that adopted either the London Convention or the London Protocol (some have adopted both and were not counted twice).

However, this low participation does not ipso facto mean that the criteria adopted with respect to sensitive marine areas would not inform the international rules and standards applicable to dumping of waste at sea, even for states that have not become parties to these agreements.[174]

### 3.4.3 Criteria

The Generic Waste Assessment Guidelines list 'important amenities, biological features and uses of the sea to be considered in determining the specific location of the dump-site area' including

(i) shoreline and bathing beaches;
(ii) areas of beauty or significant cultural or historical importance;
(iii) areas of special scientific or biological importance, such as sanctuaries;
(iv) fishing areas;
(v) spawning, nursery and recruitment areas;
(vi) migration routes;
(vii) seasonal and critical habitats;
(viii) shipping lanes;
(ix) military exclusion zones; and

---

172 London Convention (n 14) art 4 and Annex III (Article 4(2)); London Protocol (n 14) Annex II.

173 IMO, '2014 Waste Assessment Guidance Documents' <www.imo.org/en/OurWork/Environment/LCLP/Publications/wag/Pages/default.aspx> accessed 28 May 2018.

174 UNCLOS (n 13) art 210. On the mechanisms of incorporation by reference in UNCLOS, see A Boyle, 'Marine Pollution under the Law of the Sea Convention' (1985) 79 *American Journal of International Law* 347, 356; Y Lyons, 'The New Offshore Oil and Gas Installation Abandonment Wave and the International Rules on Removal and Dumping' (2014) 29 *International Journal of Marine and Coastal Law* 480, 486–87, 500–12.

(x) engineering uses of the seafloor, including mining, undersea cables, desalination or energy conversion sites.[175]

As can be seen from this list, the site-selection process is not focussed solely on ecologically sensitive areas, which concern four criteria out of ten: criteria 3, 5, 6 and 7. Other criteria concentrate on other values (economic, social, cultural, historical and scientific) and uses of the sea such as fishing, shipping, military and mining.

The criteria examined in this section for the identification of marine sensitive areas in the context of limiting impacts from maritime activities are very varied. While ecological criteria are dominant considerations, some instruments also seek to limit impacts on areas valued on another basis than their ecological value. These non-ecological criteria include social, cultural, economic, scientific and educational criteria. These two broad categories of criteria (ecological and non-ecological values) are analysed in the following section across all the families of instruments reviewed in Sections 2 and 3.

## 4 Comparison of the Sets of Scientific Criteria

The different sets of criteria use some recurring terminology but also different terminology developed in different fora to suit the objective of the family of instruments they are embedded in. To compare them, a reference set of criteria (Reference Criteria Set) has been established, seeking to include all the criteria used while avoiding repetitions. This Reference Criteria Set includes all the ecological and non-ecological criteria found in the sets of criteria discussed in Sections 2 and 3. The comparison of the sets of criteria is divided into two groups, each of which is examined separately: first the ecological criteria, which seek to identify an area for its ecological or biological attributes, and second, all the other criteria, including economic, social, cultural, scientific and educational criteria. This very diverse category of non-ecological criteria highlights other values of different marine (including coastal) areas on the basis of which international bodies considered that they should be granted a particular sensitivity status and afforded greater care.[176]

175 Generic Waste Assessment Guidelines, Section 19 in IMO, '2014 Waste Assessment Guidance Documents' (n 173). See Karen N Scott, Chapter 3 'Mind the Gap: Marine Geoengineering and the Law of the Sea'.

176 Gilman et al make the same distinction between ecological criteria and other criteria grouped under the category name of 'socio-economic and governance criteria'. Gilman et al, 'Designing Criteria Suites to Identify Discrete and Networked Sites of High Value across Manifestations of Biodiversity' (2011) 20 *Biodiversity Conservation* 3363. Although this analysis focusses on identification of areas prior to determining adequate

In addition to the ecological and non-ecological criteria discussed so far, the sets of criteria include other considerations in the selection of areas and their boundaries, some of which relate to the application of the ecological criteria and the objective of successfully securing the ecological characteristics of the area, whereas other criteria are broader. Considerations that relate primarily to the ecological criteria concern

(i) the geographic scale considered including the relevant biogeographic classification;[177]
(ii) the size of the area, to ensure for example, that it is large enough to encompass the ecological functions on the basis of which it has been identified;[178] and
(iii) the urgency to identify this area as sensitive on the basis of, for instance, the level of environmental and/or anthropogenic threats it is facing.[179]

## 4.1 *Ecological Criteria*

### 4.1.1 The Reference Criteria Set

The EBSA criteria have been used as a reference for the purpose of comparing ecological criteria for the identification of sensitive marine areas on the basis that they have been applied by the COP of the CBD to all maritime zones. Furthermore, in addition to being geographically comprehensive, these criteria apply to all marine systems (coastal, marine, deep seabed, etc.) rather than only to subsets of the marine environment, such as wetlands or critical habitats for migratory species. However, their scope is limited to the identification and conservation of biodiversity rather than the protection of the marine environment in general or of critical habitats for commercially valuable fisheries, which is why other sets of criteria include criteria that are not included in the EBSA criteria.

The Reference Criteria Set includes first the seven EBSA criteria (completed with corresponding criteria developed under other instruments in order

management measures, some instruments include management consideration as a precondition for identification, and these are included in 'practical considerations' rather than 'governance'.

177 CBD Ottawa Report (n 39.) Annex V. Also an essential consideration in the design of APEIs for the Clarion-Clipperton Zone. See Section 2.3.2.

178 For example in the application of OUV criterion 9 on ecological and biological processes. UNESCO, 'The Operational Guidelines for the Implementation of the World Heritage Convention' (n 66) para 94. Also an essential component of the design of APEIs for the Clarion-Clipperton Zone. See Section 3.3.2.

179 CMS COPs have highlighted the importance of assessing risks to identify sites of critical importance. See Section 2.4.3. The assessment of threats from shipping in the designation of Special Areas under MARPOL and PPSAs is another example.

TABLE 4.11 Correspondences between ecological criteria developed to identify sensitive marine areas in different families of legal instruments aiming to protect and conserve different components of the marine environment

| Identification criteria | EBSA[a] | Wetland of international importance[b] |
|---|---|---|
| **Uniqueness or rarity or distinctiveness** | Criterion 1<br>Representativity not included | Included in Ramsar criterion 1 with respect to wetlands; the latter also includes representativity as an alternative criteri for a wetland to qualify as a 'wetland of international impor tance'. Also criterion 2 on threatened ecological communiti as they include those with restricted geographic distributio or habitat |
| **Special importance for life history of species** | Criterion 2<br>*Refugia* not included | Includes Ramsar criterion 4 applied to wetlands although the latter also refers to the provision of refuge during advers conditions. Also includes Ramsar criteria 7 and 8 for fish |
| **Importance for threatened, endangered or declining species and/or habitats** | Criterion 3<br>'Threatened ecological communities' not mentioned | Included in Ramsar criterion 2 with respect to vulnerable, e dangered or critically endangered species ('declining specie are not mentioned in the Ramsar criterion but 'vulnerable' species are) |
| **Vulnerability, fragility, sensitivity, slow recovery of species and habitats** | Criterion 4 | Ramsar criterion 2 also refers to vulnerable species and threatened ecological communities. The language of 'fragili 'sensitivity' and 'slow recovery' is not mentioned in Ramsar criteria |
| **Biological productivity** | Criterion 5 | No direct reference. Possible indirect reference through quantitative minima for subspecies, species or families of waterbirds or fish or other wetland-dependent species bein supported by a wetland (Ramsar criteria 5, 6, 7 and 9) and criterion 8 on wetlands as important source of food for fish and other wetland-dependent species |
| **Biological diversity** | Criterion 6 | Ramsar criterion 3 at scale of biogeographic regionAlso includes Ramsar criterion 7 with respect to fish |
| **Naturalness (and integrity)** | Criterion 7 | Not an express Ramsar criterion but implied in most criteri Ramsar criterion 1 refers to near-natural wetlands, meaning that they continue to function…". 'continue to function in what is considered an almost natural way[e] |
| **Representativity** | Not included in EBSA criteria but in scientific criteria for network of MPAS | Defined as a typical example of a particular wetland type found in a region[f] and included in Ramsar criterion 1 |
| **Connectivity** | Not included in EBSA criteria but in scientific criteria for network of MPAS under CBD | Not included |
| **Refugia against adverse conditions or site resistance and resilience** | Not included | Included in Ramsar criterion 4 |

a CBD COP9 Decision IX/20 (n 27).

b Ramsar Guidelines (n 46).

c UNESCO, 'The Operational Guidelines for the Implementation of the World Heritage Convention' (n 66).

| \[A]rea of OUV[c] | Critical habitats for migratory species[d] |
|---|---|
| \[C]an concur with OUV criterion 7 with respect \[t]o superlative natural phenomena or areas of \[e]xceptional beauty | IOSEA Marine Turtle Site Network criterion 1.4 on rare speciesIMMA criterion 2.1 on small and resident populations and criterion 4.1 on genetic, behavioural or ecologically distinctive characteristics EAAFP: same as Ramsar |
| \[O]UV criterion 10 | Includes 'critical habitats' that CMS instruments seek to protect |
| \[O]UV criterion 10 with refers to threatened species \[o]f OUV | CMS-listed species meet the definition of threatened, endangered or declining species. So this criterion is met even when not stated |
| \[N]o express reference in selection criteria for \[i]dentification of OUV but implied in criterion 10 \[a]nd can be part of the demonstration of existing \[a]dverse effects needed to obtain World Heritage \[s]ite status | Endangered status/fragility/habitat sensitivity of the species is assumed. EEAFP: Vulnerability, sensitivity or fragility of the habitat is not an express criterion for selection but threatened ecological communities can be EEAFP Flyway |
| \[N]ot included | No direct reference but possible (unconfirmed) overlap with criteria on aggregations and feeding areas |
| \[I]ncludes OUV criteria 9 and 10 | Diversity of listed or target species/sub-species, abundance or aggregations are conditions for an area to qualify as a 'critical habitat' |
| \[C]alled 'integrity' in this family of instruments. \[A]lso a condition for the designation of an OUV as \[a] World Heritage site. Can overlap with criterion 10. | Not direct reference but implied if considered as 'integrity' as defined under Ramsar (sufficient integrity of ecosystem functions can be assumed to be necessary for a site to qualify as a 'critical habitat') |
| \[P]ossibly included in OUV criterion 10 | Criterion 4.1 of IOSEA Marine Turtle Site Network criteriaCriterion 1 of EEAFP (from Ramsar criteria) |
| \[N]ot included | Criterion 4.2 of IOSEA Marine Turtle Site Network criteria. Also 4.3. Also a guiding principle in CMS COP12 Res 12.11Also implicit component in IMMA criteria |
| \[N]o explicit reference but possible inclusion in \[c]riterion 10 | Criterion 1.5 of IOSEA Marine Turtle Site Network criteria: Resistance and resilienceImplicit in EEAFP (from Ramsar criteria) Not in IMMA |

d Two sets of criteria are included here in addition to general provisions of the CMS and of COPs resolutions. Those developed to identify IMMAs and those developed for inclusion in the IOSEA Marine Turtle Site Network. Section 2.4.

e Ramsar Guidelines (n 46) para 117.

f Ibid para 114.

TABLE 4.12 Correspondences of ecological criteria developed to identify sensitive marine areas in different families of legal instruments aiming to regulate uses at sea

| Identification criteria | Special areas under MARPOL | PSSAS |
|---|---|---|
| **Uniqueness or rarity or distinctiveness** | Criterion 2.4 | Criteria 1.1 and 1.11 |
| **Special importance for life history of species** | Criteria 2.3 and 2.4 | Criteria 1.2 and 1.7 |
| **Importance for threatened, endangered or declining species and/or habitats** | Criterion 2.1 | Criterion 1.2. Also criterion 1.1 |
| **Vulnerability, fragility, sensitivity, slow recovery of species and habitats** | Criteria 2.1 and 2.4 | Criterion 1.10. May also overlap w. criterion 1.3 |
| **Biological productivity** | Criterion 2.2 | Criterion 1.6 |
| **Biological diversity** | No explicit reference. Possible overlap with criterion 2.4 (critical habitats for marine resources) | Criterion 1.5 |
| **Naturalness (and integrity)** | Not included | Criterion 1.8 (naturalness) and criterion 1.9 (integrity) |
| **Representativity** | Not included | Criterion 1.4 with high threshold. Must be an 'outstanding/illustrative ex'. Overlap with criterion 1.11 representative biogeographic types |
| **Connectivity** | Not included | Not included |
| **Refugia against adverse conditions OR site resistance and resilience** | Not express reference. May be indirectly included in criterion 2.4, where such areas are of critical importance for the support of marine ecosystems such as LMEs | No express reference. Possible overlap with criterion 1.2 |

| VMES | APEIs in the CCZ | LC/LP | OPRC convention |
|---|---|---|---|
| Criterion 1 | No explicit reference | No explicit equivalence. Possible overlap with criterion 3 | May be included in criteria 2.1 and 2.2 |
| Criterion 2 | No explicit reference | Criteria 5, 6 and 7 | Criteria 2.1 and 2.2 |
| Criteria 1 and 2 | No explicit reference | Implicit inclusion in criteria 3, 5, 6 and 7 | Criteria 2.1 and 2.2 |
| Criteria 3 and 4 | No explicit reference | No explicit equivalent. Possible overlap with criteria 3, 5 and 7 | Criteria 1, 2.1, 2.2 and 3 |
| No explicit reference but overlap with criterion 2 on functional significance of habitat. | No explicit reference | Overlap with criterion 4 | Criteria 1 and 3 |
| No explicit reference but possible overlap with criterion 5 aimed at structural complexity | No explicit reference but included in guiding principles | Criterion 3 | Criteria 2.1 and 2.2 |
| Not included but integrity of VMES' ecological functions is implied condition | An indicator for CCZ APEIs network design | Not included | Not included |
| Not included | An indicator for CCZ APEIs network design | Not included | Not included |
| Not included | Not included | Not included | Not included |
| No express reference. Possible overlap with criterion 5 | Not included | Not included | Not included |

to reflect different emphasis and use of language).[180] Second, the Reference Criteria Set also includes three additional criteria that are not included in the EBSA criteria but are included in other families of instruments.

The resulting list of all criteria in the Reference Criteria Set is therefore as follows:

(i) Uniqueness or rarity
(ii) Special importance for life history of species
(iii) Threatened or endangered or declining species or habitats
(iv) Vulnerability, fragility, sensitivity, slow recovery
(v) Biological productivity
(vi) Biological diversity
(vii) Naturalness
(viii) Representativity
(ix) Connectivity
(x) *Refugia* or site resistance or resilience

Each criterion is examined across all sets of criteria that have been legally adopted in order to identify commonalities and differences and summarised in Tables 4.11 and 4.12. An additional set of criteria has been added, due to their being abundantly referred to by international bodies and in the scientific literature: the criteria developed by the IUCN to identify Key Biodiversity Areas (KBAs).[181]

Of note, while APEIs adopted for the CCZ are included in the comparative analysis, the fact that they do not explicitly refer to several of the usual criteria should be considered carefully. As discussed in Section 3.3, these APEIs were designed as a first step or pilot and did not include criteria such as those adopted for other sensitive areas. It is hoped that criteria will soon be adopted following current studies to that effect in the context of deep seabed mining operations on the mid-Atlantic Ridge.[182]

180 Or emphasised more in other instruments such as distinctiveness, including genetic distinctiveness as a sub-criteria or uniqueness and rarity.

181 IUCN (2016) 'Global Standard for the Identification of Key Biodiversity Areas' (Version 1.0 first edn, IUCN 23 March 2016), <https://portals.iucn.org/library/sites/library/files/documents/2016-048.pdf> accessed 28 May 2018.

182 CL Van Dover, 'Scientific Rationale and International Obligations for Protection of Active Hydrothermal Vent Ecosystems from Deep-Sea Mining' (2018) 90 *Marine Policy*. 20. CL Van Dover et al, *Environmental Management of Deep-Sea Chemosynthetic Ecosystems: Justification of and Considerations for a Spatially-Based Approach* (ISA Technical Study No. 9, ISA 2011) <www.isa.org.jm/sites/default/files/files/documents/tstudy9.pdf> accessed 28 May 2018. See also the work of the SEMPIA 'Workshop: Towards the Development of a Strategic Environmental Management Plan for deep Seabed Mineral Exploration and

### 4.1.2 Comparison of Ecological Criteria Adopted in the Context of International Legal Instruments

Tables 4.11 and 4.12 present a comparison of the sets of ecological criteria discussed in Sections 2 and 3 respectively (Table 4.11 compares criteria across legal instruments aiming to protect and conserve different components of the marine environment; Table 4.12 compares criteria across legal instruments aiming to regulate uses of the sea). Figure 4.1 shows comparable, similar and overlapping criteria in green and criteria of the Reference Criteria Set for which there are no equivalent criteria in the 13 sets of criteria reviewed in beige.

The dominant observation from the analysis of correspondences between sets of criteria is that they are consistent, mutually supportive and mostly overlapping. This observation is also consistent with the scientific literature and statements by the bodies of the international fora where such criteria have been adopted.[183]

FIGURE 4.1 Comparison of ecological criteria.
In dark green: expressly similar, comparable or overlapping criteria; in bright green: implicitly similar, comparable or overlapping criteria; in beige: no comparable criteria.

---

Exploitation in the Mid-Atlantic RIDGE' (Horta, Portugal, 1–3 June 2015) <https://webgate.ec.europa.eu/maritimeforum/en/node/3718> accessed 28 May 2018.

183 Dunn et al (n 27); P Dearden and KN Topelko (2005) 'Establishing Criteria for the Identification of Ecologically and Biologically Significant Areas on the High Seas' (Background Paper prepared for Fisheries and Oceans Canada) <www.cbd.int/doc/meetings/mar/ewsebm-01/other/ewsebm-01-dearden-en.pdf> accessed 28 May 2018.

Criteria 1 to 7 of the Reference Criteria Set are all comparable or equivalent or overlap with criteria of other sets. The nature of this equivalence, comparability or overlap can take different forms. The most obvious case of equivalence or comparability is when criteria from different sets use identical, similar or equivalent language. For example, the notion of uniqueness or rarity, which is comparable to an 'exceptional' area and can be extended to a distinctive area (i.e. different, not representative) or to geographically restricted biodiversity and irreplaceable sites. Each of these criteria points to a different degree of 'rarity'. Another typical example is the criteria of naturalness and of integrity. Criteria can also be equivalent or overlap when one of the criteria is more general or to the contrary a subset of the other one. This is the case of criteria 2 and 3 of the Reference Criteria Set on special importance for life history of species and importance for threatened, endangered or declining species and/or habitats. Similarly, criteria that aim to identify different critical habitats of endangered species will meet both criteria 2 and 3.

Criterion 4 on vulnerability, fragility, sensitivity and slow recovery of species and habitats is an inclusive criterion that also points to different but comparable criteria that may be worded primarily to include only inherent fragility or to take into account theoretical vulnerability to some risks or consider exposure and vulnerability to specific risks. Some criteria can also be used to identify a feature that may be used as an indicator or proxy in the application of another criterion. For example areas of high aggregations of species in wetlands can be an indicator of high biological productivity and biodiversity. Another example concerns areas of geomorphological importance, criterion 8 of OUV criteria, which is not an express criterion in the Reference Criteria Set. However, geomorphological features may correspond to areas of high biodiversity, unique or rare ecosystems or communities, vulnerable habitats (e.g. seamounts) or biological productivity and as such be covered by the Reference Criteria Set or even be more directly used as a proxy under these criteria.[184]

184 Johnson et al highlight that several EBSAs including the Orange Shelf Edge EBSA, the East Broken Ridge Guyot EBSA, the Ua Puakaoa Seamounts EBSA and the Shelf Break Front EBSA were described based on the geomorphic features and its associated effects on productivity in the water column. DE Johnson et al, 'Reviewing the EBSA process: Improving on Success' (2018) 88 *Marine Policy* 75, 80. Examples of marine world heritage sites that are listed as meeting criterion 8 of OUV criteria include Australia Great Barrier Reef, Shark Bay in Western Australia, Macquarie and Fraser Islands, The Galapagos Islands, The Gulf of Porto, Papahānaumokuākea, Ha Long Bay and Phong Nha-Ke Bang National Park.

Analysis of the equivalence, comparability or overlap of each set of criteria with the Reference Criteria Set shows that

(i) most of the criteria in the Reference Criteria Set can be found in the 13 sets of criteria considered except for APEIs (which can be explained by the nature of this set of criteria);

(ii) the few criteria that do not have an explicit equivalent in the Reference Criteria Set are likely to have an implicit equivalent or overlap in practice when the criteria are being applied;[185] and

(iii) the alignment between the Reference Criteria Set and the set of criteria reviewed is generally balanced.[186]

Criteria 8 on representativity, 9 on connectivity and 10 on *refugia* show more limited overlaps, with criterion 9 on connectivity showing the least overlap, despite its being generally accepted as a guiding principle in MPA network design. By contrast, criteria 8 and 10 can explicitly or implicitly be found in more than half of the sets of criteria considered.

This analysis goes beyond confirming the consistency between different criteria sets for ecological criteria to identify marine sensitive areas acknowledged by many. It also suggests that they are mutually reinforcing. Given their global vetting and acceptance by states that are members of the different bodies that adopted or acknowledged these criteria, the question arises of their legal value and the extent to which this mutual supportiveness leads to the creation of a globally accepted standard, practice or procedure under UNCLOS, if not of a legal norm.

However, the fact that these ecological criteria are mutually reinforcing does not mean that they are necessarily sufficient to ensure protection and preservation of the marine environment and especially sensitive marine

185 These include criteria of a practical nature or that are not strictly ecological and/or concern assessment of threats to the sensitive area considered or expected management measures in response to the sensitivity identified, for instance the criterion on vessel traffic characteristics in the context of the establishment of special areas under MARPOL.

186 The only three exceptions to this are the OUV criteria and EEAFP criteria, where one criterion addresses most of the criteria of the Reference Criteria Set and the APEI criteria for the same reasons as highlighted above. The reasons for the imbalance in the spread of application of the OUV and EEAFP criteria are different. The overall purpose of the OUV criteria is to identify exceptional cultural and natural features whether geophysical or biological. The protection of biodiversity is a subset; criteria that deal with this subset are those where overlap can be identified. EEAFP criteria are based on Ramsar criteria, which are all included in its criterion 1, which is the reason for the over-representation of this criterion.

areas.[187] In any case, improvement of the criteria may be addressed in the context of the review process of these criteria where they exist.

## 4.2 *Non-Ecological Criteria*

### 4.2.1 Comparison Methodology

Non-ecological criteria included in the 13 sets of criteria reviewed can be categorised as follows:

(i) Economic value, including marine and coastal areas that are source of food and livelihood at local or national level (C11)
(ii) Social value (non-monetary) such as recreational sea uses and aesthetic value (C12)
(iii) Educational value for education or outreach at local or national level (C13)
(iv) Cultural or traditional use (C14)
(v) Cultural, religious or spiritual significance, such as areas that are associated with traditions, ideas, beliefs or outstanding universal significance (C15)
(vi) Scientific research value (including historical value), such as sites used for regulator monitoring or research (C16)

As for ecological criteria, non-ecological criteria can also be assorted or completed with other considerations or pragmatic considerations in the identification of an area under different instruments in relation to legal, institutional, administrative or financial measures or social and political feasibility. Equivalent criteria in each set of criteria, where they exist, are set out in Table 4.13 below for each non-ecological criterion, and other pragmatic considerations and summarised in Figure 4.2 below.

One of the first observations that stand out from both Table 4.13 and Figure 4.2 is the very low numbers of non-ecological criteria in the sets of criteria considered in contrast with the ecological criteria. The average number of ecological criteria in the 15 sets of criteria is 11.7, against 6.3 for the non-ecological

187 In addition to a need for more adequate mechanisms to prepare and adapt to impacts from climate change discussed above in the context of the criterion on refuges against adverse conditions, many publications and reports identify weaknesses. Of note, a study conducted by the Global Ocean Biodiversity Initiative (GOBI) for the CMS highlights a need for more research in the spatial and temporal distributions of marine migratory species, including considerations of representativity and connectedness in the EBSA process. The report reviews the application of the EBSA process to migratory species and indicates a strong overlap of criteria as migratory species played a principal role in the description of approximately 10% of EBSAs identification and a contributory role in 80% of EBSAs identification. CY Kot et al (n 113) 51.

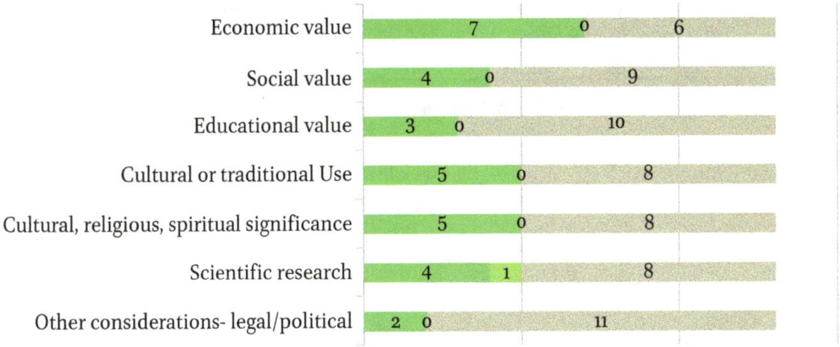


FIGURE 4.2 Comparison of non-ecological criteria.
In dark green: expressly similar, comparable or overlapping criteria; in bright green: implicitly similar, comparable or overlapping criteria; in beige: no comparable criteria.

criteria. Furthermore, five sets of criteria include no non-ecological criteria,[188] and three sets of criteria include only one non-ecological criterion,[189] the economic criterion relating to biological productivity and support for living resources. This criterion is therefore both an ecological criterion and one that relates to the economic value of an area, based on which is also included in the non-ecological criteria.

### 4.2.2 Comparison

Of the 13 sets of criteria presented in Table 4.13, 8 sets include no purely non-ecological criteria, being a criterion that does not overlap with ecological criteria. This can be explained by the purpose and subject matter of the instrument for which they were adopted. Three different situations can be distinguished:

(i) conservation-driven instruments, which include at the most management measures that are limited to monitoring, assessment of impacts or generally worded provisions towards sustainable use of resources;[190]

(ii) criteria developed in the context of specific sea-use regimes that concern marine areas removed from local populations: fishing in deep sea

---

188 Sets of criteria for EEAFP, IMMAS, VMES, APEIS and KBAS.

189 Sets of criteria for EBSAS, RAMSAR wetlands and Special Areas under MARPOL.

190 The CBD and Ramsar Convention are good examples of specific management measures being limited to monitoring and assessment of impacts. Sets of criteria developed for migratory species (EAFFP and IMMAS) are limited to purely ecological site identification criteria.

TABLE 4.13 Correspondences of non-ecological criteria developed to identify sensitive marine areas in different families of legal instruments

| Instruments | non-ecological criteria and considerations | | |
|---|---|---|---|
| | **Economic value (C11)** | **Social value (C12)** | **Educational value (C13)** |
| | Productive area—Source of food for living resources<br>Essential for local population, including artisanal/subsistence fisheries<br>Benefits local/national economy | Non-monetary benefits to society:<br>– recreational sea uses,<br>– aesthetics/natural beauty<br>– clean water/less pollution | |
| EBSA | C1—Biological Productivity | – | Not express |
| RAMSAR | C8—Food source for fish stocks | Not express. Valued once listed[a] | Not express |
| OUV | – | – | – |
| IOSEA | C3.6—Ancillary benefits for communities (monetary) | C3.6—Ancillary benefits for communities (non-monetary) | C3.3—Opportunity for educational and outreach activities |
| EEAFP | – | – | – |
| IMMAS | – | – | – |

| **Cultural or traditional use (C14)** | **Cultural, religious, spiritual significance (C15)** | **Scientific research value including historical value (C16)** | **Other considerations** |
|---|---|---|---|
| Traditional or representative sea-use | Associated with traditions, ideas, beliefs or outstanding universal significance | Site used for regular monitoring or research | Legal, institution, financial considerations or social/ political feasibility |
| – | – | Not express | – |
| Not express. Valued once listed | Not express. Valued once listed[b] | Not express | – |
| C5—Outstanding ex of sea use which is representative of a culture(s) or human interaction with the environment | C6—Associated with living traditions, ideas, beliefs, artistic or literacy works of outstanding universal significance | Implied—C8 major stages in earth history | Additional condition to gain status of World Heritage Site: Adequate protection and management system |
| C3.1—Traditional beliefs/ practices of cultural, religious and/or spiritual significance (in relation to marine turtles) | C3.1—Traditional beliefs/ practices of cultural, religious and/or spiritual significance (in relation to marine turtles) | C2.4—Site used for research and monitoring | C2.1—Legal framework<br>C2.2—Measures taken<br>C2.3—Participatory work with locals for compliance and enforcement<br>C2.5—Long-term financial and human resources available |
| – | – | – | – |
| – | – | – | – |

TABLE 4.13 Correspondences of non-ecological criteria developed to identify sensitive marine areas in different families of legal instruments (*cont.*)

| Instruments | non-ecological criteria and considerations | | |
|---|---|---|---|
| | **Economic value (C11)** | **Social value (C12)** | **Educational value (C13)** |
| | Productive area—Source of food for living resources<br>Essential for local population, including artisanal/subsistence fisheries<br>Benefits local/national economy | Non-monetary benefits to society:<br>– recreational sea uses,<br>– aesthetics/natural beauty<br>– clean water/less pollution | |
| Special Areas under MARPOL | C2.2—High natural productivity<br>C2.3—Important for important marine species<br>C2.4—Critical habitats for marine resources | –<br>Although mentioned in 2013 Guidelines | – |
| PSSAS | C1.6—Productivity<br>C2.1—Social or economic dependency<br>C2.2—Human dependency | C2.1—Social or economic dependency | C3.3—Education |
| OPRC Convention | C3—Human uses | C3—Human uses | C3—Human uses |
| VME | – | – | – |
| APEIS | – | – | – |
| LC/LP | C4—Fishing areas<br>C7—Shipping lanes<br>C10—Engineering: mining, cables, etc. | C1—Recreational uses<br>C2—Areas of beauty or significant cultural or historical value | No express equivalent |
| KBAS | – | – | – |

a Ramsar Strategic Framework and Guidelines highlight that socio-economic and cultural values must be considered along with the maintenance of the ecological character of the listed Ramsar site (n 47) paras 35, 94–95.

b Ibid.

| Cultural or traditional use (C14) | Cultural, religious, spiritual significance (C15) | Scientific research value including historical value (C16) | Other considerations |
|---|---|---|---|
| Traditional or representative sea-use | Associated with traditions, ideas, beliefs or outstanding universal significance | Site used for regular monitoring or research | Legal, institution, financial considerations or social/ political feasibility |
| – | – | – | – |
| Although cultural value is mentioned in 2013 Guidelines | Although cultural value is mentioned in 2013 Guidelines | Although scientific value is mentioned in 2013 Guidelines | |
| C2.3—Cultural heritage | C2.3—Cultural heritage | C3.1—Research C3.2—Baseline for monitoring studies | – |
| C3—Human uses | C3—Human uses | C3—Human uses | – |
| – | – | – | |
| – | – | – | |
| C2—Areas of beauty or significant cultural or historical value | C2—Areas of beauty or significant cultural or historical value | C3—Scientific importance | – |
| – | – | – | – |

ecosystems (VMEs) and deep seabed mining in the Clarion-Clipperton Zone (APEIs); and
(iii) criteria developed for the implementation of MARPOL Special Areas, which are based on the text of MARPOL as adopted in the late 1970s.[191]

The remaining five sets of criteria, which all include three or more non-ecological criteria, share an important characteristic. They are all designed to not only identify marine areas that present certain characteristics of sensitivity or value, but also have specific regulation of sea uses taken with respect to these areas. Three different situations can be distinguished:

(i) The OUV criteria are a first step towards the designation of a site as a World Heritage Site. Although the World Heritage Convention is a conservation-driven convention, it includes a strict set of mechanisms to ensure protection and conservation of the site under the purview of the World Heritage Commission.[192]
(ii) Criteria in the context of the regulation of activities are concerned about not only the environmental impact of activities but also their impact on other legitimate uses of the sea. Such criteria are the PSSA criteria as well as those developed in the context of dump-site selection for the disposal of waste at sea under the London Convention and the London Protocol and for oil sensitivity mapping under OPRC Convention.
(iii) Criteria designed to guide the designation of sites to establish a network of MPAs made of critical habitat-sites for marine turtles. They presuppose domestic laws, regulations and measures to protect and preserve the marine environment within them. They present a national (as well as regional) perspective on MPAs that cannot be limited to ecological criteria.

There is therefore no clear trend demonstrating broad acceptance of non-ecological criteria. Economic value of an area is the only criterion included implicitly or explicitly in more than half of the sets of criteria reviewed. Cultural or traditional uses and cultural, religious or spiritual significance are included in seven sets. Scientific research is included in six sets.

It seems that this difference between ecological and non-ecological criteria may partly be linked to a fundamental difference in states' approaches to the environment compared to other values of the oceans. States have the sovereign right to exploit their natural resources, but it is subject to their obligation to

191 Section 3.1.1.
192 Section 2.3.

protect and preserve the marine environment.[193] States' freedom to fish on the high seas is also subject to their obligation to protect and preserve the marine environment.[194] All activities at sea are subject to states' obligation to protect and preserve the marine environment. However, some of the non-ecological criteria may also correspond to other obligations under UNCLOS, other treaties and general international law. For example, marine scientific research may be seen as both a right and an obligation in the context of the obligation to monitor the effects of activities that may adversely impact the marine environment[195] and the obligation to monitor the status of sensitive areas provided in different international instruments.

## 5 Conclusion

The analysis of 13 sets of criteria has shown the systematic presence of most of the 10 ecological criteria of the Reference Criteria Set, with connectivity and *refugia* being less generally used. However, the absence of connectivity from the set of identification criteria for independent sites does not mean that connectivity is excluded from the guidelines for the identification and selection of sites in the context of the creation of a network. With respect to the absence of an express equivalent criterion for *refugia*, it may be linked to the perception of the need for such *refugia* in the context of the subject matter of each instrument and perhaps the fact that the perception of the need is linked to developing science on the understanding of the steps needed to adapt to the effects of climate change in the marine environment.[196] The inclusion of non-ecological criteria is different and appears to be based primarily on the underlying objective of the instruments being implemented with the criteria and whether it is primarily focussed on ecological values or is more holistic.[197]

193 UNCLOS (n 13) art 193.
194 Ibid art 192.
195 UNCLOS (n 13) arts 204 and 206.
196 Johnson, Ferreira and Kenchington (n 160).
197 GC Ray, 'Marine Protected Areas: Past Legacies and Future Consequences. "You Can't Know Where You're Going Unless You Know Where You've Been"' (2015) 25 *Aquatic Conservation: Marine and Freshwater Ecosystems* 1; S Wells et al, 'Building the Future of MPAs—Lessons From History' (2016) 26 *Aquatic Conservation: Marine and Freshwater Ecosystems* 101.

This study may have legal implications that deserve further attention. The eight ecological criteria that are generally found across sets of criteria aimed at identifying sensitive marine areas within and beyond national jurisdiction may serve as a general guide or global standard in the protection and preservation of the marine environment beyond national jurisdiction. Although they may not be directly binding, their global acceptance by states, substantive mutual supportiveness and adoption by the relevant competent authorities under UNCLOS give them a particular legal status and authority.[198]

Furthermore international courts and tribunals have clearly confirmed that states' obligations under UNCLOS must be interpreted in light of the general corpus of international law,[199] including the obligation to protect and preserve rare or fragile ecosystems as well as the habitat of depleted, threatened or endangered species.[200] They also judged that states have an obligation to act with due diligence to discharge their obligation to protect and preserve the marine environment.[201] Generally accepted ecological criteria may therefore serve as

198 Pavoni discusses the importance of mutual supportiveness between rules developed under different instruments (in competing regimes) and the way in which they may have normative significance and serve to progressively consolidate general principles of international law. R Pavoni, 'Mutual-supportiveness as a Principle of Interpretation and Law-Making: A Watershed for the WTO-and-Competing-Regimes' Debates' (2010) 21(3) *European Journal of International Law* 549. Brunée discusses the influential role placed by Conference of Parties to different Multilateral Environmental Agreements (MEAs). J Brunée, 'COPing with Consent: Law-Making Under Multilateral Environmental Agreements' (2002) 15 *Leiden Journal of International Law* 1. See also a discussion of convergence in international, as a means to limit fragmentation. M Andenas and E Bjorge, *A Farewell to Fragmentation: Reassertion and Convergence in International Law. Studies on International Courts and Tribunals* (Cambridge University Press 2015).

199 As recalled in the Pulp Mills Case, a treaty shall be 'interpreted in good faith in accordance with the ordinary meaning to be given to the terms of the treaty in their context and in light of its object and purpose. That interpretation will also take into account, together with the context, any relevant rules of international law applicable in the relations between the parties'. The decision further states that these rules of international law include 'rules of general international law or contained in multilateral conventions to which the two States are parties'. *Pulp Mills on the River Uruguay* (*Argentina v Uruguay*) (Judgment of 20 April 2010) ICJ Reports 2010, 14, para 66. A Boyle, 'Further Development of the Law of the Sea Convention: Mechanisms for Change' (2005) 54 *International & Comparative Law Quarterly* 563.

200 UNCLOS (n 13) art 194(5); South China Sea Arbitration (n 17).

201 The obligation of due diligence is an obligation to 'deploy adequate means, to exercise best possible efforts, to do the utmost to obtain this result'. It entails not only the adoption of appropriate rules and measures but also a certain level of vigilance in their enforcement and the exercise of administrative control applicable to public or private operators, such as the monitoring of their activities. See for instance *Request for an Advisory Opinion*

a minimum standard. Should an area qualify as a sensitive area under these criteria, particular attention would be expected by states to monitor activities that may adversely affect them especially in case of a risk of irreparable harm on depleted, threatened or endangered species or their habitats or on rare or fragile ecosystems and habitats.

---

*Submitted by the Sub-Regional Fisheries Commissions* (*SFRC*) (Advisory Opinion of 2 April 2015) ITLOS Reports 2015, 4, paras 116–24.

CHAPTER 5

# The Use of IMO Instruments for Marine Conservation on the High Seas

*Aldo Chircop*

## Abstract

Chapter 5 explores the use of IMO regulatory and management tools to protect sensitive marine areas and species, including special areas designated under the International Convention for the Prevention of Pollution from Ships, 1973/78 (MARPOL), particularly sensitive sea areas (PSSAs) and routeing and reporting. It discusses how these instruments have been applied and their potential for use in high seas areas to address conservation needs. The conclusion is there is sufficient legal and institutional basis for MARPOL special areas, PSSAs, routeing measures and reporting measures to protect ecosystems and species in high seas areas, but likely with some adjustments.

## 1 Introduction

Marine transportation is a crucial ocean use that services 80 per cent of international trade.[1] The international shipping industry responsible for the carriage of that trade is largely owned, managed, operated and regulated at the global level. The industry relies on complex and extensive webs of global and regional shipping routes located both offshore and inshore. They are mostly longstanding traditional routes and are usually located without any prior consideration for environmental sensitivities and modern integrated coastal and ocean management. Accordingly, international shipping has a substantial footprint on ocean space and its ecosystems. The impacts include operational and accidental marine and air pollution, transposition of exotic species from one ecosystem to another through ballast water and hull biofouling, collisions with marine mammals, surface and underwater noise impacts on marine life, and groundings that cause major damage to very sensitive coastal systems.

1 World Bank Group and United Nations, *The Potential of the Blue Economy* (World Bank 2017) 21.

 | DOI:10.1163/9789004373303_006

Under the 1982 United Nations Convention on the Law of the Sea (UNCLOS), all states have a duty to protect and preserve the marine environment, both directly and through international organisations.[2] The International Maritime Organization (IMO) is the competent international organisation that facilitates the ability of states to fulfil this fundamental duty in the international shipping industry. Since the 1960s the mandate of the IMO has evolved from a predominant focus on maritime safety to include the environmental impacts of international shipping. The IMO's environmental portfolio spans, among other issues, pollution prevention, places of refuge for ships in need of assistance, greenhouse gas emissions and other air emissions, ballast water management, harmful antifouling systems, biofouling, protection of sensitive marine areas and measures for particular species, underwater noise, ship recycling, and related safety management and seafarer training. In recent years, the protection of particularly sensitive areas of the marine environment from international shipping has encompassed a growing mix of hard and soft regulatory tools originally designed for the regulation and management of maritime safety. These tools have now been used also to help safeguard environmental values. The use of these tools draws authority from the IMO's own constitutive instrument, UNCLOS and international maritime conventions. Increasingly, the use of international maritime tools is coordinated with or complements other tools for marine conservation available under other multilateral environmental instruments.[3]

This chapter discusses the established and growing IMO experience with the use of legal and management tools adopted for the protection of sensitive marine areas and species, in particular (a) special areas designated under the International Convention for the Prevention of Pollution from Ships, 1973/78 (MARPOL),[4] (b) particularly sensitive sea areas (PSSAs) under the authority

2 United Nations Convention on the Law of the Sea (adopted 10 December 1982, entered into force 16 November 1994) 1833 UNTS 3 (UNCLOS) art 192.

3 For example, the Convention on Biological Diversity (adopted 5 June 1992, entered into force 29 December 1993) 31 ILM 818 (CBD). Article 8 of the CBD provides a duty for in situ conservation, which is frequently discharged through the establishment of marine protected areas and may be further supported through the designation of particularly sensitive sea areas (PSSAs) through the IMO.

4 International Convention for the Prevention of Pollution from Ships (adopted 2 November 1973, entered into force 2 October 1983) 1340 UNTS 184, as amended by the Protocol Relating to the International Convention for the Prevention of Pollution from Ships of 1973 (adopted 17 February 1978, entered into force 2 October 1983) 1340 UNTS 61, as amended (MARPOL). Annexes I and II entered into force on the same date as the convention.

of the Convention on the International Maritime Organization, 1948[5] and associated protective measures (APMs) under various instruments,[6] and (c) permanent or seasonal routeing and reporting measures to avoid or minimise negative impacts on particular species.[7] The chapter further explores the extent to which these instruments have been applied or have the potential for application to high seas areas to facilitate contemporary marine conservation needs, as well as pollution prevention and the role of the IMO.

## 2 The Scope of the IMO Environmental Mandate

To appreciate how the IMO mandate has responded to new demands in the governance of international shipping, including to contribute to conservation values on the high seas, one must consider the evolution of the role of IMO to address new societal needs since its inception. When first established and during its early years, the IMO's mandate was focussed predominantly on maritime safety and technical aspects of shipping, although some tasks were also assigned to it by the International Convention for the Prevention of Pollution of the Sea by Oil, 1954 (OILPOL).[8] The limited mandate at the time can be explained in part by the initial reticence of its members to confer on the IMO more than a consultative status. Also, in the post-Second World War and reconstruction period, marine environmental degradation was not a priority. International concern was first triggered in the 1950s–1960s by prominent individuals, such as Rachel Carson[9] and Jacques-Yves Cousteau,[10] who through

5 Convention on the Inter-Governmental Maritime Consultative Organization (adopted 6 March 1948, entered into force 17 March 1958) 289 UNTS 48, as amended and renamed Convention on the International Maritime Organization, 9 UNTS 61 (IMO Convention).

6 For example, special areas under MARPOL Annexes (n 4), and routeing measures under the International Convention for the Safety of Life at Sea (adopted 1 November 1974, entered into force 25 May 1980) 1184 UNTS 2, as amended (SOLAS) Ch V.

7 Ibid Ch V, regs 10 and 11.

8 International Convention for the Prevention of Pollution of the Sea by Oil (adopted 12 May 1954, entered into force 26 July 1958) 327 UNTS 3 (OILPOL), now superseded by MARPOL (n 4) art 9(1).

9 Rachel Carson, *The Sea Around Us* (Oxford University Press 1951); Rachel Carson, *Silent Spring* (Houghton Miffin 1962).

10 Cousteau was probably the first to launch a series in the era of mass television introducing a global public to ocean diversity and threats to its well-being. 'Jacques Cousteau Centennial: What He Did, Why He Matters' *National Geographic News* (11 June 2010) <http://news.nationalgeographic.com/news/2010/06/100611-jacques-cousteau-100th-anniversary-birthday-legacy-google> accessed 24 May 2018. His classic work was *The Silent World: A Story of Undersea Discovery and Adventure* (Harper & Brothers Publishers 1953).

print and visual media provoked much public attention on environmental and ocean issues.

Attention to the marine environment in the IMO was triggered in a seismic shift by a massive casualty and the role of the media in communicating its impacts. The Torrey Canyon in 1967 casualty off the southwestern Atlantic coast of the United Kingdom was a major milestone in the institutional growth of the IMO and international maritime regulation. The Marine Environment Protection Committee (MEPC) and Legal Committee were established as a result.[11] Several new conventions in international public and private maritime law were adopted, almost all concerning marine pollution. The United Nations Stockholm Conference on the Human Environment in 1972[12] and the Third United Nations Conference on the Law of the Sea in 1973–82 further influenced the evolution of the IMO mission, in particular by cementing its role as the competent international organisation with respect to international shipping.[13]

The evolving mandate has enabled the IMO to better respond to major drivers of regulation, such as passenger safety, technological change, new commercial practices, maritime casualties, marine security and climate change concerns. Although MARPOL was adopted in 1973, it had minimal support in its early years. It was the Amoco Cadiz casualty that triggered the amending protocol, which generated the support needed to bring that instrument into force. Subsequent casualties producing massive environmental impacts similarly triggered the IMO into further action. The public perception was that in Torrey Canyon and other casualties resulting in major oil spills such as Amoco Cadiz, Exxon Valdez, Erika and Prestige, the polluters were not held fully accountable. A common view was that regulatory initiatives were reactive rather than anticipatory. In the public eye, vessel-source pollution was a major threat that did not appear to be sufficiently addressed. While IMO's environmental mandate today is still primarily dominated by the regulation of the various

11 Aldo Chircop, 'The International Maritime Organization' in Donald Rothwell et al (eds), *The Oxford Handbook on the Law of the Sea* (Oxford University Press 2015) 417, 424–25.

12 United Nations Declaration on the Human Environment, in particular principles 6 (discharge of toxic substances into the environment), 7 (marine pollution) and 22 (liability and compensation of pollution victims). United Nations Conference on the Human Environment, 'United Nations Declaration on the Human Environment' in 'Report of the United Nations Conference on the Human Environment' (5–16 June 1972) UN Doc A/CONF/48/14/Rev.1, 3. See also 'Annex III General Principles for Assessment and Control of Marine Pollution', Ibid 73.

13 IMO, 'Implications of the United Nations Convention on the Law of the Sea for the International Maritime Organization: Study by the Secretariat of the IMO' IMO Doc LEG/MISC.7 (19 January 2012).

types of vessel-source pollution (e.g., oil, noxious substances, sewage, garbage), it has expanded to address other threats posed by ships, such as the introduction of exotic species in alien environments, harmful antifouling systems (such as tributyltin), atmospheric emissions harmful to public health, safe and sustainable recycling of ships, and greenhouse gas emissions.

In addition to the evolving functions, the IMO's contemporary environmental mandate has a global scope, affecting marine areas under national jurisdiction and on the high seas, and draws from varied sources. The core mandate is set out in its constitutive instrument in general and specific thematic terms.[14] The general mandate is the provision of 'machinery for co-operation among Governments in the field of governmental regulation and practices relating to technical matters of all kinds affecting shipping engaged in international trade' irrespective of the theme.[15] The IMO facilitates the ability of member states to fulfil their duty to cooperate for the protection and preservation of the marine environment, either directly or through international organisations.

The IMO's thematic mandate is 'to encourage and facilitate the general adoption of the highest practicable standards in matters concerning maritime safety, efficiency of navigation and prevention and control of marine pollution from ships'.[16] While the theme might suggest an environmental mandate limited to vessel-source pollution and without other environmental impacts of shipping, the functions of the IMO Assembly include the wider power to recommend to members

> adoption of regulations and guidelines concerning maritime safety, the prevention and control of marine pollution from ships and other matters concerning the effect of shipping on the marine environment assigned to the Organization by or under international instruments, or amendments to such regulations and guidelines which have been referred to it.[17]

The international instruments concerned are UNCLOS and maritime conventions and could conceivably include any new instruments, including a future agreement on the conservation and sustainable use of marine biological diversity of areas beyond national jurisdiction.[18]

UNCLOS sets out implicit environmental roles for the IMO as the competent international organisation with respect to international shipping in several

14 IMO Convention (n 5) art 1(a).
15 Ibid.
16 Ibid.
17 Ibid art 15(j).
18 See Robin Warner, Chapter 6 'Marine Living Resources beyond National Jurisdiction'.

provisions.[19] The roles tend to be with reference to international conventions, regulations, generally accepted rules, and standards and procedures for international shipping and which have the effect of nourishing UNCLOS rights and duties. These provisions preponderantly aim to prevent, reduce and control pollution from ships,[20] including dumping from ships.[21] In prescribing responsibilities of the flag state, UNCLOS sets out measures to be adopted, which are to conform to generally accepted international regulations, procedures and practices.[22] These comprise measures necessary to ensure safety at sea including for the master and crew to be conversant in and to observe applicable international regulations on safety of life at sea, prevention of collisions, and prevention, reduction and control of marine pollution.[23] As will be discussed below, safety and collision prevention instruments have been used to source mandatory and voluntary APMs for PSSAs and other routeing and reporting measures aimed to prevent animal strikes.

The IMO interprets its thematic mandate in the broader terms of sustainable shipping.[24] The IMO's marine environmental mandate is further nourished by the international maritime conventions, which assign such roles to it. In addition to MARPOL, the IMO has responsibilities conferred in other conventions pertinent to all marine spaces, including ballast water management,[25] anti-fouling systems[26] and oil pollution preparedness and response.[27] By its own count, the IMO has adopted over 160 codes, recommendations, guidelines and other environmental matters in shipping.[28]

---

19 See Division for Ocean Affairs and the Law of the Sea, '"Competent or Relevant International Organizations" under the United Nations Convention on the Law of the Sea' (1996) 31 *Law of the Sea Bulletin* 79.

20 UNCLOS (n 2) arts 211(1), (2), (5) and (7); 217(1)–(4); 218(1); 219; 220(1)–(3) and (7); 226(1); 228.

21 Ibid arts 210(4) and (6), 216.

22 Ibid art 94(5).

23 Ibid art 94(4)(c).

24 Strategic Plan for the Organization (2016–2021) IMO Res A.1097(29) (25 November 2015).

25 International Convention for the Control and Management of Ships' Ballast Water and Sediments (adopted 13 February 2004, entered into force 8 September 2017) IMO Doc BWM/CONF/36 (16 February 2004).

26 International Convention on the Control of Harmful Anti-fouling Systems on Ships (adopted 5 October 2001, entered into force 17 September 2008) Can TS 2010 No 15.

27 International Convention on Oil Pollution Preparedness, Response and Co-operation (adopted 30 November 1990, entered into force 13 May 1995) 30 ILM 733. This was supplemented by the Protocol on Preparedness, Response and Co-operation to Pollution Incidents by Hazardous and Noxious Substances (adopted 15 March 2000, entered into force 14 June 2007) ATS 2007 4.

28 IMO, 'List of Codes, Recommendations, Guidelines and other Environment-related Non-mandatory Instruments Note by the Secretariat' IMO Doc MEPC 70/INF.7 (10 August 2016).

The diversity of ocean uses, interests and impacts on the marine environment concerns numerous economic sectors and consequently involves the mandates of various intergovernmental organisations focussed on one or more sectors and/or marine regions. This is recognised in UNCLOS, which anticipates cooperation not only among states, and states and international organisations, but also between and among international organisations.[29] In its mandate, the IMO is expected to respond to issues 'concerning the effect of shipping on the marine environment' referred to it by other United Nations (UN) organs and agencies.[30] It is common for the IMO Secretariat to disseminate information on developments in other international regimes through information circulars to the IMO's membership. The IMO has also actively cooperated on specific environmental initiatives with other UN organisations on global and regional initiatives.[31]

Since its inception, the IMO has demonstrated a high degree of resiliency and adaptability to new demands placed upon it by the international community. It remains relevant and well placed to respond to emerging new demands in the governance of shipping, including on the high seas, both in terms of exercising its regulatory mandate and cooperating with other international organisations. The discussion now turns to how the IMO is addressing the needs of sensitive marine areas and species and the circumstances for the adoption of MARPOL special areas, PSSAs and routeing measures in response to particular environmental threats, before subsequent discussion on applications to high seas areas.

## 3 MARPOL Special Areas

### 3.1 *The Scope of MARPOL*

The leitmotif of MARPOL is ships and their environmental performance with respect to the services they provide or receive. The MARPOL and its six annexes

29 For example, with respect to cooperation in the development and transfer of marine technology (UNCLOS (n 2) art 278); power of the Council on behalf of the International Seabed Authority to enter into agreements with the UN and other international organisations (Ibid art 162(2)(f)).

30 IMO Convention (n 5) art 1(d).

31 For example, at the global level, IMO has cooperated with the Basel Convention Secretariat with respect to the development of the Hong Kong Convention. At the regional level, IMO has cooperated with the United Nations Environment Programme within the framework of the Regional Seas Programme and in some cases by establishing regional centres, such as the Regional Marine Pollution Emergency Response Centre for the Mediterranean Sea.

TABLE 5.1 Status of MARPOL

| Instrument | Date adopted | Date in force | Years to entry into force | Number of state parties | Percentage of global tonnage |
|---|---|---|---|---|---|
| MARPOL (by Protocol) | 17 February 1978 | 2 October 1983 | 6 | 156 | 99.42 |
| Annexes I & II | 17 February 1978 | 2 October 1983 | 6 | 156 | 99.42 |
| Annex III | 17 February 1978 | 1 July 1992 | 14 | 148 | 98.81 |
| Annex IV | 17 February 1978 | 27 September 2003 | 25 | 142 | 96.54 |
| Annex V | 17 February 1978 | 31 December 1988 | 10 | 153 | 98.97 |
| Protocol, 1997, Annex VI | 26 September 1997 | 19 May 2005 | 8 | 91 | 96.89 |

SOURCE: ADAPTED FROM IMO, 'STIATUS OF TREATIES AS OF 24 MAY 2018'[a]
a IMO, 'Status of IMO Treaties' (n 37).

command a large number of state parties representing the vast majority of global tonnage (Table 5.1). The regulatory mission is focussed on *ships* of state parties.[32] The standards, prescribed actions and prohibitions principally concern ships, and while ports provide important services, the flag state has a duty to establish sanctions for its vessels that commit violations wherever the infraction occurs.[33] With respect to particular discharges, the rules define distances from the nearest land.[34] Where violations occur within the jurisdiction of a coastal state, that state has a similar duty to enforce or to inform the flag

32 MARPOL (n 4) art 3(1). The only exception concerns government ships on non-commercial service. However, the state of service has a duty to ensure 'that such ships act in a manner consistent, as far as is reasonable and practicable with the present Convention'. Ibid art 3(3).

33 Ibid art 4(1).

34 Generally referring to the baseline from which the territorial sea is measured. There is a variation with respect to the northeast coast of Australia. Ibid Annex I, Ch I, reg 1(10). The rule is replicated in other annexes.

state of the infraction.[35] Also, in the course of their normal trading activities and irrespective of actual violations, port states may inspect visiting ships to determine whether a ship has violated a rule.[36] In practice, MARPOL is enforced on the basis of the 'no more favourable treatment' principle, with the consequence that ships of non-parties are affected in port state control (PSC) inspections.[37] Consequently, the MARPOL mission is defined in largely functional rather than spatial terms, so that the regulatory reach includes the operations of ships on the high seas.

MARPOL regulation can be described as 'layered'. Regulations are based on universal standards for all ships (e.g., garbage management and disposal), standards applicable to specific classes of ships (e.g., double hulls for oil tankers), and standards for particular marine areas that are higher than the general norms. The last category of standards includes special areas and emission control areas, and other defined spaces short of formal technical designation as special areas. In some special areas, certain requirements may be even higher than what is otherwise applicable in the special area concerned.[38]

### 3.2 *Special Area Elements*

#### 3.2.1 Concept

The idea that certain areas of the marine environment should be subject to zero discharge rules was first set out in OILPOL.[39] Prohibited zones for tankers were designated and defined in terms of 50 nautical miles (NM) from the coast and other distance limits for specific areas,[40] and also with a similar rule with

35 Ibid art 4(2). Where the coastal state informs the flag state and defers to its jurisdiction, the flag state has a duty to promptly inform the coastal state of the action it has taken. Ibid art 4(3).

36 Ibid art 6(2).

37 MARPOL (n 4) art 5(4). The prospect of MARPOL being enforced against the ships of non-parties is likely remote, but not impossible. The IMO has 172 member states representing 97.28 per cent of world tonnage. IMO, 'Status of IMO Treaties (14 May 2018)' <www.imo.org/en/About/Conventions/StatusOfConventions/Documents/Status%20-%202018.pdf> accessed 24 May 2018.

38 For example, MARPOL (n 4) Annex II, Ch 9, reg 43 concerns special requirements for the use or carriage of oils in the Antarctic area and prohibits the carriage, ballast or fuel use of heavy oils, except for certain ships.

39 OILPOL (n 8) Annex A.

40 For example for the Adriatic (30 NM and with provision for a further extension by 20 NM, with an exception near the island of Vis), North Sea (100 NM from the coasts of Belgium, Denmark, Federal Republic of Germany, Netherlands and United Kingdom, and up to the limit of Norway's 50 mile zone), northeast Atlantic (defined by geographical coordinates), and Australian zone (150 miles with some exceptions). Ibid Annex A, reg 1.

variable distances for non-tankers.[41] State parties were able to propose a reduction or expansion of the no-discharge zone off their coasts up to 100 NM.[42] However, the concept of a special area that was designated and justified on the basis of objective scientific criteria and traffic patterns and whose limits were set by the collective of states, rather than by individual states, was first introduced in the original MARPOL. When adopted in 1973, MARPOL designated the Mediterranean Sea, Baltic Sea, Black Sea, Red Sea areas (including the Gulfs of Suez and Aqaba) and 'Gulfs' area as special areas for oil (Annex I) and garbage (Annex V), and the Baltic and Black Seas for noxious liquid substances in bulk (Annex II). At the time there were no designations or special area rules in Annexes III and IV.

Following the adoption of the original MARPOL, a series of tanker casualties starting with the SS Argo Merchant off Nantucket in 1976 and until Amoco Cadiz in 1978 re-energised regulatory efforts on oil pollution, moving the US President Jimmy Carter to call upon the IMO to convene an intergovernmental conference to consider amendments to MARPOL and the International Convention for the Safety of Life at Sea, 1974 (SOLAS).[43] The International Conference on Tanker Safety and Pollution Prevention adopted the 1978 amendments to MARPOL and SOLAS, and 19 resolutions, the ninth of which addressed the need for the IMO to study how specific ocean areas could receive a reasonable degree of protection because of their particular sensitivity, an idea that reinforced both the rationale for special areas and the development of future PSSA guidelines.[44] Although many more years would pass before the idea was implemented, it would form the basis of the first guidelines for special areas and PSSAs proposed by MEPC 31 and adopted in 1991 by the Assembly pursuant to its mandate.[45]

As a novel instrument, the 1991 guidelines were formulated in a manner not only to guide proposals for special areas and PSSAs, but also to provide an integrated context and educate proponents on the range of threats posed

---

41 Ibid Annex A, reg 2.

42 Ibid Annex A, reg 3.

43 SOLAS (n 6).

44 International Conference on Tanker Safety and Pollution Prevention (London, 6–17 February 1978), Final Act of the Conference with Attachments including the Protocol of 1978 relating to the International Convention for the Safety of Life at Sea, 1974 and the Protocol of 1978 relating to the International Convention for the Prevention of Pollution from Ships, 1973 (Intergovernmental Maritime Consultative Organization 1978) (Tanker Safety Conference) res 9.

45 Guidelines for the Designation of Special Areas and the Identification of Particularly Sensitive Sea Areas, IMO Res A.720(17) (6 November 1991). The Guidelines were adopted under the authority of the IMO Convention (n 5) art 15(j).

by shipping, the tools available in the IMO arsenal to address the threats (e.g., special areas, PSSA designation, routeing measures), the role of marine protected areas at the national and international levels, and other international conventions that provide additional protection and tools. Subsequent versions of the guidelines largely excised the educational text and focussed more on the actual requirements and guidance on the procedure for proposals for special areas and PSSAs. Amendments adopted in 1999 clarified procedures for PSSA identification and adoption of APMs,[46] and amendments in 2001 adopted new guidelines for both areas for the purpose of simplifying proposal procedures.[47] In 2005 the guidelines for PSSA designation were further amended and in the process revoked the 1991 and 1999 resolutions and superseded the PSSA guidelines contained in Annex II of the 2001 resolution.[48]

More recently, the adoption of amendments to MARPOL Annex IV (sewage) necessitated consequential amendment of the special area guidelines to provide for the designation of such areas for the prevention of sewage pollution from passenger ships.[49] This latest iteration is the 2013 Guidelines for the Designation of Special Areas under MARPOL (2013 Guidelines).[50]

### 3.2.2 Definition

MARPOL defines special area as 'a sea area where for recognized technical reasons in relation to its oceanographical and ecological condition and to the particular character of its traffic the adoption of special mandatory methods for the prevention of sea pollution by oil is required'.[51] These are marine spaces that require a higher level of protection than the normal mandatory pollution prevention measures for ships on international voyages and affect waste management on board, at-sea discharge rates and concentrations, port reception facilities for on-board waste, and in trading areas. As Table 5.2 demonstrates,

46 Procedures for the Identification of Particularly Sensitive Sea Areas and the Adoption of Associated Protective Measures and Amendments to the Guidelines contained in Resolution A.720(17), IMO Res A.885(21) (25 November 1999).

47 Guidelines for the Designation of Special Areas under MARPOL 73/78 and Guidelines for the Identification and Designation of Particularly Sensitive Sea Areas, IMO Res A.927(22) (29 November 2001).

48 Revised Guidelines for the Identification and Designation of Particularly Sensitive Sea Areas, IMO Res A.982(24) (1 December 2005).

49 Consequential Amendments to the Revised Guidelines for the Identification and Designation of Particularly Sensitive Sea Areas (Resolution A.982(24)) Note by the Secretariat, IMO Doc MEPC 67/10 (22 July 2014).

50 2013 Guidelines for the Designation of Special Areas under MARPOL, IMO Res A.1087(28) (21 February 2014). They revoke the 2005 Revised PSSA Guidelines.

51 MARPOL (n 4) Annex I, Ch I, reg 1(11). The same definition is reproduced in Annex IV, Ch I, reg 1(6) and Annex V, Ch I, reg 1(14).

TABLE 5.2 MARPOL special areas and emission control areas

| Annexes | Special areas and emission control areas | Date adopted | Date in force | Date effective |
|---|---|---|---|---|
| I: Oil (10) | Mediterranean Sea | 2 November 1973 | 2 October 1983 | 2 October 1983 |
| | Baltic Sea | 2 November 1973 | 2 October 1983 | 2 October 1983 |
| | Black Sea | 2 November 1973 | 2 October 1983 | 2 October 1983 |
| | Red Sea | 2 November 1973 | 2 October 1983 | Not in effect |
| | 'Gulfs' area | 2 November 1973 | 2 October 1983 | 1 August 2008 |
| | Gulf of Aden | 1 December 1987 | 1 April 1989 | Not in effect |
| | Antarctic area | 16 November 1990 | 17 March 1992 | 17 March 1992 |
| | North West European waters | 25 September 1997 | 1 February 1999 | 01 August 1999 |
| | Oman area (Arabian Sea) | 15 October 2004 | 1 January 2007 | Not in effect |
| | Southern South African waters | 13 October 2006 | 1 March 2008 | 1 August 2008 |
| II: Noxious liquid substances (1) | Antarctic area | 30 October 1992 | 1 July 1994 | 1 July 1994 |
| IV: Sewage (1) | Baltic Sea | 15 July 2011 | 1 January 2013 | Not in effect |
| V: Garbage (8) | Mediterranean Sea | 2 November 1973 | 31 December 1988 | 1 May 2009 |
| | Baltic Sea | 2 November 1973 | 31 December 1988 | 1 October 1989 |
| | Black Sea | 2 November 1973 | 31 December 1988 | Not in effect |
| | Red Sea | 2 November 1973 | 31 December 1988 | Not in effect |
| | 'Gulfs' area | 2 November 1973 | 31 December 1988 | 1 August 2008 |
| | North Sea | 17 October 1989 | 18 February 1991 | 18 February 1991 |
| | Antarctic area | 16 November 1990 | 17 March 1992 | 17 March 1992 |
| | Wider Caribbean region | 4 July 1991 | 4 April 1993 | 1 May 2011 |
| VI: Air pollution (6 in 3 regions) | Baltic Sea ($SO_x$) | 26 September 1997 | 19 May 2005 | 19 May 2006 |
| | North Sea ($SO_x$) | 22 July 2005 | 22 November 2006 | 22 November 2007 |
| | North American ($SO_x$/PM) | 26 March 2010 | 1 August 2011 | 1 August 2012 |
| | North American ($NO_x$) | 26 March 2010 | 1 August 2011 | 1 January 2016* |
| | US Caribbean Sea ($SO_x$/PM) | 26 July 2011 | 1 January 2013 | 1 January 2014 |
| | US Caribbean Sea ($NO_x$) | 26 July 2011 | 1 January 2013 | 1 January 2016* |

Legend: (*): for ships built after this date; PM: particulate matter; $NO_x$: nitrogen oxides; $SO_x$: sulphur oxides

SOURCE: ADAPTED FROM IMO, 'SPECIAL AREAS UNDER MARPOL'[a]

a IMO, 'Special Areas under MARPOL' <www.imo.org/en/OurWork/Environment/SpecialAreasUnderMARPOL/Pages/Default.aspx> (accessed 24 May 2018).

special areas have been designated under Annexes I (oily wastes), II (noxious liquid substances),[52] IV (sewage) and V (garbage). Special areas are designated in marine spaces where bordering states commit to the provision of reception facilities in their ports for the wastes that may not be discharged at sea and become effective on notification to the IMO that such facilities are in place.[53] Special discharge measures may also be adopted in the absence of special area designation. For example, the IMO adopted high standards for on-board waste management and at-sea discharge in polar waters under MARPOL Annexes I, II, IV and V and the Polar Code's environmental provisions, without designating special areas in Arctic and Antarctic waters.[54] This approach enabled the introduction of high protective standards without the contingent requirement for the provision of port reception facilities in regions where such infrastructure is unavailable.

Table 5.2 also shows designated emission control areas (ECAs), which are analogous to special areas and are designated under Annex VI (air pollution from ships). The ECAs are not central to the discussion in this paper and are discussed briefly. The ECA is defined as an area 'where the adoption of special mandatory measures for emissions from ships is required to prevent, reduce and control air pollution from NOx or SOx and particulate matter or all three types of emissions and their attendant adverse impacts on human health and the environment'.[55] Special areas and ECAs are adopted through amendments to MARPOL and on the basis of proposals submitted by member states in accordance with guidelines adopted for this purpose.[56]

Despite the broad base of support, not all MARPOL state parties are necessarily parties to Annexes III to VI, since only the first two annexes are mandatory. In theory this means that states not parties to Annexes IV and V are not bound by the more restrictive sewage and garbage special areas discharge rules. However, from a practical perspective, vessels flying the flags of non-parties to Annexes IV and V are still subject to port inspections (e.g., for the Garbage Record Book) under the MARPOL's 'no more favourable treatment'

52 The definition of special area is not present in Annex II. Technically, Antarctic waters are not designated as a special area under this annex and the pertinent rule prescribes zero discharge of noxious liquid substances. Ibid Annex II, Ch V, reg 13(8).

53 For example, with respect to the Red Sea Area, 'Gulfs' Area, Gulf of Aden Area and Oman Area of the Arabian Sea in Annex I. MARPOL (n 4) Annex I, Ch VI, reg 38(6).

54 Ibid Annex I, Ch XI, reg 47; Annex II, Ch X, reg 22; Annex IV, Ch VII, reg 18; Annex V, Ch III, reg 14. International Code for Ships Operating in Polar Waters (Polar Code) IMO Res MEPC.265(68) (15 May 2015) Part IIA, Chs 1, 2, 4 and 5.

55 MARPOL (n 4) Annex VI, Ch I, reg 2(8).

56 IMO Res A.1087(28) (n 50).

principle and as enforced in the PSC Memoranda of Understanding (MoU).[57] Garbage is particularly significant, because recent amendments to Annex V prohibit disposal at sea of most types of garbage with very few exceptions.[58]

### 3.2.3 Criteria

The criteria for the designation of special areas have remained constant since first proposed in the original MARPOL, and consist of three clusters concerning oceanographic conditions, ecological conditions and vessel traffic characteristics (Table 5.3), although additional information may be submitted.[59] The definition does not set limits to the size of the proposed area, and generally size depends on collective consideration of the criteria and the values to be protected. In general, however, the areas have encompassed entire seas or semi-enclosed seas (e.g., Baltic Sea, Black Sea, Mediterranean Sea), or exclusive economic zones (EEZs) (Northwest European Waters), or parts of EEZs (Southern South African waters).[60]

The oceanographic conditions must be such that they may lead to an increase in the concentration of harmful substances in the waters or sediments of the proposed area.[61] Examples of the conditions that may lead to such an outcome are listed in Table 5.3 (column 1) and are proposed as inclusive rather than exclusive. For example, the restricted circulation of the waters of the Baltic, Black and Mediterranean Seas was a factor in designating them as special areas. The long residence time of waters also had repercussions for the assimilative capacity for wastes discharged into the marine environment.

The vulnerability of the proposed area's ecology is such that it demands protection from harmful substances to produce particular conservation outcomes.[62] The listed outcomes are not set out in inclusive terms, as in the case of oceanographic conditions, but are sufficiently broad to encompass a wide range of values. When the Mediterranean Sea was first included as a special area,

---

57 Europe and the North Atlantic (Paris MoU); Asia and the Pacific (Tokyo MoU); Latin America (Acuerdo de Viña del Mar); Caribbean (Caribbean MoU); West and Central Africa (Abuja MoU); the Black Sea region (Black Sea MoU); the Mediterranean (Mediterranean MoU); the Indian Ocean (Indian Ocean MoU) and the Persian Gulf (Riyadh MoU). For links to each MoU, see IMO, 'Port State Control' <www.imo.org/blast/mainframe.asp?topic_id=159> accessed 24 May 2018.

58 Amendments effective as of 1 January 2013. The exemptions are food waste, animal carcasses, cargo residues contained in wash water, and environmental friendly cleaning agents.

59 IMO Res A.1087(28) (n 50) para 2.3.

60 Ibid para 2.2.

61 Ibid para 2.4.

62 Ibid para 2.5.

TABLE 5.3 MARPOL special area criteria

| Oceanographic conditions | Ecological conditions | Vessel traffic characteristics |
| --- | --- | --- |
| The area possesses oceanographic conditions that may cause the concentration or retention of harmful substances in the waters or sediments of the area, including<br>– particular circulation patterns (e.g. convergence zones and gyres) or temperature and salinity stratification;<br>– long residence time caused by low flushing rates;<br>– extreme ice state; and<br>– adverse wind conditions. | The area possesses ecological conditions indicating that protection of the area from harmful substances is needed to preserve<br>– depleted, threatened or endangered marine species;<br>– areas of high natural productivity (such as fronts, upwelling areas, gyres);<br>– spawning, breeding and nursery areas for important marine species and areas representing migratory routes for seabirds and marine mammals;<br>– rare or fragile ecosystems such as coral reefs, mangroves, seagrass beds and wetlands; and<br>– critical habitats for marine resources including fish stocks and/or areas of critical importance for the support of large marine ecosystems. | The sea area is used by ships to an extent that the discharge of harmful substances by ships when operating in accordance with the requirements of MARPOL for areas other than special areas would be unacceptable in the light of the existing oceanographic and ecological conditions in the area. |

SOURCE: ADAPTED FROM IMO RES A.1087(28)[a]

a Ibid annex. See Youna Lyons, Chapter 4 'Identifying Sensitive Marine Areas in the High Seas: A Review of the Scientific Criteria adopted under International Law'.

the long list of ecological concerns included several species of the threatened or endangered lists (e.g., Mediterranean monk seal, green turtles), spawning and breeding grounds for important resources (e.g., Bluefin tuna), seagrass beds (e.g., *posidonia oceanica*) and wetlands, among others.

Vessel traffic conditions are required to be quantitatively and qualitatively such that the normal discharge standards under the various MARPOL annexes would not be sufficient to protect the ecological values in the proposed area's oceanographic conditions.[63] This was clearly the case in the Baltic,

63 Ibid para 2.6.

Mediterranean and North Sea, which constitute major international navigation routes for virtually all forms of marine transportation. The volumes of shipping are so high that the assimilative capacities of the seas concerned are under substantial stress.

Oceanographic, ecological and vessel traffic characteristics may be further buttressed by other important factors, such as the threats posed by the sheer volume of shipping to vital coastal and inshore amenities, such as economic activities like tourism, aquaculture, fishing, renewable energy generation and production of fresh water by desalination.[64] For example, coastal tourism in the Mediterranean is a major revenue generator for many of the sea's coastal states.

The 2013 Guidelines recognise that marine environmental degradation is exacerbated by cumulative impacts of pollutants from various sources, such as pollution from land-based sources, dumping and atmospheric pathways.[65] The potential cumulative impact of the various sources may also be considered, especially where proponents indicate that action is being taken with respect to other sources as well. In this regard, proposals for special areas are strengthened where the proponents demonstrate that coastal and/or ocean management regimes are in place and used to address environmental degradation.[66] Although this was not an argument for the Mediterranean when it was first included as a special area on the adoption of MARPOL in 1973, by 1976 it had adopted a treaty-based management regime.[67]

### 3.2.4 Procedure

Designated special areas constitute new rules and standards in MARPOL and consequently their adoption engages the formal amendment procedure of the convention.[68] In practice, this entails use of the tacit acceptance procedure.[69] Fulfilling the requirements of the submission and the related burden of proof rests with the proponent states. Proposals are submitted to the MEPC

64 Ibid para 2.8.

65 Ibid para 2.9.

66 Ibid para 2.10.

67 Barcelona Convention for Protection of the Mediterranean Sea Against Pollution (adopted 16 February 1976, entered into force 12 February 1978) 1102 UNTS 27, amended on 10 June 1995 and retitled Convention for the Protection of the Marine Environment and the Coastal Region of the Mediterranean, amendments in force in 2004.

68 MARPOL (n 4) art 16.

69 Ibid. Amendment of MARPOL depends on whether the proposed amendment applies to the convention itself, an annex or an appendix to an annex. The tacit acceptance rule provides for a procedure for a technical amendment (e.g., to an annex) adopted directly by the MEPC to be deemed to have been accepted by state parties and for its entry into force.

and should be accompanied by suggested draft amendments to MARPOL and a background document prepared by the proponent state(s).[70] The background document should contain the following:

- definition and geographical coordinates of the area, including charts;
- type of special area proposed;
- general description of the area, including
  - oceanography,
  - ecological characteristics,
  - social and economic value,
  - scientific and cultural significance,
  - environmental pressures from ship-generated pollution,
  - other environmental pressures, and
  - measures already taken to protect the area;
- annexes with more detailed material or references to readily available documentation;
- analysis of how the proposed sea area fulfils special area criteria; and
- information on the availability of adequate reception facilities in the proposed area.[71]

In summary, the proponent states are required to establish why the normal rules and standards are insufficient to address the needs of the proposed area and why those needs should be addressed by making exceptions. Most importantly, they have to demonstrate that the region adjacent to the proposed special area has the necessary reception facilities for the wastes concerned. This is a critical requirement, because while lack of information on reception areas might not, per se, deny the designation, it might delay its effectivity. The effectivity of a special area depends on the proponent states communicating the actual availability of reception areas to the IMO, and at which point all other member states are so informed. Thus states in the special areas have a duty to provide adequate reception facilities for oily residues and mixtures,[72] noxious liquid substances,[73] sewage[74] and garbage.[75] As Table 5.2 indicates, not all adopted special areas are actually in effect, even though some were designated when MARPOL was first adopted in 1973. It is ironic that from time to time within the IMO, concern has been expressed regarding the potential proliferation of special areas.[76]

70 IMO Res A.1087(28) (n 50) paras 3.2 and 3.3.
71 Ibid para 3.3.
72 MARPOL (n 4) Annex I, Ch VI, reg 38.
73 Ibid Annex II, Ch VIII, reg 18.
74 Ibid Annex IV, Ch IV, regs 12–13.
75 Ibid Annex V, Ch I, reg 8.
76 IMO Res A.1087(28) (n 50) para 2.2.

## 4 Particularly Sensitive Sea Areas

### 4.1 *PSSA Elements*

#### 4.1.1 Concept

As noted earlier, the genesis of the PSSA concept in the IMO can be traced back to the 1978 International Conference on Tanker Safety and Pollution Prevention, when the IMO was called upon to consider how to provide protection for sensitive marine areas and also to provide an inventory of such areas.[77] The history has been traced and discussed in the literature.[78] Tasked with following up, the MEPC commenced discussions in 1986 and continued them through 1990. The Great Barrier Reef, which would become the first PSSA, was the type of sensitive system that MEPC had in mind. Given that whatever additional protection was likely to be an intrusion into the freedom of navigation, an appropriate procedure and predetermined criteria for submissions to enable collective review through an IMO structure were necessary. What emerged was the first generation of guidelines adopted in 1991 by an IMO Assembly resolution.[79] The Assembly tasked the MEPC to maintain the guidelines under review and in doing so amended them in 1999,[80] 2001,[81] 2005[82] and with Secretariat-proposed minor consequential amendments in 2015.[83] A new

77 Tanker Safety Conference (n 44) res 9.

78 The early history of PSSAs is discussed by Kristina Gjerde and David Freestone 'Particularly Sensitive Sea Areas—An Important Environmental Concept at a Turning Point' (1994) 9 *International Journal of Maritime & Coastal Law* 431; Gerard Peet 'Particularly Sensitive Sea Areas—A Documentary History' (1994) 9 *International Journal of Maritime & Coastal Law* 469. For subsequent treatment, see Markus J Kachel, *Particularly Sensitive Sea Areas: The IMO's Role in Protecting Vulnerable Marine Areas* (Springer 2008); Julian Roberts *Marine Environment Protection and Biodiversity Conservation: The Application and Future Development of the IMO's Particularly Sensitive Sea Concept* (Springer 2010); Aldo Chircop, 'The Designation of Particularly Sensitive Sea Areas: A New Layer in the Regime for Marine Environment Protection from International Shipping' in Aldo Chircop et al (eds), *The Future of Ocean Regime Building: Essays in Tribute to Douglas M Johnston* (Nijhoff 2009) 573.

79 Guidelines for the Designation of Special Areas and the Identification of Particularly Sensitive Sea Areas (n 45).

80 Procedures for the Identification of Particularly Sensitive Sea Areas and the Adoption of Associated Protective Measures and Amendments to the Guidelines contained in Resolution A.720(17) (n 46).

81 Guidelines for the Designation of Special Areas under MARPOL 73/78 and Guidelines for the Identification and Designation of Particularly Sensitive Sea Areas (n 47). The MEPC also approved the Guidance Document for Submission of PSSA Proposals to IMO, 'Identification and Protection of Special Areas and Particularly Sensitive Sea Areas' IMO Doc MEPC 48/7/1 (30 April 2002) subsequently revised in 2005 (n 48).

82 The current version is IMO Res A.982(24) (n 48).

83 Amendments to the Revised Guidelines for the Identification and Designation of Particularly Sensitive Sea Areas (Resolution A.982(24)) Note by the Secretariat (n 49), IMO Res

Guidance Document for Submission of PSSA Proposals to IMO[84] and Proposal Review Form were adopted in 2006.[85] The 2005 Revised PSSA Guidelines constitute the current guidance for the preparation, submission and consideration of new proposals. They perform three functions:

- provide guidance to IMO member governments in the formulation and submission of applications for designation of PSSAs;
- ensure that in the process all interests—those of the coastal state, flag state and the environmental and shipping communities—are thoroughly considered on the basis of relevant scientific, technical, economic and environmental information regarding the area at risk of damage from international shipping activities and the APMs to prevent, reduce or eliminate that risk; and
- provide for the assessment of such applications by IMO.[86]

When discussion first started, the initial view was that IMO already provided options for the protection of sensitive areas, whether at a regional or subregional basis through MARPOL special areas or more surgically for example through routeing measures under SOLAS. But it was clear that the PSSA concept, although related, was different from MARPOL special areas, which more directly focussed on pollution prevention and suggested an even higher level of protection in response to the vulnerabilities of particular areas of the marine environment. The area that demanded additional protection could need special area designation but likely needed other forms of protection as well. MARPOL special areas were creatures of a conventional instrument, whereas the PSSA was not, and rather drew from the IMO's general mandate. Also, special areas designated under UNCLOS Article 211(6) were considered different. Under the latter, a special area designation was likely geographically narrower and the proposed special measures would be mandatory. APMs under a PSSA could be simply recommended. However, an opinion provided by the United Nations Division for Ocean Affairs and the Law of the Sea was that the PSSA Guidelines are derived from Article 211(6), although conceding that they are

MEPC.267(68) (15 May 2015) 'Report of the Marine Environment Protection Committee (MEPC) on its Sixty-Eighth Session' IMO Doc MEPC 68/21 (29 May 2015) annex 13.

84 IMO, 'Guidance Document for Submission of PSSA Proposals to IMO' IMO Doc MEPC.1/Circ.510 (10 May 2006) (IMO Doc MEPC.1/Circ.510).

85 IMO, 'Report of the MEPC on its Fifty-Fifth Session' IMO Doc MEPC 55/23 (16 October 2006) (IMO Doc MEPC 55/23) Annex 20. The adoption of the proposal was the result of a United States initiative. 'Particularly Sensitive Sea Area Proposal Review Form Submitted by the United States' IMO Doc MEPC 55/8 (16 June 2006).

86 IMO Res A.982(24) (n 48) para 1.

'far more detailed and "liberal" in their approach'.[87] However, the IMO's actual authority for PSSA-making lies in its own convention and other maritime conventions, and that an Article 211(6) mandatory measure could simply be an APM.[88]

The evolution of the guidelines from one version to another was not always smooth. In particular, the 2005 revisions were contentious, because they came in the wake of the difficult discussion on the Western European Waters designation and the subsequent proposal for a Baltic Sea-wide PSSA, as will be discussed further below. Not all states felt that changes to the guidelines were needed.[89] The idea of a moratorium on designations pending review of the guidelines was floated but not adopted.[90]

### 4.1.2 Definition

A PSSA is defined as a marine area 'that needs special protection through action by the IMO, because of its significance for recognized ecological, socio-economic, or scientific attributes where such attributes may be vulnerable to damage by international shipping activities'.[91] When the IMO designates a PSSA, it is accompanied by APMs aimed at preventing, reducing or eliminating the threat or identified vulnerability.[92] Table 5.4 sets out the PSSAs adopted to date.

PSSAs can also be adopted in principle, pending completion of internal IMO assessment of the proposal and final designation. At the time of writing, the most recent PSSA to be designated is the Tubbataha Reefs Natural Park in the Philippines.[93] New proposals or expressions of intentions for PSSAs have been

---

87 IMO, 'Comments made by the Division for Ocean Affairs and the Law of the Sea of the United Nations (DOALOS) in connection with Issues Raised in Document LEG 87/16/1' IMO Doc LEG 87/WP.3 (13 October 2003) (IMO Doc LEG 87/WP.3).

88 IMO Convention (n 5) arts 1(a) and 15(j). Further, according to DOALOS, the PSSA 'Guidelines conform to the requirements of article 237 [of the UNCLOS], as a subsequent agreement adopted by IMO Assembly in furtherance of the general principles set forth in the Convention, in particular, those in Part XII'. IMO Doc LEG 87/WP.3 (n 87).

89 IMO, 'Report of the MEPC on its Fifty-First Session' IMO Doc MEPC 51/22 (22 April 2004) 29–32.

90 Led by Norway in the MEPC, the view that prevailed was that a moratorium would have had to be an Assembly decision and was not within the MEPC's mandate. Ibid 32.

91 IMO Res A.982(24) (n 48) para 1.2.

92 Ibid.

93 Designation of Tubbataha Reefs Natural Park as a Particularly Sensitive Sea Area, IMO Res MEPC.294(71) (7 July 2017).

TABLE 5.4 PSSAs designated by IMO

| PSSA | Year | Proponents |
|---|---|---|
| Great Barrier Reef (GBR) | 1990 | Australia |
| Sabana-Camaguey Archipelago | 1997 | Cuba |
| Wadden Sea | 2002 | Denmark, Germany, Netherlands |
| Florida Keys sea area | 2002 | United States |
| Malpelo Island | 2002 | Colombia |
| Paracas National Reserve | 2003 | Peru |
| Western European Waters | 2004 | Belgium, France, Ireland, Portugal, Spain and United Kingdom |
| Torres Strait (GBR PSSA extension) | 2005 | Australia, Papua New Guinea |
| Canary Islands | 2005 | Spain |
| Galapagos Archipelago | 2005 | Ecuador |
| Baltic Sea (excluding Russian waters) | 2005 | Denmark, Estonia, Finland, Germany, Latvia, Lithuania, Poland, Sweden |
| Papahānaumokuākea Marine National Monument (Hawaii) | 2008 | United States |
| The Strait of Bonifacio | 2011 | France, Italy |
| Saba Bank, North-eastern Caribbean | 2012 | Netherlands |
| Southwest part of Coral Sea (extension to GBR PSSA) | 2015 | Australia |
| The Jomard Entrance | 2016 | Papua New Guinea |
| Tubbataha Reefs Natural Park | 2017 | Philippines |

SOURCE: IMO, 'PARTICULARLY SENSITIVE SEA AREAS'[a]

a IMO, 'Particularly Sensitive Sea Areas' <www.imo.org/en/OurWork/Environment/PSSAs/Pages/Default.aspx> accessed 24 May 2018.

submitted by Malaysia,[94] Mauritania,[95] and Viet Nam[96] and are at different stages.

94 'Protection of Pulau Kukup (Kukup Island) and Tanjung Piai (Cape Piai) Submitted by Malaysia' IMO Doc MEPC 68/INF.22 (6 March 2015).

95 'Re EEZ waters adjacent to the Banc d'Arguin National Park of Mauritania. Protection of Banc d'Arguin National Park World Heritage Site and an Adjacent Sea Area Submitted by Mauritania' IMO Doc MEPC 69/INF.19 (11 February 2016).

96 'Protection of Ha Long Bay (Quang Ninh)—Cat Ba (Hai Phong) Marine Protected Area Submitted by Viet Nam' IMO Doc MEPC 69/INF.12 (29 January 2016).

#### 4.1.3 Criteria

The designation and adoption of APMs have to satisfy three integral components, each with criteria to be satisfied. First, the proposed area has to have particular attributes that attest to its sensitivity. Second, the proposed area must be vulnerable to international shipping activities. Third, the protection of the area demands the use of tools that are within the IMO's competence.[97]

(a) Attributes of the proposed area

The scientific criteria to be addressed under the first component are broader than those for MARPOL special areas. Whereas for the latter the scientific criteria are oceanographic and ecological, in the case of PSSAs they are (a) ecological, (b) social, cultural and economic, and (c) scientific and educational (Table 5.5).

The proponent state needs to demonstrate only 1 of the 17 criteria.[98] This minimal requirement was the subject of much discussion in the 2005 amendments, as it was argued that any marine space could satisfy at least one

TABLE 5.5 PSSA sensitivity criteria

| Ecological cluster | Social, cultural and economic cluster | Scientific and educational cluster |
|---|---|---|
| (1) Uniqueness or rarity<br>(2) Critical habitat<br>(3) Dependency<br>(4) Representativeness<br>(5) Diversity<br>(6) Productivity<br>(7) Spawning or breeding grounds<br>(8) Naturalness<br>(9) Integrity<br>(10) Vulnerability<br>(11) Bio-geographical importance | (12) Economic benefit<br>(13) Recreation<br>(14) Human dependency | (15) Research<br>(16) Baseline and monitoring studies<br>(17) Education |

SOURCE: ADAPTED FROM IMO RES A.982(24)[a]

a (n 48). See Youna Lyons, Chapter 4 'Identifying Sensitive Marine Areas in the High Seas: A Review of the Scientific Criteria adopted under International Law'.

97 IMO Res A.982(24) (n 48) para 1.5.

98 Ibid para 4.4.

criterion, and scientific assessment could be difficult.[99] Nor was there agreement to prioritise criteria or to emphasise local ecosystems,[100] given that larger ecosystems should also be candidates, as was the case with the Great Barrier Reef. The view that prevailed was that the criteria should be broad enough to encompass a broad range of areas on a precautionary basis.[101]

In practice, the PSSAs designated to date (Table 5.6) have tended to be special places that satisfied more than one criterion in Table 5.5. The geography of the PSSA has been varied, such as: coral reef systems (Great Barrier Reef[102] and Coral Sea Extension,[103] Sabana-Camaguey,[104] Florida Keys sea area,[105] Saba Bank;[106] Tubbataha Reefs Natural Park[107]); tidal wetlands (Wadden Sea[108]);

99 The United States felt that scientific recognition was important and that 'the criteria are not as clear and meaningful as they should or could be, thus making it difficult to conduct a meaningful review of proposed PSSAs'. 'Proposed Amendments to Assembly Resolution A.927(22) to Strengthen and Clarify the Guidelines for the Identification and Designation of Particularly Sensitive Sea Areas Submitted by the United States' IMO Doc MEPC 52/8 (9 July 2004) (IMO Doc MEPC 52/8).

100 The Russian Federation proposed giving priority to ecological attributes. 'Proposed Amendments to Guidelines for the Identification and Designation of Particularly Sensitive Sea Areas (Annex 2 of IMO Resolution A.927(22)) Submitted by the Russian Federation' IMO Doc MEPC 52/8/1(6 August 2004).

101 'Proposed Amendments to Assembly Resolution A.927(22) to Strengthen and Clarify the Guidelines for the Identification and Designation of Particularly Sensitive Sea Areas (PSSAs)—Comments on MEPC 52/8 Submitted by WWF' IMO Doc MEPC 52/8/4 (18 August 2004) (IMO Doc MEPC 52/8/4).

102 Identification of the Great Barrier Reef Region as a Particularly Sensitive Sea Area, IMO Res MEPC.44(30) (16 November 1990) in 'Report of the MEPC on its Thirtieth Session' IMO Doc MEPC 30/44, Annex 17.

103 Designation of the South-west Coral Sea as an Extension of the Great Barrier Reef and Torres Strait Particularly Sensitive Sea Area, IMO Res MEPC.268(68) (15 May 2015) Report of the MEPC on its Sixth-Eighth Session, IMO Doc MEPC 68/21(29 May 2015) Annex 14.

104 Identification of the Archipelago of Sabana-Camaguey as a Particularly Sensitive Sea Area, IMO Res MEPC.74(40) (25 September 1997) in 'Report of the MEPC on its Forth Session' IMO Doc MEPC 40/21 (27 October 1997) Annex 3.

105 Identification of the Sea Area around the Florida Keys as Particularly Sensitive Sea Area, IMO Res MEPC.98(47) (8 March 2002) in 'Report of the MEPC on its Forth-Seventh Session' IMO Doc MEPC 47/20 (18 March 2002) Annex 5.

106 Designation of the Saba Bank as a Particularly Sensitive Sea Area, IMO Res MEPC.226(64) (5 October 2012) in 'Report of the MEPC on its sixty-Fourth Session' IMO Doc MEPC 64/23 (11 October 2012) Annex 15.

107 Designation of Tubbataha Reefs Natural Park as a Particularly Sensitive Sea Area, IMO Res MEPC.294(17) (7 July 2017) in 'Report of the MEPC on its Seventy-First Session' IMO Doc MEPC 71/17 (24 July 2017) Annex 18.

108 Identification of the Wadden Sea as a Particularly Sensitive Sea Area, IMO Res MEPC.101(48) (11 October 2002) in 'Report of the MEPC on its Forty-Eighth Session' IMO Doc MEPC 48/21 (24 October 2002) Annex 5.

TABLE 5.6 PSSA attributes

| PSSA | Geography and sensitivity | International/national status |
|---|---|---|
| Great Barrier Reef | – Vast complex of reefs and islands; largest barrier reef in the world<br>– Important for bird life, fishing grounds and tourism | – World Heritage List site<br>– Marine protected area |
| Sabana-Camaguey Archipelago | – Coastal archipelago and barrier reef<br>– Coral reefs, cays, mangroves, seagrass beds<br>– Migratory birds<br>– Crocodiles, dolphins, manatees<br>– Fishing and tourism | – Marine protected area |
| Wadden Sea | – Transboundary area shared by three states<br>– Wetlands, salt marshes, dynamic tidal ecosystem<br>– Habitats for birds, seals, fish, plants | – Marine protected area in three states<br>– EU Bird Directive protection<br>– EU Habitats Directive protection for dunes<br>– Biosphere Reserve |
| Florida Keys sea area | – Third largest barrier coral reef system<br>– Mangroves<br>– Tourism and small fishery | – National marine sanctuary |
| Malpelo Island | – Water surrounding island<br>– Biologically very productive; biogeographically significant; home to a unique Pacific coral reef community; whales and large fish<br>– Bird life | – Marine protected area |
| Paracas National Reserve | – Great biodiversity including migratory birds sea lions and fish<br>– Pre-Incan relics | – National reserve<br>– Ramsar Convention site |

TABLE 5.6 PSSA attributes (*cont.*)

| PSSA | Geography and sensitivity | International/national status |
|---|---|---|
| Western European Waters | – EEZs and territorial seas<br>– Birds, fish and underwater life<br>– Fisheries and tourism | – MARPOL Annex I special area designation |
| Torres Strait (GBR PSSA extension) | – International strait<br>– Same as Great Barrier Reef | – Same as Great Barrier Reef |
| Canary Islands | – Archipelago<br>– Rich biodiversity, including whales and dolphins<br>– Tourism | – EU Habitats Directive (macaronesian biogeographical region)<br>– Biosphere Reserve (La Palma, El Hierro)<br>– Canaries Natural Spaces Network |
| Galapagos | – Archipelago<br>– Unique location for study evolution<br>– Many endemic species of animals and birds<br>– Tourism | – World Heritage List site<br>– Marine protected area |
| Baltic Sea | – Semi-enclosed sea with unique mixture of salt and fresh water<br>– Beaches important for birds<br>– Many threatened species | – MARPOL Annexes I, IV, V and VI special areas and emission control area |
| Papahānaumokuākea Marine National Monument (Hawaii) | – Archipelago (islands), atolls, reefs and EEZ<br>– Migratory species<br>– Important for native Hawaiian culture | – World Heritage List site<br>– Marine national monument |
| The Strait of Bonifacio | – International strait<br>– Marine mammals and birds | – Marine protected area in two states<br>– Within the Mediterranean MARPOL Annexes I, IV and V special areas |

| PSSA | Geography and sensitivity | International/national status |
|---|---|---|
| Saba Bank, North-eastern Caribbean | – Rock and coral bank or atoll<br>– Diversity of algae, seaweed and fish | – Marine protected area<br>– IUCN Red List species |
| Southwest part of Coral Sea (extension to GBR PSSA) | – Coral reefs with high biodiversity<br>– Indigenous cultural and social significance to coastal communities | – Marine reserve |
| Jomard Entrance | – International strait<br>– Part of Louisiade archipelago, fringing reefs and part of coral triangle<br>– Turtle nesting grounds<br>– Important for bird life | – Critical habitat for rare turtles on IUCN Red List<br>– CITES Appendix I and II species |
| Tubbataha Reefs Natural Park | – Route for international navigation through archipelagic waters<br>– Coral reef complex<br>– Turtle foraging area<br>– Habitat for seabirds and fish | – World Heritage List site<br>– Ramsar Convention site<br>– Marine protected area (natural park)<br>– No-take zone |

SOURCE: ADAPTED IN PART FROM IMO, 'PARTICULARLY SENSITIVE SEA AREAS'[a]

a IMO, 'Particularly Sensitive Sea Areas' <http://pssa.imo.org/#/intro> accessed 24 May 2018.

islands and archipelagos (Malpelo Island waters,[109] Paracas National Reserve,[110] Canary Islands,[111] Galapagos[112]); straits (Bonifacio,[113] Torres Strait Extension of

109 Identification of the Sea Area around Malpelo Island as a Particularly Sensitive Sea Area, IMO Res MEPC.97(47) (8 March 2002) in 'Report of the MEPC on its Forty-Seventh Session' IMO Doc MEPC 47/20 (18 March 2002) Annex 4.

110 Designation of the Marine Area of the Paracas National Reserve as a Particularly Sensitive Sea Area, IMO Res MEPC.106(49) (18 July 2003) in Report of the MEPC on its Forty-Ninth Session' IMO Doc MEPC 49/22 (8 August 2003) Annex 12.

111 Designation of the Canary Islands as a Particularly Sensitive Sea Area, IMO Res MEPC.134(53) (22 July 2005) in 'Report of the MEPC on its Fifty-Third Session' IMO Doc MEPC 53/24 (25 July 2005) annex 22.

112 Designation of the Galapagos Archipelago as a Particularly Sensitive Sea Area, IMO Res MEPC.135(53) (22 July 2005) in IMO Doc MEPC 53/24 (n 111), Annex 23.

113 Designation of the Strait of Bonifacio as a Particularly Sensitive Sea Area, IMO Res MEPC.204(62) (15 July 2011) in 'Report of the MEPC on its Sixty-Second Session' IMO Doc MEPC 62/24 (16 July 2011) Annex 22.

the Great Barrier Reef PSSA,[114] Jomard[115]); EEZs (Western European Waters,[116] Papahānaumokuākea Marine National Monument[117]); and a semi-enclosed sea (Baltic[118]). All were already under international and/or national protection, thus providing additional justification for protection from international shipping. Most have tended to be coastal ecosystems, but offshore ecosystems have also been included.

(b) Vulnerability

The PSSA proposals must demonstrate that the areas are at risk from international shipping activities, entailing consideration of two sets of risk factors (Table 5.7).[119] The first column lists vessel characteristics, and the second column lists factors that may exacerbate the vessel risks. Factors that assist with ascertaining the overall risk and the necessary additional protective measures are additional evidence of threats and potential damage posed by shipping to the area, history of groundings, casualties and spills, any potential shifting of risks from the proposed PSSA area to other areas, other environmental stressors, and measures in place and their impact.[120]

Table 5.8 provides examples of some of the major threats identified in PSSA submissions. In the Western European Waters PSSA proposal, the proponents demonstrated that the area was a busy multiple-use area and had major international navigation routes that included all vessel types.[121] The area had suffered some of the worst maritime casualties resulting in massive oil pollution.

114 Designation of the Torres Strait as an Extension of the Great Barrier Reef Particularly Sensitive Sea Area, IMO Res MEPC.133(53) (22 July 2005) in IMO Doc MEPC 53/24 (n 111), Annex 21.

115 Designation of the Jomard Entrance as a Particularly Sensitive Sea Area, IMO Res MEPC.283(70) (28 October 2016) in 'Report of the MEPC on its Seventieth Session' IMO Doc MEPC 70/18 (11 November 2016) Annex 12. See also background information in: 'Protection of Jomard Entrance, Louisiade Archipelago Submitted by Papua New Guinea' IMO Doc MEPC 69/INF.14 (4 February 2016); 'Designation of the Jomard Entrance as a PSSA Submitted by Papua New Guinea' MEPC 70/8 (11 July 2016).

116 Designation of the Western European Waters as a Particularly Sensitive Sea Area, IMO Res MEPC.121(52) (15 October 2004) in 'Report of the MEPC on its Fifty-Second Session' IMO Doc MEPC 52/24 (18 October 2004) Annex 10.

117 Designation of the Papahānaumokuākea Marine National Monument as a Particularly Sensitive Sea Area, IMO Res MEPC.171(57) (4 April 2008) in 'Report of the MEPC on its Fifty-Seventh Session' IMO Doc MEPC 57/21 (7 April 2008) Annex 12.

118 Designation of the Baltic Sea Area as a Particularly Sensitive Sea Area, IMO Res MEPC.136(53) (22 July 2005) in IMO Doc MEPC 53/24 (n 111), Annex 24.

119 IMO Res A.982(24) (n 48) para 5.1.

120 Ibid.

121 'Designation of a Western European Particularly Sensitive Sea Area Submitted by Belgium, France, Ireland, Portugal, Spain and the United Kingdom' IMO Doc MEPC 49/8/1 (11 April 2003).

TABLE 5.7 Risk factors

| Vessel traffic characteristics | Natural factors |
|---|---|
| – Operational factors: Types of maritime activities (e.g., small fishing boats, small pleasure craft, oil and gas rigs) in the proposed area that by their presence may reduce the safety of navigation.<br>– Vessel types: Types of vessels passing through or adjacent to the area (e.g. high-speed vessels, large tankers, or bulk carriers with small under-keel clearance).<br>– Traffic characteristics: Volume or concentration of traffic, vessel interaction, distance offshore or other dangers to navigation are such as to involve greater risk of collision or grounding.<br>– Harmful substances carried: Type and quantity of substances on board, whether cargo, fuel or stores, that would be harmful if released into the sea. | – Hydrographical: Water depth, bottom and coastline topography, lack of proximate safe anchorages and other factors that call for increased navigational caution.<br>– Meteorological: Prevailing weather, wind strength and direction, atmospheric visibility and other factors that increase the risk of collision and grounding and also the risk of damage to the sea area from discharges.<br>– Oceanographic: Tidal streams, ocean currents, ice, and other factors that increase the risk of collision and grounding and also the risk of damage to the sea area from to the sea area from discharges. |

SOURCE: ADAPTED FROM IMO RES A.982(24)[a]
a (n 48).

The area was also subject to bad weather, which was in fact a factor in some casualties such as Torrey Canyon, Amoco Cadiz and Prestige. There was infrastructure in place to respond to threats, but the heightened risk posed by the oil tanker trade in that environment necessitated a mandatory reporting system. In the case of the Florida Keys sea area, the heavily used sea lanes, danger of groundings and anchoring threatened physical damage to the coral reefs.[122] While the risks in the case of Malpelo Island were on a different order

122 'Designation of the Marine Area around the Florida Keys as a Particularly Sensitive Sea Area Submitted by the United States' IMO Doc MEPC 46/6/2 (19 January 2001) (IMO Doc MEPC 46/6/2) 9–10.

TABLE 5.8 Threats posed and APMs adopted

| PSSA | Shipping threats | APMs |
|---|---|---|
| Great Barrier Reef | – Risks of ship groundings, pollution and noise | – Recommended pilotage for large ships and carriers of hazardous materials<br>– Mandatory reporting<br>– Two-way route<br>– Area to be avoided |
| Sabana-Camaguey Archipelago | – Dense shipping traffic | – Area to be avoided (with traffic separation schemes outside PSSA) |
| Wadden Sea | – Dense shipping traffic | – Mandatory deep-water route |
| Florida Keys sea area | – Dense shipping traffic<br>– Risk of ship groundings<br>– Anchoring | – Areas to be avoided (new and amendment of existing)<br>– Mandatory no-anchoring areas |
| Malpelo Island | – Illegal fishing<br>– Diving activities from cruise ships<br>– Dense shipping traffic<br>– Risk of ship groundings | – Area to be avoided for large ships and fishing vessels |
| Paracas National Reserve | – Increasing shipping trend near reserve<br>– Accidents and operational pollution in the area<br>– Risks of ship collisions and groundings | – Area to be avoided for ships carrying hydrocarbons<br>– Traffic separation schemes |
| Western European Waters | – Dense shipping traffic<br>– History of major maritime casualties and oil pollution | – Mandatory ship reporting system for oil tankers |
| Torres Strait (GBR PSSA extension) | – Same as Great Barrier Reef | – Recommended pilotage<br>– Two-way shipping route |
| Canary Islands | – Oil tanker route | – Traffic separation schemes (including inshore traffic zones and precautionary areas) |

| PSSA | Shipping threats | APMS |
|---|---|---|
| | | – Areas to be avoided<br>– Mandatory ship reporting for oil tankers |
| Galapagos Archipelago | – Risks of ship groundings<br>– Cruise shipping | – Area to be avoided for oil tankers and ships carrying hazardous cargoes<br>– Mandatory ship reporting system for all ships<br>– Traffic separation schemes |
| Baltic Sea | – Dense shipping traffic | – Traffic separation schemes (including inshore traffic zones, precautionary areas) in three areas<br>– Deep-water route<br>– Areas to be avoided<br>– Mandatory ship reporting system<br>– Pilotage rules |
| Papahānaumokuākea Marine National Monument (Hawaii) | – Dense shipping traffic | – Areas to be avoided<br>– Ship reporting system (mandatory for ships passing through PSSA to or from a US port; voluntary for all others) |
| The Strait of Bonifacio | – Dense shipping traffic | – Mandatory ship reporting for all ships<br>– Recommendations for navigation and use of two-way route<br>– Recommended pilotage |
| Saba Bank, North-eastern Caribbean | – Oil tanker traffic<br>– Risks of collisions | – Area to be avoided<br>– Mandatory no-anchoring area |
| Southwest part of Coral Sea (extension to GBR PSSA) | – Risks of collisions and groundings | – Two-way routes<br>– Areas to be avoided |

TABLE 5.8 Threats posed and APMs adopted (*cont.*)

| PSSA | Shipping threats | APMs |
|---|---|---|
| The Jomard Entrance | – Increasing traffic<br>– Risks of groundings, collisions and spills<br>– Fishing<br>– Marine debris | – Two-way routes<br>– Precautionary area |
| Tubbataha Reefs Natural Park | – Substantial shipping traffic<br>– Oil tanker traffic<br>– Risk of pollution and groundings | – Area to be avoided for ships of more than 150 gross tonnage<br>– Recommended pilotage for large ships and vessels carrying hazardous material |

SOURCE: ADAPTED IN PART FROM IMO, 'PARTICULARLY SENSITIVE SEA AREAS'[a]
a IMO, 'Particularly Sensitive Sea Areas' <http://pssa.imo.org/#/intro> accessed 24 May 2018.

of magnitude than in Western European Waters and the Florida Keys sea area, the threat posed by the user factors (illegal fishing, diving activities from cruise ships, major international traffic) posed relatively high risks to the waters of the island.[123]

An important consideration in assessing vulnerability is determining the geographical scope of the area with respect to the threats posed by international shipping. While PSSA size is proposed by submitting states, the MEPC decides on the appropriate size commensurate with the evidenced threats.[124] A challenge is to maintain a reasonable balance between a coastal state's concerns and international navigation rights. The 2005 Revised PSSA Guidelines do not set a size limit, leaving the geographical scope to be determined as part of the response to the evidenced threats. Moreover, the proposal may include a buffer zone.[125]

123 IMO, 'Additional Information for the Designation of Malpelo Island as a "Particularly Sensitive Sea Area"' IMO Doc MEPC 46/6/3 (16 February 2001).

124 IMO Res A.982(24) (n 48) para 8.2.3.

125 The Galapagos Submission included an area to be avoided to function as 'a separation zone or contiguous zone for the proposed PSSA' justified on the basis of currents, logistical concerns in tackling spills, and correspondence of buffer area to the existing national marine protected area. 'Designation of the Galapagos Archipelago as a Particularly Sensitive Sea Area Submitted by Ecuador' IMO Doc MEPC 51/8/2 (24 December 2003) 20–21.

Size of a PSSA has always been an issue. While on the one hand the size should be efficient in addressing the problem, on the other hand, and without any limits, it could encompass entire EEZs, and enclosed and semi-enclosed seas with substantial restrictive impacts on the freedom of navigation. In reality, the Great Barrier Reef PSSA had a very large geographical footprint that was never disputed, indeed geographically much larger than the Baltic Sea PSSA submitted by proponent states,[126] and the first semi-enclosed sea to be proposed. However, the Baltic Sea PSSA covered an entire sea with numerous jurisdictions, unlike Australia's sole jurisdiction. Moreover, the Baltic Sea PSSA was already designated as a MARPOL special area for Annex I, and some states argued that the proposal did not provide added value to that status.[127] The Russian Federation, major flag states Liberia and Panama, and industry organisations with consultative status expressed concerns over the proposed size. Concern was further expressed that there would be proliferation of PSSAs without sufficient justification. This was one of the relatively few occasions when consensus at this level of rule-making could not be achieved and in the end the designation proceeded without Russian waters.

Similar criticism was levied at the Western European Waters submission of Belgium, France, Ireland, Portugal, Spain and the United Kingdom.[128] This time the proposed area was larger than the Baltic Sea and was comparable to the Great Barrier Reef[129] and accompanied by proposals to ban single hull traffic and introduction of mandatory reporting for tankers. The area had experienced many maritime casualties with catastrophic pollution. Numerous concerns were raised, including consistency with the freedom of navigation protected by UNCLOS, the risk of pushing tanker traffic into open ocean and closer to other states[130] and the stress on search and rescue capabilities.[131]

---

126 'Designation of the Baltic Sea Area as a Particularly Sensitive Sea Area Submitted by Denmark, Estonia, Finland, Germany, Latvia, Lithuania, Poland and Sweden' IMO Doc MEPC 51/8/1 (19 December 2003).

127 IMO, Russian position, in IMO Doc MEPC 51/22 (n 89) 34–35 and annex 8.

128 'Designation of a Western European Particularly Sensitive Sea Area Submitted by Belgium, France, Ireland, Portugal, Spain and the United Kingdom' IMO Doc MEPC 49/8/1 (11 April 2003).

129 'Designation of a Western European Particularly Sensitive Sea Area Submitted by Liberia, Panama, the Russian Federation, BIMCO, ICS, INTERCARGO, INTERTANKO and IPTA' IMO Doc LEG 87/16/1 (12 September 2003).

130 The northern limit was adjusted to respond to this criticism.

131 IMO, 'Report of the MEPC on its Forty-First Session' IMO Doc MEPC 49/22 (8 August 2003) 47.

(c) Measures within IMO competence

While a PSSA submission is under consideration, the assessment process includes the APMs proposed to address the identified threats. Proposed compliance and control procedures are also included in the submission.[132] APMs may consist of mandatory directions and/or recommended actions for shipping in the area adopted through the IMO and as permitted by maritime conventions and guidelines.[133] The 2005 Revised PSSA Guidelines provide a non-exclusive list of three major types of APMs, namely (a) special area and emission control area designations for pollution prevention, (b) ships' routeing and reporting systems, and (c) other measures developed specifically for the PSSA.[134] The first category consists of measures that are exclusively available under MARPOL as discussed above. In practice, the Baltic and West European Waters were already designated as Annex I special areas before PSSA designation.

The second category consists of measures available exclusively under the authority of Chapter V of SOLAS[135] and are the most commonly adopted measures (Table 5.9). A routeing system is defined as 'any system of one or more routes or routeing measures aimed at reducing the risk of casualties; it includes traffic separation schemes, two-way routes, recommended tracks, areas to be avoided, no-anchoring areas, inshore traffic zones, roundabouts, precautionary areas and deep water routes'.[136] A ship reporting system is defined as one '... that has been established by a Government or Governments after it has been accepted by the Organization as complying with all requirements

132 IMO Res A.982(24) (n 48) para 7.9. See also IMO Doc MEPC.1/Circ.510 (n 84) para 3.5.3.

133 Potential measures could include 'designated anchorage areas and methods; no anchorage zones; closure of routes to certain types of vessels or cargoes; speed restrictions; compulsory pilotage or tug escort to ensure safe navigation in or near PSSAs; prohibitions or restrictions on cargo transfer; required submission of pre-filed passage plans and adherence to time schedules; special under-keel clearance restrictions; regulation of offshore bunkering; prohibition of intentional discharges, including ballast waters; seasonal closures to protect migrating marine mammals'. IMO, 'Additional Protection for Particularly Sensitive Sea Areas (PSSAs)' IMO Doc MEPC 46/6/1 (19 January 2001) (IMO Doc MEPC 46/6/1) para. 2.4.10. See also 'Draft Guidance Document on Associated Protective Measures for Particularly Sensitive Sea Areas Submitted by WWF' IMO Doc MEPC 49/8/2 (8 May 2003). One interesting idea advanced in the latter document is air pollution emission limitations, which could require exclusive use of high-quality low sulphur fuels in or near PSSAs.

134 IMO Res A.982(24) (n 48) para 6.

135 SOLAS (n 6) Ch V, reg 10 (routeing systems) and reg 11 (reporting systems).

136 General Provisions on Ships' Routeing, IMO Res A.572(14) (20 November 1985) para 2.1.1. Specific methods are set out in Part 4 of this document.

TABLE 5.9 Routeing and reporting systems

| Routeing measures | Reporting measures |
| --- | --- |
| – Areas to be avoided<br>– Deep-water routes<br>– Established direction of traffic flow<br>– Inshore traffic zones<br>– Precautionary areas<br>– Recommended direction of traffic flow<br>– Recommended routes<br>– Recommended tracks<br>– Roundabouts<br>– Traffic lanes<br>– Traffic separation schemes<br>– Two-way routes<br>– No-anchoring area?<br>– Separation zone or line? | – Radar surveillance<br>– Interactive ship/shore reporting |

SOURCES: IMO RES A.572(14)A AND IMO DOC MSC 64/221/ADD.1[b]
a IMO Res A.572(14) (n 136).
b IMO Doc MSC 64/221/Add.1 (n 137).

of SOLAS regulation V/8-I, except paragraph (e) thereof'.[137] It implies interaction between a ship and a shore-based facility and can be mandatory or recommended.[138] There are also weather routeing measures.[139] The general purpose is to improve navigation safety in areas of dense traffic and restricted sea-room, because of various factors, including obstructions, bathymetry and meteorological conditions.[140] The specific objective of each routeing system

137 Guidelines and Criteria for Ship Reporting Systems, IMO Res MSC.43(64) (9 December 1994) in 'Report of the MSC on its Sixty-Forth Session' IMO Doc MSC 64/221/Add.1 (11 October 2012) (IMO Doc MSC 64/221/Add.1).

138 The Sub-committee on Navigation, Communications and Search and Rescue, through its Navigation Working Group, is currently considering the benefits of automated ship-to-shore systems for mandatory reporting systems adopted by the IMO in order to reduce verbal reporting. A test conducted cooperatively by Brazil, Norway and Singapore demonstrated benefits. IMO, 'Report to the MSC' IMO Doc NCSR 4/29 (24 March 2017) 22.

139 Recommendation on Weather Routeing, IMO Res A.528(13) (17 November 1983).

140 IMO Res A.572(14) (n 136) para 1.1.

depends on 'the particular hazardous circumstances which it is intended to alleviate'.[141]

The third is an inclusive category permitting the fashioning of new or unique measures, such as pilotage and towage. Where such new measures are proposed, it is particularly important for the proponent to explain the legal basis for the proposed measure and that it falls within the IMO's mandate to adopt. The power to adopt new measures cannot be presumed to be encompassed by an existing convention, and where it is not, the proposal may well have to seek amendment of a convention to provide the necessary authority. APMs may be accompanied by different degrees of authority. They may consist of mandatory or recommended actions, whether for all ships or particular ships. Special mandatory measures may also be adopted under Article 211(6) of UNCLOS.[142]

The IMO practice of APM assessment suggests that proposed measures are evaluated by necessity, legality, proportionality, efficiency and effectiveness. The measure must be needed to safeguard evidenced values and is directly related to the threat addressed. Proponents must explain the legal basis for the proposed measures. In addition to mandatory reporting, the original Western European PSSA submission proposed a ban on single-hull tanker navigation in the PSSA.[143] The ban effectively concerned design and construction standards and potentially violated Article 211(6)(c).[144] It was eventually withdrawn and the only APM proposed was a mandatory reporting system.[145] The proposed measure(s) must be proportionate to the threat, and if there is a choice, the least restrictive is favoured.[146] In its submission on the Florida Keys sea area, the United States suggested that a no-anchoring area may be preferable to an area to be avoided when the problem concerns anchoring rather than navigation.[147] Compulsory pilotage might not be proportionate, as was evidenced

141 Ibid para 1.2.

142 IMO Res A.982(24) (n 48) para 7.5.2.3(iii).

143 'Designation of a Western European Particularly Sensitive Sea Area Submitted by Belgium, France, Ireland, Portugal, Spain and the United Kingdom' IMO Doc MEPC 49/8/1 (11 April 2003).

144 'Such additional laws and regulations may relate to discharges or navigational practices but shall not require foreign vessels to observe design, construction, manning or equipment standards other than generally accepted international rules and standards'. UNCLOS (n 2) art 211(6)(c).

145 'Mandatory Ship Reporting System for the Western European Particularly Sensitive Sea Area Submitted by Belgium, France, Ireland, Portugal, Spain and the United Kingdom' IMO Doc NAV 50/3/6 (2 April 2004).

146 IMO Doc MEPC 46/6/1 (n 133) para 2.1.3.

147 IMO Doc MEPC 52/8 (n 99) 2; IMO Doc MEPC 46/6/2 (n 122).

in discussions of the Baltic Sea PSSA[148] and the initial proposal for the Torres Strait extension.[149] The Baltic Sea PSSA was originally designated in principle, and the hope was that compulsory pilotage would be one of the measures considered.[150] Both the Russian Federation and United States argued against compulsory pilotage during amendment of the 2005 Revised PSSA Guidelines.[151] The proposed measure must be one that can be implemented and does not unnecessarily impede navigation safety and efficiency.[152] The measure must also be expected to be effective in removing or mitigating the threat. Finally, APMs are not static and may be reviewed and changed periodically on submissions by proponent states.

PSSAs and APMs are annotated on navigation charts maintained in accordance with International Hydrographic Organization symbols and methods.[153] As of the effective date following IMO adoption, the coastal state may legislate and enforce APMs in accordance with powers conferred and constraints set by UNCLOS. Flag states have a responsibility to ensure compliance by their ships.[154]

#### 4.1.4 Procedure

Proponent states are responsible for ensuring that substantive requirements set out in the 2005 Revised PSSA Guidelines as amended are complied with. This has not always been the case. Difficulties encountered in some submissions

148 'Designation of the Baltic Sea Area as a Particularly Sensitive Sea Area Submitted by Denmark, Estonia, Finland, Germany, Latvia, Lithuania, Poland and Sweden' IMO Doc MEPC 51/8/1 (19 December 2003). The proposal indicated a number of protective measures in place, namely compulsory reporting in some areas and traffic surveillance, routeing systems in certain areas for big draught ships, local pilotage and the MARPOL 73/78 special area and emission control area.

149 IMO, 'Outcome of NAV 52 related to PSSAs Note by the Secretariat' IMO Doc MEPC 55/8/2/Add.1 (7 September 2006).

150 'Designation of the Baltic Sea Area as a Particularly Sensitive Sea Area Submitted by Denmark, Estonia, Finland, Germany, Latvia, Lithuania, Poland and Sweden' IMO Doc MEPC 51/8/1 (19 December 2003). The new measures proposed for consideration were common monitoring system based on compulsory reporting and traffic surveillance, routeing systems, escort and escorting tugs, pilotage and areas to be avoided. See also 'HELCOM Expert Working Groups Submitted by Denmark, Germany and Sweden' IMO Doc NAV 50/INF.5 (29 April 2004).

151 IMO Doc MEPC 52/8/1 (n 100). IMO Doc MEPC 52/8 (n 99). See also the explanation for omission of reference to compulsory pilotage in 'Report of the Correspondence Group Submitted by the United States' IMO Doc MEPC 53/8/2 (15 April 2005).

152 IMO Res A.982(24) (n 48) para 7.6.

153 Ibid para 9.1.

154 Ibid para 9.3.

included the incomplete nature of the required information and the lack of linkages between the identified threats from international shipping, environmental vulnerabilities, reasonable foreseeability of actual or potential harm, size of the PSSA and the requested APMs, forcing the MEPC to request additional information before further consideration.[155] This was the case with Colombia's original submission to designate Malpelo Island as a PSSA.[156] Ecuador did not satisfy the review committee on one particular requested measure, because of the paucity of shipping in the Galapagos Area.[157] There has been significant inconsistency in the manner information has been organised in submissions, despite the recommendations in the Guidance Document.[158]

The issue was flagged on a number of occasions and necessitated the revision of the Guidance Document and adoption of the Proposal Review Form in 2006.[159] The form introduced a rigorous questionnaire aimed at encouraging a more careful and detailed submission of key information to facilitate the assessment of the proposal. Even so, proposals submitted since continue to vary in detail and justification, although the recent review of the Philippine proposal suggests that the form appears to be improving the process.[160] The Technical Group reviewing the Philippine proposal to ensure compliance with the 2005 Revised PSSA Guidelines addressed the elements in a systematic manner and concluded that the proposal to designate the Tubbataha Reefs Natural Park as a PSSA complied with the Guidelines. The Technical Group did not identify gaps or appear to need any substantial new information.

The procedure commences with a formal submission by the proponent state(s) to the MEPC.[161] The MEPC considers the proposal against the criteria set out in the Guidelines on a case by case basis and establishes a

155 See concerns expressed by the US during deliberations to produce the 2005 Revised PSSA Guidelines (IMO Res A.982(24) (n 48)). IMO Doc MEPC 52/8 (n 99) Annex 7–8.

156 The information gap was raised during consideration at the MEPC's 44th Session, 6–13 March 2000. Colombia responded by providing the requested information in IMO, 'Additional Information for the Designation of Malpelo Island as a Particularly Sensitive Sea Area' IMO Doc MEPC 46/6/3 (16 February 2001).

157 This concerned a proposed traffic separation scheme that the NAV Sub-committee (at the time) did not consider as the most appropriate measure. Instead, the Sub-committee agreed that recommended tracks mandatory as a condition of port entry constituted the most appropriate measure. IMO, 'Outcomes of NAV 53 and MSC 83 on PSSAs Note by the Secretariat' IMO Doc MEPC 57/7 (4 December 2007).

158 IMO Doc MEPC.1/Circ.510 (n 84)

159 IMO Doc MEPC 55/23 (n 85).

160 IMO, 'Identification and Protection of Special Areas and PSSAs, Report of the Technical Group on PSSAs' IMO Doc MEPC 69/WP.11 (21 April 2016).

161 IMO Res A.982(24) (n 48) para 8.3.

multidisciplinary (environmental, scientific, maritime, legal) technical group to undertake an in-depth assessment. The proponents present the proposal to the Technical Group and are asked questions by members. On completion of their assessment, the Technical Group presents a report to the MEPC indicating the outcome of the assessment. If satisfied, the MEPC may designate the PSSA in principle and consider further the proposed APMs and/or request consideration by the Assembly, another IMO committee or sub-committee as appropriate.

The technical appraisal of routeing measure(s) as the proposed APMs is undertaken by the Sub-Committee on Navigation, Communication and Search and Rescue (NCSR). At this stage, a proposal for APMs is required to be submitted in accordance with a prescribed template.[162] The proposed APMs have to be described in technical detail, setting out the purpose and justification (demonstrated need, reasons, history of incidents, classes of ships, and impacts on navigation). The proposed area of application is to be described and charts submitted. If other states have a common interest, there has to be indication of consultations undertaken and any agreements that may have been concluded. Joint submissions are encouraged. Evidence of traffic considerations, alternative routeing, other ocean uses, state of hydrography, etc., is to be submitted. Marine environmental considerations have also to be reported on, including factors that affect shipping in the vicinity of the proposed area (e.g., weather, currents, tides, ice). The explanations have to include how the requested measure will protect the marine environment and limitations to sea-room. Whether mandatory or recommended, they have to be justified with reference to what is essential to navigation safety and environment protection. The proponent also is to indicate compliance and enforcement actions that will be undertaken. Other pertinent information, such as consultations with other stakeholders and a proposed date of implementation have to be indicated.

If a mandatory reporting system is proposed, a separate template applies.[163] Objectives, demonstrated need and classes of ships affected have to be set out. The proposal requires contextual information, such as hydrographical, meteorological and environmental considerations, state of hydrography, existing

162 The template was developed by the Secretariat in response to a request from NCSR Subcommittee. IMO, 'Model Document Template for Ships' Routeing Proposals Note by the Secretariat' IMO Doc NCSR 2/3 (9 September 2014). The template is also available at IMO, 'Ships' Routeing' <www.imo.org/en/OurWork/Safety/Navigation/Pages/ShipsRouteing.aspx> accessed 24 May 2018.

163 Also developed by the Secretariat is IMO, 'Model Document Template for Ship Reporting System Proposals Note by the Secretariat' IMO Doc NCSR 2/3/1 (25 September 2014). The template is also available at IMO, 'Ships' Routeing' (n 162).

and proposed aids to navigation, and detailed characteristics of ship traffic. If mandatory reporting is for environmental reasons, there has to be explanation of how it is reasonably expected to significantly prevent or reduce the risk of pollution or other damage to the marine environment. The reporting system is to have clear geographical scope and represented on charts. Where other states have a common interest, the consultations undertaken and agreements reached have to be described and any other related agreements mentioned. In such cases, states are encouraged to make joint submissions. The details of the actual reporting system are to be described (e.g. format, content of reports, times and geographical positions for submitting reports, authority to whom reports should be sent and available services). Also to be described are information services to ships and procedures, communications required for the system, applicable rules and regulations, shore-based facilities to support the operation of the system, current measures, and contingencies should shore-based communication facilities fail. As in routeing measures, the proponent is expected to explain how failure of compliance will be addressed. Finally, the proposed date of implementation is to be indicated.

The committee or sub-committee ascertains APM proposal compliance with the rules and procedures of the legal instrument concerned. Reference to the Legal Committee may be necessary if the APM addresses questions of legal interpretation.[164] MEPC designation of the PSSA occurs after the APMs are approved. Alternatively, it is conceivable that the proposed APMs are not approved, with the consequence that the MEPC has the discretion 'to reject the PSSA application entirely or request that the proposing Member Government submit new proposals' for APMs, or amend their submission, or submit new information. The decision is accompanied by reasons.

Previously under the 2001 Guidelines, there was significantly more flexibility for submissions, with the effect that PSSA proposals could be submitted without proposing APMs. All that was required was an indication of the types of measures contemplated and an expectation that at least one APM would be proposed within two years of the designation in principle, failing which the designation would lapse.[165] This was the case with the Baltic Sea PSSA, consideration of which required paring down of subsequent ideas for APMs. As was pointed out by the United States Delegation, the practice did not permit

164 For example, the Western European Waters PSSA proposal raised an issue with respect to compliance with the UNCLOS Article 211. See 'Designation of a Western European Particularly Sensitive Sea Area Submitted by Liberia, Panama, the Russian Federation, BIMCO, ICS, INTERCARGO, INTERTANKO and IPTA' IMO Doc LEG 87/16/1 (13 September 2003).

165 IMO Doc MEPC.1/Circ.510 (n 84) Part 3.5.

a holistic consideration of the proposal from inception,[166] whereas others saw the designation in principle as an application of the precautionary approach until substantive action could be taken.[167] The 2005 Revised PSSA Guidelines now enable a more integrated consideration of all the elements of a submission to ensure that the proponent indicates how proposed APMs 'provide the needed protection from the threats of damage posed by international maritime activities occurring in and around the area'.[168]

Curiously, there is no required review of PSSAs (or MARPOL special areas) subsequent to their adoption to determine how well they functioned, their continued need and any lessons learned. The 2005 Revised PSSA Guidelines provide that

> IMO should provide a forum for the review and re-evaluation of any associated protective measure adopted, as necessary, taking into account pertinent comments, reports, and observations of the associated protective measures. Member Governments which have ships operating in the area of the designated PSSA are encouraged to bring any concerns with the associated protective measures to IMO so that any necessary adjustments may be made. Member Governments that originally submitted the application for designation with the associated protective measures, should also bring any concerns and proposals for additional measures or modifications to any associated protective measure or the PSSA itself to IMO.[169]

This has never happened and raises an important question concerning accountability post-designation. It is unclear whether all proponent states have acted consistently with the protection and conservation needs of PSSAs. Some PSSAs (e.g., Papahānaumokuākea, Great Barrier Reef, Baltic Sea, Western European Waters, Galapagos) have had routeing and reporting measures reviewed, including any additional measures designated separately. Others do not appear to have had APMs revisited in the IMO (e.g., Malpelo, Paracas, Sabana-Camaguey).

More importantly, there is a dearth of evidence or lack of clarity on whether PSSAs and MARPOL special areas have improved environmental conditions

166 IMO Doc MEPC 52/8 (n 99) 3.

167 IMO Doc MEPC 52/8/4 (n 101) 4.

168 Ibid para 7.5.2. Despite its concern with designations-in-principle, the United States obtained such designation for the Papahānaumokuākea Marine National Monument (Hawaii) PSSA at the 56th MEPC Session, before its final PSSA designation at the 57th MEPC Session in 2008.

169 IMO Res A.982(24) (n 48) para 8.4.

and, if so, to what extent and in what ways. There have been past calls for a reporting and review requirement. In 2013, the World Wildlife Fund and the International Union for Conservation of Nature called for a review of the Great Barrier Reef PSSA and its APMs and in so proposing suggested that other PSSAs should be reviewed as well.[170] More recently in MEPC 70 in 2016, the Russian Federation submitted a proposal at the IMO for this purpose.[171] While recalling past calls for some form of review process, the Russian Federation has proposed amending the 2005 Revised PSSA Guidelines 'to provide for the reporting and monitoring of changes in the environmental status of sea areas with the status of PSSA or special sea area' on a five-year basis.[172] The regular report to be submitted by proponents would describe change in the environmental status of MARPOL special areas and PSSAs. This new requirement would serve to raise public awareness of the social agenda of the industry and to motivate shipowners to comply with the regulations for the prevention of pollution from ships.

## 5 Routeing and Reporting Measures Other Than in PSSAs

UNCLOS encourages states to promote the adoption of routeing measures to minimise threats of accidents through either diplomatic conferences or the IMO.[173] The IMO has exclusive competence for the development of rules for ships' routeing systems under the authority of SOLAS[174] and, as discussed earlier, has adopted the General Provisions on Ships' Routeing and related guidance notes.[175]

While SOLAS is the most important maritime safety instrument, it also plays an important role in the protection of the marine environment. Among other, SOLAS provides a framework for the adoption of regulations, rules and standards that ensure the seaworthiness and responsible operation of ships. In this

170 'The Need to Evaluate the Effectiveness of PSSAs and their APMs, A Case Study of the Great Barrier Reef Submitted by the World Wildlife Fund and the International Union for Conservation of Nature' IMO Doc MEPC 65/9 (8 March 2013).

171 'The Need to Evaluate the Status Effectiveness of Special Areas and Particularly Sensitive Sea Areas, Submitted by the Russian Federation' IMO Doc MEPC 70/8/1 (15 August 2016).

172 Ibid.

173 UNCLOS (n 2) art 211(1).

174 SOLAS (n 6) ch V, reg 10(2).

175 IMO Res A.572(14) (n 136). Adopted measures are published by the IMO in IMO, *Ships' Routeing* (IMO 2015), updated periodically. See also IMO, 'Guidance Note on the Preparation of Proposals on Ships Routeing Systems and Ship Reporting Systems' IMO Doc MSC/Circ.1060 (6 January 2003) (Guidance Note).

respect proper construction, equipping and operation of ships are important so that human life at sea, maritime trade and the marine environment are protected. Apart from that indirect and preventative function, SOLAS also provides tools that are more directly deployed for marine environment protection.

In addition to MARPOL special areas and PSSAs, another important package of tools available to achieve specific conservation goals are routeing and reporting measures under Chapter V of SOLAS. These have already been discussed with respect to APMs for designated PSSAs. Routeing and reporting measures, as well as traffic separation schemes pursuant to Rule 10 of the Convention on the International Regulations for Preventing Collisions at Sea, 1972 (COLREG),[176] have also long been used to promote maritime safety and environment protection, but outside the framework of MARPOL or other marine environmental instruments. In considering proposals for new or amending of routeing measures for environmental purposes, the IMO considers whether

1. the proposed routeing system can reasonably be expected to significantly prevent or reduce the risk of pollution or other damage to the marine environment of the area concerned;
2. given the overall size of the area to be protected, or the aggregate number of environmentally sensitive areas established or identified in the geographical region concerned, the use of routeing systems—particularly areas to be avoided—could have the effect of unreasonably limiting the sea area available for navigation; and
3. the proposed routeing system meets the requirements of these General Provisions.[177]

The IMO has been requested to adopt routeing and reporting measures to achieve a specific marine conservation goal separately from PSSA designations, when such a designation was not considered appropriate by the proponent(s).

The need to protect marine mammals and other animals, such as dolphins, from ship strikes and other threats has been long advocated by the International Whaling Commission (IWC).[178] The IWC Scientific Committee identified possible mitigation measures to minimise and manage the risk (Table 5.10), some of which have benefitted from IMO routeing and reporting measures.

176 Convention on the International Regulations for Preventing Collisions at Sea (adopted 20 October 1972, entered into force 15 July 1977) 1050 UNTS 16. Regulation 10 provides for traffic separation schemes.

177 Guidance Note (n 136) para 3.5.

178 For example, see the International Whaling Commission, 'IWC Draft Strategic Plan to Mitigate Impacts of Ship Strikes on Cetacean Populations, 2017–2020' <https://iwc.int/private/downloads/dr1UJzeCuNpAWs9Xf9caBw/IWC_Strategic_Plan_on_Ship_Strikes_Working_Group_FINAL.pdf> accessed on 24 May 2018.

TABLE 5.10 Ship strike mitigation measures

| Measure | Situation to which it might be applied | Implementation process (and observations) | Examples |
|---|---|---|---|
| **Keeping vessels away from whales** | | | |
| Permanent routing measures through TSS, ATBA or port approach routes | Long-term patterns of whale distribution are sufficiently predictable and well understood to enable a robust analysis of the risk reduction that might be achieved | Implemented through IMO or national regulation if within territorial sea. Proposals should follow the IMO process including data on the problem, the risk reduction achieved and implications for shipping. (Generally well respected by industry.) | Bay of Fundy, Canada; Boston, United States; California, United States; Panama; Cabo de Gata, Spain |
| Seasonal routing measures | Similar requirements to permanent routing but applicable where there are strong seasonal patterns in whale distribution | As above. | Roseway Basin, Canada; Great South Channel, United States |
| Recommended (voluntary) routes | Similar requirements to permanent routing through TSS or ABTA but not mandatory | Implemented by IMO or coastal State as a non-mandatory measure. | Peninsula Valdez, Argentina; Hauraki Gulf, New Zealand; Glacier Bay, United States; Ports on U.S. east coast |
| Short-term (days—weeks) and Dynamic routing measures | Implemented in response to short- term observations of whale aggregations or known high risk areas. Need almost real-time reporting systems that can identify such aggregations | Voluntary measures that need to be communicated to mariners. (Can be difficult to encourage compliance.) | DMAs off U.S. east coast; Gibraltar Strait |

| Measure | Situation to which it might be applied | Implementation process | Examples (and observations) |
|---|---|---|---|
| **Slowing vessels down** | | | |
| Permanent speed restriction zones | Long-term patterns of whale distribution are predictable and well understood but routing measures are not practicable | Can be voluntary or mandatory if implemented in national waters. | East coast of United States (mandatory); Glacier Bay, United States; Hauraki Gulf, New Zealand |
| Seasonal speed restriction zones | As above but applicable where there are strong seasonal patterns in distribution | As above. | Panama; California, United States; Peninsula Valdez, Argentina |
| Dynamic Management Areas for speed restrictions | Implemented in response to short- term observations of whale aggregations or known high risk areas. Need reporting systems that can identify such aggregations | Voluntary measures that need to be communicated to mariners. (Can be difficult to encourage compliance.) | United States east coast |
| **Avoidance manoeuvres** | | | |
| Real-time alerting tools to warn vessels of the presence of whales or aggregations that allow vessels to alter course or slow down | A rapid reporting network of whale sightings or acoustic detections alerts all vessels transiting an area to the locations of whales so that they can alter course or slow down | Individually designed and implemented reporting systems. | REPCET, ACCOBAMS, Mediterranean Sea; WhaleAlert, Boston, United States |
| Observations from the vessel that allow avoiding action to be taken | Only effective for vessels capable of rapid maneuvers to avoid whale sightings (e.g. vessels of a few thousand GT or less) | Additional dedicated observers, education and outreach to mariners. | Many initiatives |

SOURCE: IMO DOC MEPC 69/10/3[a]

a 'Information on Recent Outcomes regarding Minimising Ship Strikes to Cetaceans Submitted by the International Whaling Commission' IMO Doc MEPC 69/10/3 (12 February 2016).

For example, already in 1998 the United States submitted a proposal to the IMO for a mandatory ship reporting system for the protection of endangered North Atlantic Right Whales off its north and southeastern coasts, which was adopted later that year.[179] In 2007 Canada took an analogous step for the protection of the same species off Southeast Nova Scotia, by requesting a new recommended seasonal route aimed at sensitising traffic to the aggregation of animals at a particular time of the year.[180] More recently, the use of COLREG to effect changes to the shipping lanes in the Santa Barbara Channel in California helped reduce ship strike risk to blue whales.[181] Also under COLREG, in 2014 a traffic separation scheme was adopted to minimise overlaps between shipping routes and humpback whale migration routes off the Pacific coast of Panama.[182] These measures were based on scientific evidence that demonstrated how adjustments to routes or traffic schemes in the areas helped avoid conflicts with concentrations of whales. In addition to adopting appropriate measures, the IMO adopted, following a United States initiative, a guidance document to help minimise ship strikes of cetaceans.[183]

Proposals to the IMO to adopt new routeing systems or amend existing systems for specific environmental purposes, such as the protection of marine mammals, must also comply with applicable maritime conventions and the Routeing Guidelines and are submitted to the NCSR Sub-Committee in the prescribed format, as in the case of APMs for PSSAs. The same procedure described earlier applies. As a standing technical committee, the NCSR Subcommittee evaluates the proposals and recommends their adoption to the Maritime Safety Committee (MSC).

## 6 Application of IMO Instruments to High Seas Areas

Having identified and discussed the role of major IMO instruments employed to address international shipping impacts on sensitive marine areas and species,

179 Description of the Mandatory Ship Reporting Systems for Protection of Endangered North Atlantic Right Whales in Sea Areas Off the Northeastern and Southeastern Coasts of the United States, MSC.85(70) (7 December 1998).

180 Establishment of New Recommended Seasonal Area To Be Avoided in Roseway Basin, South of Nova Scotia, MSC 83/28/Add.3 (2 November 2007).

181 IMO, 'New and Amended Existing Traffic Separation Schemes' IMO Doc COLREG.2/Circ.64 (4 December 2012) annex 6.

182 IMO, 'New Traffic Separation Schemes' IMO Doc COLREG.2/Circ.65 (23 May 2014).

183 IMO, 'Guidance Document for Minimizing the Risk of Ship Strikes with Cetaceans' IMO Doc MEPC.1/Circ.674 (31 July 2009).

the chapter now turns to consider the extent to which these instruments have been applied or have the potential for application to high seas areas. As indicated earlier, MARPOL is an instrument that regulates pollution prevention from ships on international voyages and irrespective of their location. At the same time, MARPOL special areas, PSSAs and routeing and reporting measures are area-based management tools. In the discussion so far, consideration of the coastal, port and flag state jurisdictional context in UNCLOS has not been central in the treatment of IMO instruments. In particular, PSSAs and routeing measures were discussed against the implicit backdrop of coastal state jurisdiction. Consideration of area-based management tools for high seas areas requires explicit consideration of jurisdiction, because proponents of such tools cannot rely on coastal state powers and have limited options to address compliance and enforcement. High sea areas commence from the outer limit of the EEZ and will include the water column above the extended continental shelf and international seabed area.[184] Arguably, a coastal state may request an area-based measure from the IMO to address sensitive ecosystems and species of the extended continental shelf within the rights it enjoys.[185]

Jurisdiction on the high seas is thus an important consideration. With very limited exceptions, coastal states may not exercise jurisdiction over high seas areas or foreign vessels in those areas.[186] Rather it is the flag state that enjoys exclusive jurisdiction over its ships.[187] When an international regime so permits, port states might be in a position to exercise jurisdiction when a ship suspected of an enforceable violation is voluntarily in their ports.

From an IMO perspective, an additional set of considerations relates to the extent it is desirable to impose additional requirements on the exercise of the freedom of navigation. While navigation on the high seas is always subject to international maritime safety, pollution and security rules and standards adopted by the IMO, the imposition of additional requirements for designated high seas areas will have to cross a high threshold to demonstrate the requisite

184 UNCLOS (n 2) arts 86 (high seas), 78 (saving high sea area status of waters above the continental shelf, subject to art 76) and 135 (saving of high sea area status of waters above the international seabed area).

185 However, under the UNCLOS, 'The exercise of the rights of the coastal State over the continental shelf must not infringe or result in any unjustifiable interference with navigation and other rights and freedoms of other States as provided for in this Convention'. Ibid art 78(2).

186 The exceptions are few, such as seizure of pirate ships (UNCLOS (n 2) art 105) and measures to avoid pollution arising from maritime casualties (ibid art 221). However, these exceptions permit functional rather than spatial jurisdiction similar to maritime zones under national jurisdiction.

187 UNCLOS (n 2) art 94(1).

authority and compelling need. As mentioned earlier, in considering the potential range of measures available to address threats to sensitive areas, the least restrictive are favoured if the desired effect can be achieved efficiently.

### 6.1 *Application of* MARPOL *Special Areas to High Seas Areas?*

In principle, MARPOL special areas and emission control areas may be designated for high seas areas so long as the criteria set out in the applicable annexes and the 2013 Special Area Guidelines are satisfied. The definition of special area in Annexes I, IV and V refers to 'sea area' without qualifying whether it has to be within national jurisdiction. The 2013 Guidelines suggest that a special area 'may encompass the maritime zones of several States, or even an entire enclosed or semi-enclosed sea area',[188] implying that high seas areas are not included. However, the 2013 Guidelines, although adopted by the Assembly, ought not to be read to constrain the interpretation and application of the parent convention. In some regions special areas have in fact included high seas areas. The designation of the entire Mediterranean Sea as a special area for the purposes of Annexes I (oil) and V (garbage) occurred when much of the regional sea was not encompassed by EEZs, and remains so today.[189] The Wider Caribbean region, similarly designated for Annex V purposes, includes high sea areas.[190] The Antarctic area, defined for Annex I (oil) as 'the sea area south of latitude 60°S'[191] largely comprises high sea areas. It is also designated as a special area for the purposes of Annexes II (noxious liquid substances)[192] and V.[193] While Arctic waters as defined in MARPOL and the Polar Code do not constitute a formal special area, the extent of protection is comparable to that for special areas.

188 IMO Res A.1087(28) (n 50) para 2.2.

189 MARPOL (n 4) annex I, ch I, reg 1(11). The region is defined as 'the Mediterranean Sea area means the Mediterranean Sea proper including the gulfs and seas therein with the boundary between the Mediterranean and the Black Sea constituted by the 41° N parallel and bounded to the west by the Straits of Gibraltar at the meridian of 005°36' W'. The definition is replicated in Annexes III, IV and V.

190 Ibid Annex V, Ch I, reg 1(14). The provision defines the region as follows: 'The Wider Caribbean Region means the Gulf of Mexico and Caribbean Sea proper including the bays and seas therein and that portion of the Atlantic Ocean within the boundary constituted by the 30° N parallel from Florida eastward to 77°30′ W meridian, thence a rhumb line to the intersection of 20° N parallel and 59° W meridian, thence a rhumb line to the intersection of 7°20′ N parallel and 50° W meridian, thence a rhumb line drawn southwesterly to the eastern boundary of French Guiana'.

191 Ibid Annex I, Ch I, reg 1(7).

192 Ibid Annex II, Ch V, reg 13(8).

193 Ibid Annex V, Ch I, reg 1(7).

There are hurdles that will need to be overcome. The first criterion, namely oceanographic conditions, may pose challenges. Proponents would need to demonstrate that the oceanography of the proposed high sea area is such as to cause the concentration or retention of harmful substances in the waters or sediments. Given that most high seas areas are high-energy environments, this criterion may be difficult to meet. In general, there are high seas areas that can be described as possessing particular circulation patterns, and more specifically convergence zones and gyres. The Sargasso Sea in the North Atlantic Ocean is an example of a mid-ocean gyre and has been described as an anticyclonic gyre whose physical boundaries are defined in terms of the Gulf Stream to the west, the North Atlantic Current to the north, the Canaries Current to the east, and the North Equatorial Current to the south.[194] The presence of the free-floating pelagic plant *sargassum* is one of its unique features.

The proposed high sea area would need to meet ecological conditions that justify protection from harmful substances or other vessel impacts. The proposal will turn on the facts. While the criteria in the 2013 Special Area Guidelines appear to provide as examples elements that are characteristic of coastal systems (e.g., spawning, breeding and nursery areas, coral reefs, mangroves, seagrass beds and wetlands), they appear to leave some room to interpret elements to be found in the ecology of the high seas (e.g., upwellings, gyres, migratory routes for seabirds and marine mammals, areas of critical importance for the support of large marine ecosystems).

A difficult challenge will be to demonstrate that vessel traffic is to such an extent the general discharge standards under the various MARPOL annexes are unacceptable in the light of the existing oceanographic and ecological conditions of a proposed high sea area. The reason for the difficulty is that the general discharge standards have been increased to a point that the marine environment is expected to assimilate the permitted wastes. Proponents would need to demonstrate that even with such standards in a high-energy environment, there is still likelihood of concentration of permitted wastes as to affect ecological attributes. This may be possible in the case of the Sargasso Sea, while relatively free of pollution, but where numerous species rely on the *sargassum*

194 AH Knap, 'Overview of the Importance of the Sargasso Sea and this Report' in MW Lomas *et al* (eds), *Oceanography of the Sargasso Sea: Overview of Scientific Studies Sargasso Sea* (Alliance Science Report Series, No 5, Sargasso Sea Alliance and Government of Bermuda 2011) 5 <www.sargassoseacommission.org/storage/documents/No5_Oceanography_HI.pdf> accessed 24 May 2018. See also David Freestone and Viva Harris, 'Particularly Sensitive Sea Areas Beyond National Jurisdiction: Time to Chart a New Course?' In Myron H. Nordquist, John Norton Moore and Ronán Long (eds), *International Marine Economy: Law and Policy* (Brill 2017) 322.

and 'tar and plastic can amalgamate with the *Sargassum* community'.[195] Moreover, there are important semi-enclosed seas that have not yet been designated special areas and which are likely to meet the requirements. With its heavy international traffic and extremely sensitive ecosystems, the South China Sea is a prime candidate for virtually all MARPOL special area and emission control area designations.[196] Coastal states of the region would need to make a collaborative submission, although at this time such an initiative is not likely politically realistic. As for any other area of the high seas, the proponent(s) can be any state party to MARPOL. Since the adoption of the special area is by an amendment to the applicable annex (usually on the basis of consensus), the applicant serves as the trigger, but the rule is one that is supported by the other state parties.

Any new high seas special areas will benefit from the existing system. MARPOL already has an advanced port inspection rule, and rules and standards are enforced in all of the PSC MoUs around the world. Shipowners will have an incentive to comply, irrespective of the flags under which they register their ships.

### 6.2 *Application of PSSAs to High Sea Areas?*

Unlike MARPOL special areas, at this time there are no PSSAs designated for high seas areas. The Arctic Marine Shipping Assessment Report of the Arctic Council recommended consideration of PSSAs in the region without suggesting limiting the designations to areas within national jurisdiction.[197] In theory, there is nothing in the IMO Convention and the 2005 Revised PSSA Guidelines that suggests PSSAs may not be designated for high seas areas. The original guidelines, and as subsequently amended, although containing scientific and

---

195 Knap (n 194) 5.

196 In general on these potential ideas, see Vu Hai Dang, *Marine Protected Areas Network in the South China Sea Charting a Course for Future Cooperation* (Brill 2014); Aldo Chircop, 'Regional Cooperation in Marine Environmental Protection in the South China Sea: A Reflection on New Directions for Marine Conservation' (2010) 41 *Ocean Development & International Law* 1.

197 Arctic Council, 'Arctic Marine Shipping Assessment 2009 Report' (Arctic Council 2009) <https://pame.is/images/03_Projects/AMSA/AMSA_2009_report/AMSA_2009_Report_2nd_print.pdf> accessed 24 May 2018. Recommendation D (Specially Designated Arctic Marine Areas) provides the following: 'That the Arctic states should, taking into account the special characteristics of the Arctic marine environment, explore the need for internationally designated areas for the purpose of environmental protection in regions of the Arctic Ocean. This could be done through the use of appropriate tools, such as "Special Areas" or Particularly Sensitive Sea Areas (PSSA) designation through the IMO and consistent with the existing international legal framework in the Arctic'. Ibid 10.

technical parameters that are as geographically neutral as possible, were developed and updated while learning from sensitive areas within national jurisdiction. The early rules were modelled on the experiences of the Great Barrier Reef and the Sabana-Camaguey archipelago. Thus the Guidelines (and Proposal Review Form) might need to be interpreted or adapted in some respects to enable the submission of proposals for high seas areas with different ecosystems.

High seas PSSA proposals will need to meet high ecological thresholds, although unlike MARPOL special areas, they do not need to meet oceanographic criteria as well. The emphasis is on demonstrating ecosystemic, habitat, species and critical functions, and their fragility and biogeographic importance. More of an issue are the social, cultural and economic criteria which are usually satisfied as a result of close geographical proximity and functional dependence of coastal communities on the sensitive area. The more distant the proposed high seas PSSA is from the coast, the more difficult it will be to demonstrate the nexus. However, only one criterion has to be met, so that as long as an ecological criterion is satisfied, the proponent is likely dispensed from further meeting other criteria.

Another potential constraint in an application is disclosure of action at the national level. Under the Guidelines the proponent state is expected to indicate what steps have been taken to protect the area (usually within its jurisdiction) with respect to its vessels, port entry conditions, and foreign vessels in the area consistently with international law. Given that the proposed area on the high seas is beyond its jurisdiction, it may be constrained in meeting this expectation. Proponent states are also queried on whether the proposed area has also been designated as a World Heritage Site or Biosphere Reserve, and whether that area is subject to international, regional or national conservation action or agreements. Again, this expectation is usually met with respect to areas within national jurisdiction but may also be met in part with respect to high seas areas where conservation areas may be in place. In recent years, the Northwest Atlantic Fisheries Organization has closed 20 areas identified as vulnerable to bottom contact gears.[198] These include seamounts, deep sea coral areas, and high sponge concentrations located in whole or in part in high seas areas. The Council of the International Seabed Authority has designated areas of particular environmental interest in the Clarion-Clipperton Fracture Zone and mandated the Legal and Technical Commission 'to make

198 Northwest Atlantic Fisheries Organization, 'NAFO Conservation and Enforcement Measures' FC Doc 17-01 (2017) <www.nafo.int/Portals/0/PDFs/fc/2017/CEM-2017-web.pdf> accessed 24 May 2018, art 17.

recommendations, where appropriate, to the Council relating to the network of areas of particular environmental interest, on the basis of the results of workshops with a view to redefining, where necessary, the details of the size, location and number of required areas of particular environmental interest'.[199] But perhaps a more substantial challenge for the proponent state(s) is to demonstrate the nexus between sensitivity-cum-vulnerability of a proposed high seas area and international shipping.[200]

As indicated earlier, the IMO Convention empowers the IMO to respond to requests concerning impacts on the marine environment communicated to it by other international organisations.[201] Although unprecedented, there is nothing to suggest that the IMO would not be in a position to consider a proposal for a measure(s) in high seas areas where the membership of a requesting organisation authorises that organisation to do so. International organisations potentially in such a position because of their mandate include the International Seabed Authority, International Whaling Commission and regional fishery management organisations. The guidance normally applicable to states would need to be applied to such international organisations mutatis mutandis.

In principle, the IMO is not limited to adopting routeing and reporting measures limited in application to areas within jurisdiction. The real test is the functionality and efficiency of the APM in meeting the proposed high seas area's needs. A real limitation is special mandatory measures pursuant to Article 211(6) of UNCLOS, which clearly are limited to areas within the EEZ. While proposals for special areas may be advanced by MARPOL state parties, proposals for high seas PSSAs can be submitted by any IMO member state.

### 6.3 *Routeing and Reporting Measures for High Seas Areas?*

As mentioned above with respect to PSSAs, the IMO may adopt routeing and reporting measures for virtually any maritime traffic area. This includes the purpose of protecting sensitive marine areas and species. The limitations are functional rather than spatial. The idea of routeing on the high seas was probably first addressed by IMO in the International Convention for the Safety of Life at Sea, 1960.[202] That convention recommended the practice of recognised routes across the North Atlantic, but the actual selection of routes was left to

---

199 International Seabed Authority, 'Decision of the Council Relating to An Environmental Management Plan for the Clarion-Clipperton Zone' ISBA/18/C/22 (26 July 2012).

200 Especially where other industrial activities might be already taking place.

201 IMO Convention (n 5) art 1(d).

202 International Convention for the Safety of Life at Sea, 1960 (adopted 17 June 1960, entered into force 26 May 1965) ATS 1968 No 7, Ch V, reg 8.

carriers. State parties could require carriers to give public notice of such routes and any changes and could even require them to use those routes and as far as practicable to avoid the Grand Banks during the fishing season and areas endangered by ice. Therefore the idea of designating routes on the high seas, this time by the IMO, is not unprecedented.

Also not new is the idea that ships on the high seas may be subject to some form of reporting. SOLAS places no geographical limitations for reporting systems.[203] In addition, there are other reporting communication requirements on ships, irrespective of location. For example, SOLAS requires the master to communicate danger messages when the vessel encounters particular conditions that are hazardous for navigation (e.g., derelicts, presence of ice, storms, etc.).[204] For certain classes of ships, SOLAS has requirements for the long-range identification and tracking of ships, including a requirement to be fitted with an automatic identification system (AIS).[205]

While routeing and reporting measures to protect sensitive high seas areas and species may be adopted, assuming compelling need is demonstrated, and the respective guidelines complied with, it is not clear who is responsible for management of the measure and enforcement. The General Provisions on Ships' Routeing assume that the proponent government(s) will implement the routeing system.[206] How this can be accomplished in a high seas area (perhaps other than the extended continental shelf) and with what legal authority consistent with UNCLOS are unclear. Moreover, that government will not possess enforcement authority, unless requested or authorised by the flag state. While through AIS data it will be possible to determine the location of a vessel, it is unclear what state (other than the flag state) would be in a position to take the necessary enforcement action.

## 7 Conclusion

In summary, this study has demonstrated that there is sufficient legal and institutional basis for the application of MARPOL special areas, PSSAs, routeing measures and reporting measures to ecosystems and species in high seas areas requiring protection and conservation. Since many of the existing frameworks

203 SOLAS (n 6) ch V, reg 11.

204 Ibid Ch V, reg 31.

205 Ibid Ch V, reg 19-1. This requirement applies to passenger ships, including high-speed passenger craft, cargo ships, including high-speed craft, of 300 gross tonnage and more, and mobile offshore drilling units.

206 IMO Res A.572(14) (n 136) paras 3.5 and what follows.

and practices for such measures have developed primarily with areas within national jurisdiction in mind, there will likely need to be adjustments to facilitate proposals for their application to particular high seas areas. The proposals will be expected to meet challenges, in particular with regard to how to propose conservation measures without unduly or unreasonably intruding into the freedom of navigation. While the criteria for special areas, routeing measures and reporting measures are likely sufficient, the procedures and criteria for PSSAs would likely benefit from adjustments to expressly indicate the eligibility of high seas areas, the ability of any state to make a submission, applicable conditions for submissions, and expected enforcement or compliance processes. The IMO appears to have the authority necessary to respond not only to proposals from member states, but also to requests from other international organisations.

As a concluding thought, the discourse on the potential extension of MARPOL special areas, PSSAs and other routeing and reporting measures to address marine environment protection and conservation on the high seas will likely encounter the frequent concern expressed in the IMO that the proliferation of such areas will water down the special protection they provide. The 'anti-proliferation' argument pervasive in the international shipping industry appears to reflect a concern that there should be as minimal regulatory intrusion as possible into the exercise of the freedom of the high seas, so that the designation of marine areas entailing higher standards or particular rules ought to be exceptions rather than more frequent and general practice. The flaw in this perception is that the idea of absolute freedom of the high seas has long been overtaken by the practice of regulated and responsible use of the high seas, and that the demand for enhanced protection of the marine environment, including through protected areas, can reasonably be expected to lead to greater demands for protective measures from international shipping. Moreover, it betrays insufficient appreciation for the need to mainstream the utilisation of the precautionary approach to marine environment protection from the impacts of international shipping.[207] The larger the marine protected areas (and their networks), the more likely the demand for commensurate special areas, PSSAs and other area-based measures. The idea that MARPOL special areas, PSSAs and other measures for marine conservation are rules of exception may have to give way to the need to mainstream such practices over more and potentially larger ocean spaces.

207 See in particular the concluding commentary in Freestone and Harris (n 194).

CHAPTER 6

# Conservation and Management of Marine Living Resources beyond National Jurisdiction: Filling the Gaps

*Robin Warner*


## Abstract

This chapter examines the multiple factors affecting the status of marine living resources in marine areas beyond national jurisdiction (ABNJ), focusing on the gaps and deficiencies in the international law framework regulating these resources. The United Nations process to develop an international legally binding instrument (ILBI) for the conservation and sustainable use of marine biological diversity in ABNJ represents a historic opportunity to lay the foundation for a more integrated and cross-sectoral system of oceans governance in ABNJ. This chapter explores the potential role of the ILBI in remediating gaps in the international law framework for marine living resources in ABNJ.


## 1 Introduction

The process initiated by the United Nations General Assembly (UNGA) in UNGA Resolution 69/292[1]—to develop the elements of an international legally binding instrument (ILBI) for conservation and sustainable use of marine biological diversity in areas beyond national jurisdiction (ABNJ)—has prompted wide-ranging research into existing ocean governance frameworks and their applicability to conservation and sustainable use of marine biological diversity in ABNJ. UNGA Resolution 69/292 provides that negotiations to develop the new ILBI should address the four elements of a package deal agreed by states in 2011. These elements comprise 'marine genetic resources, including questions on the sharing of benefits, measures such as area based management

1 Development of an International Legally Binding Instrument under the United Nations Convention on the Law of the Sea on the Conservation and Sustainable Use of Marine Biological Diversity of Areas Beyond National Jurisdiction, UNGA Res 69/292 (19 June 2015) 69th Session, Agenda Item 74(a), UN Doc A/Res/69/292 (UNGA Res 69/292). See also <www.un.org/en/ga/search/view_doc.asp?symbol=A/Res/69/292> accessed 24 May 2018.

 | DOI:10.1163/9789004373303_007

tools, including marine protected areas, environmental impact assessments and capacity-building and the transfer of marine technology'.[2] The Preparatory Committee meetings held in 2016 identified additional crosscutting issues for consideration, including definitions, scope of the instrument, relationship of the instrument to other instruments and frameworks, institutional arrangements, compliance, responsibility and liability, dispute settlement and final clauses.[3] UNGA Resolution 69/292 also stipulates that the process to develop the ILBI should not undermine existing relevant legal instruments and frameworks and relevant global sectoral and regional bodies.[4] These negotiations are referred to as the BBNJ (biodiversity beyond national jurisdiction) discussions.

Conservation and sustainable use of marine living resources beyond national jurisdiction has been a central focus of the BBNJ discussions in both the Working Group and Preparatory Committee meetings. Even though the BBNJ discussions are not specifically about high seas marine living resources, but rather the broader concept of the species, habitats and ecosystems encompassing marine biodiversity in ABNJ, the process leading to the ILBI has the potential to enhance the conservation and sustainable use of marine living resources beyond national jurisdiction. Discussion of marine living resources in the BBNJ process has encompassed both pelagic and benthic resources, including fish stocks, marine mammals and sedentary species, as well as their habitats and the ecosystems of which they form part. This chapter will examine the multiple factors affecting the status of marine living resources in ABNJ, focussing particularly on the gaps and deficiencies in the international law framework regulating these resources. It will then explore the potential role of the ILBI in remediating these gaps and deficiencies.

## 2 The Status of Marine Living Resources and Marine Biological Diversity beyond National Jurisdiction

In recent decades, commercial exploitation of marine living resources beyond national jurisdiction has grown in scale and intensity with the demise of some inshore fisheries and the extension of coastal state resource jurisdiction to 200

2 Ibid para 2.

3 Preparatory Committee Established by General Assembly Resolution 69/292, 'Chair's Indicative Suggestions of Clusters of Issues and Questions to Assist Further Discussions in the Informal Working Groups at the Second Session of the Preparatory Committee' <www.un.org/Depts/los/biodiversity/prepcom_files/IWGs_Indictive_Issues_and_Questions.pdf> accessed 24 May 2018.

4 UNGA Res 69/292 (n 1) para 3.

nautical miles (NM).[5] An estimated 10 to 20 per cent of the world's commercial fish catch is now derived from high seas areas.[6] Highly migratory stocks, such as tuna and marlin, have been vigorously exploited by distant water fishers in high seas areas, where until recently, catch restrictions have been rare.[7] With a large proportion of the world's fisheries under pressure from the commercial fishing industry, interest has also grown in previously undisturbed deep-sea fisheries, such as those which aggregate around seamounts.[8]

Fisheries experts estimate that most tuna stocks are close to or have reached their full level of exploitation or maximum sustainable yield in recent years. Large fleets of long-liners and purse seiners have been targeting tuna since the 1950s.[9] While quotas and size limits have been imposed by global and regional organisations managing tuna fisheries, there has been some resistance to such restrictions, and enforcement of quotas in high seas areas has been difficult.[10] The incidental catch of longline tuna fishers has been high with a large number of non-target species, such as billfish, turtles, birds, dolphins and sharks, being injured, killed and discarded through encounters with lines and hooks.[11] Some regional fisheries organisations (RFMOs) managing tuna and other highly migratory and straddling stocks are beginning to implement a precautionary approach to these fisheries imposing measures such as lower quotas, and spatial, time and gear restrictions on the fishing vessels of their member states.[12] These organisations are also beginning to examine the need for ecosystem-based management of these fisheries, which takes into account the associated and dependent species and habitats affected by fishing for

5 CJ Carr and HN Scheiber, 'Dealing with a Resource Crisis: Regulatory Regimes for Managing the World's Marine Fisheries' (2002) 21 *Stanford Environmental Law Journal* 45; MC Baker et al, *The Status of Natural Resources on the High Seas* (WWF, IUCN, WCPA 2001) 22.

6 K Gjerde and C Breide (eds), *Towards a Strategy for High Seas Marine Protected Areas: Proceedings of the IUCN, WCPA and WWF Experts Workshop on High Seas Marine Protected Areas* (Malaga, Spain, 15–17 January, IUCN 2003) vii.

7 Carr and Scheiber (n 5) 53.

8 EJ Molenaar, 'Unregulated Deep Sea Fisheries: A Need for a Multi-Level Approach' (2004) 19(3) *International Journal of Marine and Coastal Law*, 223; LA Kimball, 'Deep Sea Fisheries of the High Seas: The Management Impasse' (2004) 19(3) *International Journal of Marine and Coastal Law* 260.

9 S Earle, *A Message of the Oceans* (G.P. Putnam's Sons 1995) 182–84; Baker et al (n 5) 66.

10 Carr and Scheiber (n 5) 59; R Rayfuse, 'The Challenge of Sustainable High Seas Fisheries' in N Schrijver and Friedl Weiss (eds), *International Law and Sustainable Development. Principles and Practice* (Martinus Nijhoff 2004) 474.

11 A Byatt et al, *The Blue Planet: A Natural History of the Oceans* (BBC Worldwide Ltd 2001) 306–07; Gjerde and Breide (n 6) 6.

12 Rayfuse (n 10) 473–74.

the target species.[13] Implementation of precautionary and ecosystem-based measures in the conservation and management of high seas fisheries, however, is far from comprehensive and still in the early stages of development.[14]

Marine scientists are also concerned about the threats to deep-water demersal fish, which are concentrated on the continental slopes and around deep-sea habitats such as seamounts.[15] These include species such as grenadiers, smoothheads, orange roughy and the pelagic armourhead. Unlike many inshore and coastal fish species, they exhibit characteristics such as great longevity, low fecundity and high age at first maturity.[16] They can also be endemic to a particular deep-sea habitat.[17] These features of deep-water fish species inhibit their ability to recover quickly from overexploitation. The methods used to fish for these species, especially bottom trawling, have major impacts on their surrounding habitat and on associated and dependent ecosystems.[18] As well as crushing fragile deep-sea coral structures, bottom trawlers sweep up large numbers of non-target species. All deep-sea fish caught in a trawl including non-target species will be dead on reaching the surface as a result of damage to their skins and pressure and temperature changes.[19] In some oceanic regions, such as the Tasman Sea and the Southern Ocean, there are already examples of deep-water species such as orange roughy and Patagonian toothfish that have been commercially exploited to levels close to extinction.[20]

Fishing is one of the key drivers of biological diversity loss and ecological change in the open ocean beyond national boundaries for a number of reasons. There are much higher levels of overfishing and overfished stocks for straddling and migratory species in the open ocean in ABNJ than in national waters, with twice the rate of overfished stocks in open ocean than that within

---

13 Ibid 474–76.

14 Ibid 474–75.

15 K Probert, 'Seamounts, Sanctuaries and Sustainability: Moving Towards Deep Sea Conservation' (1999) 9 *Aquatic Conservation: Marine and Freshwater Ecosystems* 602; JA Koslow, 'Fish Stocks and Benthos of Seamounts' in Hjalmar Thiel and JA Koslow (eds), *Managing Risks to Biodiversity and the Environment on the High Seas, Including Tools such as Marine Protected Areas—Scientific Requirements and Legal Aspects* (Federal Agency for Nature Conservation 2001) 43; JM Gordon, 'Deep Water Fish and Fisheries' in Hjalmar Thiel and J Anthony Koslow (eds), *Managing Risks to Biodiversity and The Environment on the High Seas, Including Tools such as Marine Protected Areas—Scientific Requirements and Legal Aspects* (Federal Agency for Nature Conservation 2001) 31.

16 Koslow (n 15) 43; Gordon (n 15) 32.

17 Ibid.

18 Gordon (n 15) 33.

19 Ibid.

20 Koslow (n 15), 46–47.

national jurisdictions.[21] Bycatch in open-ocean fisheries and the indirect impacts of abandoned, lost or discarded fishing gear have been implicated in the severe decline of fish, sea turtle, shark, seabird and marine mammal populations.[22] Open-ocean fisheries have also been shown to reduce pelagic biological diversity and ecosystem resilience.[23] Fisheries have altered trophic relationships in open-ocean communities, generating trophic cascades that can lead to ecosystem-level impacts and regime shifts.[24] In addition, the ecological impacts of open-ocean fisheries and climate change can act synergistically to induce profound transformations of ecosystem dynamics.[25]

## 3 Governance Gaps and Deficiencies in the Conservation and Sustainable Use of Marine Living Resources beyond National Jurisdiction

The BBNJ process has brought into sharp relief a range of governance gaps for conservation and sustainable use of marine living resources in ABNJ. These gaps take different forms and have been variously categorised as regulatory, substantive and implementation gaps.[26] In terms of an overarching governance gap, a number of commentators have noted that the current legal and institutional regime for the high seas based primarily on flag state jurisdiction is inadequate to address the significant stresses on marine biological diversity beyond national jurisdiction.[27] Poor regulation of high seas fishing has been acknowledged throughout the BBNJ process as a key threat to marine biological diversity beyond national jurisdiction.[28] There is no international rule-making structure for the high seas that can hold individual states accountable for their

21 GO Crespo and DC Dunn, 'A Review of the Impacts of Fisheries on Open-ocean Ecosystems' (2017) 74(9) *ICES Journal of Marine Science* 2286.

22 Ibid 2288.

23 Ibid 2291.

24 Ibid 2289.

25 Ibid 2292.

26 KM Gjerde et al, *Regulatory and Governance Gaps in the International Regime for the Conservation and Sustainable Use of Marine Biodiversity in Areas Beyond National Jurisdiction* (IUCN 2008).

27 R Rayfuse and M Wilder, 'Sustainability, Uncertainty and Global Fisheries' in JW Handmer, TW Norton and SR Dovers (eds), *Ecology, Uncertainty and Public Policy* (Pearson Education Limited 2001) 138; Carr and Scheiber (n 5) 45.

28 RA Barnes, 'The Proposed LOSC Implementation Agreement on Areas Beyond National Jurisdiction and its Impact on International Fisheries Law' (2016) 31(4) *The International Journal of Marine and Coastal Law* 593.

failure to act in the face of actions by their fishing vessels that have adverse impacts on the marine environment beyond national jurisdiction. While the deficiencies in flag state control over fishing vessels have been remedied to a limited extent by introduction of the 2009 FAO Port State Measures Agreement[29] and trade-based measures, these measures are relatively recent and not applicable to all states. A more integrated and cross-sectoral governance structure is needed to adequately protect not only the interests of individual users but also of the international community.[30]

There are a wide variety of binding international law agreements and non-binding arrangements at global and regional levels relevant to the conservation and sustainable use of fish stocks in ABNJ. The UN Fish Stocks Agreement (UNFSA)[31] provides the first comprehensive blueprint for sustainable fisheries conservation and management in ABNJ and model provisions for cooperation between coastal states and flag states with regard to the regulation of high seas fishing fleets. The principal objective of the UNFSA is to ensure the long-term conservation and sustainable use of straddling and highly migratory fish stocks through the effective implementation of the relevant provisions of the UNCLOS. It builds on states' obligations under the UNCLOS to cooperate in the conservation of straddling and highly migratory fish stocks on the high seas and in exclusive economic zones (EEZ) and environmental protection obligations under instruments such as the Convention on Biological Diversity (CBD) and the Oceans Chapter of Agenda 21 by

- establishing a framework for regional cooperation to manage highly migratory and straddling fish stocks;[32]
- requiring fisheries management for highly migratory and straddling stocks to be based on precautionary and ecosystem based approaches;[33]
- specifying in detail the duties of flag states;[34] and

29 Agreement on Port State Measures to Prevent Deter and Eliminate Illegal, Unreported and Unregulated Fishing (adopted 22 November 2009, entered into force 5 June 2016) [2010] ATNIF 41.

30 B Kuemlangan, et al, 'Integrative Policy and Legal Instruments, Approaches and Tools: Fisheries and Biodiversity Conservation' in SM Garcia, J Rice, A Charles (eds), *Governance of Marine Fisheries and Biodiversity Conservation, Interaction and Coevolution* (Wiley Blackwell 2014) 175–78.

31 United Nations Agreement for the Implementation of the Provisions of the United Nations Convention on the Law of the Sea of 10 December 1982 Relating to the Conservation and Management of Straddling Fish Stocks and Highly Migratory Fish Stocks (adopted 8 September 1995, entered into force 11 December 2001) 2167 UNTS 3 (UNFSA).

32 Ibid art 21.

33 Ibid art 6.

34 Ibid arts 18–19.

- enhancing means for monitoring, control and enforcement of conservation and management measures for highly migratory and straddling stocks.[35]

UNFSA was primarily designed to conserve and manage straddling and highly migratory stocks in high seas areas, although two of its articles apply within national jurisdiction.[36] States parties are urged to apply the general principles on conservation and management of straddling and highly migratory stocks set out in Article 5 of the UNFSA to the conservation and management of such stocks within national jurisdiction. Arguably, UNFSA has dramatically extended the potential scope of regional fisheries management agreements, since it may be interpreted as obliging states parties and their flag vessels to comply with the management regulations of all RFMOs regardless of whether they are parties to these agreements.[37] Article 8(4) of UNFSA provides that only those states that agree to implement conservation and management measures established by RFMOs with regard to highly migratory and straddling stocks shall have access to the fishery resources to which those measures apply.

In addition to codifying all the relevant international environmental law principles for conservation and management of marine living resources, the UNFSA provides practical guidance for RFMOs on cooperative compliance and enforcement measures. In its model provision for a qualified right of reciprocal boarding and investigation of suspected illegal fishing vessels on the high seas, UNFSA legitimises the establishment of collaborative high seas compliance and enforcement schemes among states parties.[38] Article 21 prescribes the circumstances in which a flag state other than the flag state of the suspected fishing vessel may board and investigate the vessel for an alleged violation of conservation and management measures on the high seas. In any high seas area covered by an RFMO or arrangement, a state party to UNFSA that is also a member of the RFMO or arrangement may send its duly authorised inspectors to board and inspect the fishing vessels of other flag states that are party to the UNFSA and suspected of violations of the conservation and management measures of the RFMO or arrangements.[39] Where there are clear grounds for believing that a vessel has engaged in any activity contrary to the conservation and management measures of the RFMO or arrangement, the inspecting state must notify the flag state promptly, and the flag state must respond within

35 Ibid art 23.

36 Ibid arts 6 and 7.

37 Ibid art 8(3).

38 EJ Molenaar, 'Participation, Allocation and Unregulated Fishing: The Practice of Regional Fisheries Management Organizations' (2003) 18(4) *International Journal of Marine and Coastal Law* 473.

39 UNFSA (n 31) art 21(1).

three working days. The flag state may then either fulfil its obligation to investigate the alleged violation or authorise the inspecting state to investigate.[40] Where the flag state fails either to respond or to take action, the inspecting state may proceed with the investigation, including bringing the vessel to the nearest appropriate port.[41] These provisions extend the very limited circumstances provided for under the UNCLOS and customary international law in which flag vessels may be boarded by officials of other flag states on the high seas.[42] Since the entry into force of the UNFSA, a complex pattern of high seas fisheries regulation has continued to emerge, as new RFMOs have been established and existing RFMOs adapt their agreements and institutions to incorporate the UNFSA provisions.

### 3.1 *Regulatory Gaps and Deficiencies*

While RFMOs exist in the majority of high seas areas for both tuna and non-tuna fisheries, there are some significant gaps in geographic coverage including the Arctic and the Central and Southwest Atlantic.[43] In these regions, conservation and management of fishing is left to the discretion of the individual flag states with vessels fishing in these areas. Relatively few RFMOs include all the participants in a regional fishery among their members.[44] Only those member states bound by an RFMO agreement are required to apply its conservation and management measures. The fishing activities of vessels flagged to non-parties in particular regions can therefore undermine the conservation and management measures adopted by RFMO parties. Many highly migratory and straddling fish stocks fall within the scope of RFMO agreements, but some discrete high seas fish stocks, such as orange roughy and sedentary species, are not covered by these agreements.[45]

UNFSA and some RFMO agreements address environmental impact assessments (EIA) in general terms, but specific terms for EIAs are only spelled out

40 Ibid art 21(5) and (6).

41 Ibid art 21(8).

42 These circumstances are outlined in Article 110(1) of UNCLOS and include reasonable grounds for suspecting the ship is engaged in piracy, the slave trade, unauthorised broadcasting where the flag state of the warship boarding the suspected vessel has jurisdiction under Article 109 of UNCLOS, the ship is without nationality and though flying a foreign flag or refusing to show its flag, the ship is, in reality, of the same nationality as the warship.

43 R Warner, K Gjerde and D Freestone, 'Regional Governance for Fisheries and Biodiversity' in SM Garcia, J Rice, A Charles (eds), *Governance of Marine Fisheries and Biodiversity Conservation, Interaction and Coevolution* (Wiley Blackwell 2014) 212.

44 Ibid.

45 Y Takei, *Filling Regulatory Gaps in High Seas Fisheries. Discrete High Seas Fish Stocks, Deep-sea Fisheries and Vulnerable Marine Ecosystems* (Martinus Nijhoff 2013) 4.

for deep-sea bottom fisheries and new or exploratory fisheries.[46] RFMOs are also limited in their ability to deliver strategic or integrated EIAs of fishing and other activities in ABNJ.[47]

There is no formal global coordination mechanism to regularly oversee the conservation and management activities of RFMOs, individual flag states and port states and to monitor their performance against best practice standards for biological diversity conservation and sustainable use.[48] Similarly, there is no global mechanism for RFMOs to exchange information on patterns in conservation and sustainable use of high seas fisheries.[49] This hampers efforts to address global problems, such as illegal, unreported and unregulated (IUU) fishing, as vessels can move between regions focussing their effort on regions with less rigorous conservation and management measures. Communication between RFMOs and coordination of conservation and management efforts have tended to be ad hoc, with the Kobe Process standing out as the key example of a joint meeting of RFMOs.[50] There has been some progress in cross-sectoral communication between RFMOs and other global and regional organisations in regions such as the Northeast Atlantic and the South Pacific, but in other regions RFMOs are still in the early stages of development and are focussed on fisheries management issues rather than cross-sectoral collaboration.[51]

Since the early 1970s a patchwork of binding and non-binding regional seas arrangements (RSAs) has been negotiated around the globe to engage states in the collaborative protection of their coastal and marine environments. Many of the binding RSAs were initiated through the United Nations Environment Programme (UNEP) regional seas programme, while others are the result of independent agreements between regional partners.[52] RSAs now cover 18 regions and involve more than 143 states.[53] The areas of responsibility of the RSAs are predominantly limited to waters under national jurisdiction, and only a few of them make provision for consensual environmental protection measures in

46 Barnes (n 28) 610.

47 Ibid.

48 Warner, Gjerde and Freestone (n 43) 212.

49 Ibid.

50 Ibid.

51 Barnes (n 28) 615; UNEP-WCMC, 'Governance of Areas Beyond National Jurisdiction for Biodiversity Conservation and Sustainable Use: Institutional Arrangements and Cross-sectoral Cooperation in the Western Indian Ocean and the South East Pacific' (UN Environment World Conservation Monitoring Centre 2017) 71–72.

52 A Vallega, 'The Regional Seas in the 21st Century: An Overview' (2002) 45(11) *Ocean and Coastal Management* 925.

53 UN Environment, 'The Regional Seas Programme' <https://www.unenvironment.org/explore-topics/oceans-seas/what-we-do/working-regional-seas/why-does-working-regional-seas-matter> accessed 24 May 2018.

high seas enclaves surrounded by member states' waters and other high seas adjacent or proximate to member states' waters.[54] Those RSAs that do make provision for such measures in ABNJ include the Convention for the Protection of the Marine Environment of the North-East Atlantic (OSPAR Convention), which includes high seas areas in the Northeast Atlantic under its scope of responsibility and the South Pacific RSA, which includes within its mandate the high seas enclaves between the EEZs of its members.[55] The spread of RSAs has largely reflected and advanced the development of modern marine environmental protection principles, including biological diversity conservation and area-based management measures, such as marine protected areas.[56] As with the RFMOs, there are still varying rates of progress in incorporating biological diversity conservation objectives into the legal and institutional structures of some RSAs. Key factors limiting the effectiveness of RSAs in conserving and sustainably using marine biological diversity in ABNJ are their limited geographic scope and the absence of specific collaboration mechanisms between RSAs and RFMOs and other global organisations with responsibilities for ABNJ activities, such as the International Maritime Organization (IMO) and the International Seabed Authority (ISA).[57]

### 3.2 *Implementation Gaps and Deficiencies*

Reviews of RFMO practice at the global level disclose a number of factors that have limited their effectiveness in implementing fisheries conservation and management measures in an ecologically sustainable manner.[58] Although the majority of RFMOs have incorporated modern conservation principles into their agreements, the application of conservation and management measures varies widely across the different organisations.[59] More recently established

54 Warner, Gjerde and Freestone (n 43) 215.

55 Ibid.

56 T Treves, 'Regional Approaches to the Protection of the Marine Environment' in MH Nordquist, JN Moore and S Mahmoudi (eds), *The Stockholm Declaration and the Law of the Marine Environment* (Kluwer Law International 2003) 137–38.

57 Warner, Gjerde and Freestone (n 43) 217.

58 High Seas Task Force, 'Closing the Net: Stopping Illegal Fishing on the High Seas' (Governments of Australia, Canada, Chile, Namibia, New Zealand and the United Kingdom, WWF, IUCN and the Earth Institute at Columbia University 2006) <www.oecd.org/sd-roundtable/papersandpublications/39375276.pdf> accessed 24 May 2018; M Gianni et al, 'Unfinished Business: A Review of the Implementation of the Provisions of UNGA Resolutions 61/105 and 64/72 Related to the Management of Bottom Fisheries in Areas Beyond National Jurisdiction' (Deep Sea Conservation Coalition, September 2011) <http://dscc.visualcode.co.uk/resources/publications/unfinished-business-review-implementation-provisions-united-nations-general-assembly-resolutions-61105-6472-related-management-bottom-fisheries-areas-beyond-nat> accessed 24 May 2018.

59 Barnes (n 28) 601–04.

RFMOs, such as the Western and Central Pacific Fisheries Commission (WCPFC) and the South Pacific Regional Fisheries Management Organization (SPRFMO), have been at the vanguard of more progressive state practice.

The WCPFC Convention was the first comprehensive conservation and management regime for highly migratory fish stocks, such as tuna, in Pacific ABNJ.[60] The area of competence of the WCPFC covers a large swathe of the Pacific Ocean in the southern hemisphere, including a large area of high seas lying outside and between the 200 NM EEZs of its parties.[61] The WCPFC is empowered to adopt principles and measures for conservation and management of the highly migratory fish stocks in its area of competence that reflect key international environmental law principles and biological diversity conservation and sustainable use objectives.[62] The WCPFC must apply the precautionary approach, determine the impact of fishing activities on target and associated or dependent species and their environment, and adopt plans where necessary to ensure the conservation of species and the protection of habitats of special concern.[63] The conservation measures to be taken by the WCPFC also include those that protect biological diversity in the marine environment and assess the impact of fishing activities on other species belonging to the same ecosystem.[64] Since its inception, the WCPFC has introduced a wide array of conservation and management measures ranging from those addressing the level of fishing effort for highly migratory species, such as tuna and marlin, and the impacts of fishing on seabirds, turtles and cetaceans, to the prohibition of driftnet fishing and bottom fisheries closures.

Consistent with its biological diversity conservation obligations, SPRFMO has also adopted a wide range of conservation and management measures in relation to bottom fishing, the management of jack mackerel, the banning of gillnetting, the collection of detailed fishing data, inspections in port and at sea, transshipments, a Vessel Monitoring System, an IUU List, vessels without nationality, minimising the bycatch of seabirds and establishing of a compliance and monitoring scheme.[65] It has also begun to establish formal and informal links with other RFMOs in the region and beyond and has extended observer

60 Convention on the Conservation and Management of Highly Migratory Fish Stocks in the Western and Central Pacific Ocean (adopted 5 September 2000, entered into force 19 June 2004) (2001) 40 ILM 277 (WCPFC Convention).

61 Ibid art 3.

62 Ibid art 5(a) and (b).

63 Ibid art 5(c) and 6.

64 Ibid art 5(f) and (d).

65 SPRFMO, 'Conservation Measures' <www.sprfmo.int/conservation-measures> accessed 24 May 2018.

status to the Southeast Pacific RSA, Comision Permanente del Pacifico Sur (Permanent Commission for the South Pacific) (CPPS) and a range of relevant intergovernmental organisations and non-governmental organisations.[66] A recent UNEP-WCMC report notes, however, that there is still significant potential for further intra- and cross-sectoral cooperation between SPRFMO and other regional and global bodies.[67]

It has been the established practice of RFMOs to take decisions on their conservation and management measures by consensus, even when their instruments may not require this, and to allow for individual objections to conservation and management measures agreed by the majority of member states.[68] This allows objecting RFMO members to take advantage of uncertainties in scientific advice and can lead to only partially implemented conservation and management measures. Many of the RFMOs that were established prior to the conclusion of UNFSA permit member states to opt out or object to implementing conservation and management measures that have been agreed within the RFMO.[69] Even in the Commission for the Conservation of Antarctic Marine Living Resources (CCAMLR), whose instrument explicitly adopts a precautionary and ecosystem-based approach to marine living resource management, flaws in implementation are evident. Within CCAMLR there is a requirement for consensus in decisions on matters of substance, such as conservation and management measures, and an objection procedure is available for members to object later to measures for which they have voted.[70] This is in contrast to the WCPFC, which has introduced an innovative measure to overcome the weaknesses of consensus-based decision making. Where consensus fails, decisions on substantive matters are to be taken by a three-quarter majority of the members present. Decisions become binding on parties 60 days after their notification, but members voting against the decision or absent can, within 30 days of its adoption, seek a review of the decision.[71]

The effectiveness of conservation and management measures has most often been undermined by the inability of RFMOs to regulate the fishing activities of both member and non-member states. To address the problem of member states' non-compliance with conservation and management measures, some RFMOs, such as WCPFC and SPRFMO, have established reciprocal boarding and inspection schemes in their areas of responsibility where

66 UNEP-WCMC (n 51) 85.

67 Ibid 86.

68 Warner, Gjerde and Freestone (n 43) 212.

69 Ibid.

70 Ibid 213.

71 Ibid 214.

contracting parties board and inspect each other's fishing vessels on the high seas in accordance with the UNFSA model.[72] The problem of non-members' fishing vessels' activities has proven more intransigent. CCAMLR adopted a standard suite of fisheries measures until the mid-1990s, relying on flag state implementation of conservation and management measures supplemented by fisheries data reporting, at sea and in port inspections by member states of fishing vessels and their catch, tracking the movement of member states' fishing vessels through vessel monitoring systems, and notification of vessel movements. These measures proved to be inadequate to prevent increases in IUU fishing in the CCAMLR Area particularly for Patagonian toothfish from the mid-1990s. To address the problem of non-members' fishing vessels, CCAMLR introduced a catch documentation scheme, which prohibits entry into global markets of Patagonian toothfish without verified catch documents.[73]

## 4 Potential for the Prospective ILBI to Remediate Governance Gaps

With the right mix of flexible and supportive provisions tailored to the needs of particular regions and sectors, the ILBI has the potential to enhance and lift the performance of existing governance frameworks for conservation and sustainable use of marine living resources in ABNJ. At a minimum, the ILBI could contain international law principles and standards on conservation and sustainable use of marine biological diversity including marine living resources in ABNJ, which are generally applicable to all states parties. The high seas principles recommended by the International Union for Conservation of Nature (IUCN)[74] provide a widely accepted starting point for relevant international law norms and could be operationalised in other provisions of the ILBI. In particular the precautionary approach and ecosystem-based approach to ocean management

> should be operationalised for all high seas activities and should be used as a globally applicable default mechanism for existing, emerging and new activities on the high seas. This will require placing the burden of proof on those who argue that an activity will not cause significant harm

72 WCPFC Convention (n 60) art 24; Convention on the Conservation and Management of High Seas Fishery Resources in the South Pacific Ocean (adopted 14 November 2009, entered into force 24 August 2012) ATS 28, art 27.

73 Warner, Gjerde and Freestone (n 43) 213–14.

74 IUCN, '10 Principles for High Seas Governance' <www.iucn.org/downloads/10_principles_for_high_seas_governance___final.pdf> accessed 24 May 2018.

to show that this is so, and make the responsible parties liable for environmental harm.[75]

The ILBI could also provide an important focal point for global, regional and sectoral organisations with responsibilities for conservation and sustainable use of marine biological diversity in ABNJ through establishing a Conference of the Parties (COP), which meets regularly and has the power to take decisions on the different elements of the ILBI. The COP could establish regular as opposed to ad hoc cooperation mechanisms between RFMOs and other global and regional organisations, such as the IMO, ISA and regional seas organisations, and provide a collaborative forum and potentially financial support for such organisations to coordinate their efforts on conservation and sustainable use of marine biological diversity in ABNJ.

The ILBI has the potential to further coalesce global scientific information on marine biological diversity in ABNJ, by drawing on data, monitoring and assessment from a wider group of global, regional and sectoral organisations and sharing such information among its states parties and RFMOs. This broader collation of scientific information could provide environmental baselines for different regions in ABNJ against which to measure the cumulative impacts of human activities on marine biological diversity in ABNJ over time. These baselines could contribute to long-term conservation and sustainable use of marine living resources in ABNJ. The collation of this information has already begun with initiatives such as the Ecologically or Biologically Significant Marine Areas (EBSA) process undertaken by the CBD,[76] the various World Ocean Assessments undertaken by the UN[77] and the Global Ocean Biodiversity Initiative.[78] This central repository of scientific information could complement and supplement the capacities, functions and knowledge bases of RFMOs particularly in relation to biological diversity impacts, climate change and ecosystem-based management. A broader picture of marine biological diversity in ABNJ could provide a more informed basis for strategic environmental planning and biological diversity conservation action plans for RFMOs, RSAs and other

75 Ibid 7.

76 Convention on Biological Diversity, 'Background on the EBSA Process' <www.cbd.int/ebsa/about> accessed 24 May 2018. See Youna Lyons, Chapter 4 'Identifying Sensitive Marine Areas in the High Seas: A Review of the Scientific Criteria adopted under International Law'.

77 United Nations World Ocean Assessment <www.worldoceanassessment.org> accessed 24 May 2018.

78 Global Ocean Biodiversity Initiative, 'About GOBI' <http://gobi.org> accessed 24 May 2018.

global organisations, such as the ISA, across the whole of ABNJ. For example, access to combined information on shipping, fisheries and climate change impacts on a particular seamount in ABNJ could enable a suite of complementary conservation measures to be developed by the relevant sectoral and regional organisations in that area to protect spawning grounds and other important habitats for commercial fish species. A more holistic picture of marine biological diversity in ABNJ could also assist in developing ways to decrease the likelihood of bycatch through improved information on species habitat preferences in space and over time.

Current fisheries monitoring practices could be supplemented with broader ecosystem monitoring to comprehend the many population, community and ecosystem-level impacts from fisheries or other activities that account for cumulative effects, including from climate change. The system for sharing of scientific information to be established under the ILBI could provide support for data gathering, monitoring and assessment from a broader group of organisations that can feed into larger processes to understand the cumulative impacts of fisheries and other activities on marine biological diversity in ABNJ.

Area-based management measures introduced under the ILBI may enhance the recovery of fish stocks, ecosystem resilience and improve monitoring of changes. They could also provide a means to develop no-take reference zones to enhance ecosystem stability.

A globally and regionally coordinated approach to EIA processes under the ILBI could also provide a broader context for fisheries assessments. An important role for the ILBI will be to enhance cooperation and collaboration to make fisheries data and other relevant information available to relevant global and regional organisations, including the results of EIAs and strategic environmental assessments. A central repository of information on marine biological diversity in ABNJ could also highlight deficiencies in biological diversity conservation across ABNJ and initiate measures to mitigate adverse impacts on species, habitats and ecosystems not protected by existing regional and sectoral organisations. Some discrete high seas fish stocks not covered by existing RFMOs and sedentary species in ABNJ would be clear beneficiaries of all these measures.

Notwithstanding the positive potential the future ILBI represents for influencing improved conservation and sustainable use of marine living resources in ABNJ, the implementation of this objective still rests predominantly with the member states of RFMOs. The decision to take conservation and management measures that lead to sustainable use of high seas fish stocks and take into account associated and dependent ecosystems will remain with these sectoral bodies. Likewise the enforcement of these measures will depend on the resources, capacity and political will of RFMO member states and cooperating non-member states. Geographic and species gaps in RFMO coverage will only be remediated through political decisions on the part of relevant states to form the necessary regional organisations and cooperate with each other to achieve improved conservation and sustainable use of fish stocks. With widespread participation and adequate funding, the future ILBI does have real potential to disseminate more broadly the scientific information necessary to make informed decisions on conservation and sustainable use of marine living resources in ABNJ and to forge links between relevant global sectoral and regional organisations in achieving that objective.

## 5 Conclusion

The interdependence of the open-ocean and deep-sea environments underscores the need for legal and institutional arrangements that allow for integrated protection of the marine environment beyond national jurisdiction and the establishment of connections between global and regional bodies with regulatory competence in these areas. The rise in human activities beyond the offshore zones of coastal states poses actual and potential threats to the physical characteristics and biological diversity of the open-ocean and deep-sea environments. Arbitrary human intrusions into this largely unexplored marine domain have the potential to harm the intricate links between complex marine ecosystems and to erode components of marine biological diversity. Protection of this vast area from the range of established and emerging threats posed by intensifying human activity requires concentrated global, regional and sectoral investment in coordinating and extending the applicable environmental protection regimes.

This chapter has identified a range of governance gaps and deficiencies in global and regional arrangements regulating the conservation and sustainable use of marine living resources in ABNJ. The development of an international legally binding instrument for the conservation and sustainable use of marine biological diversity in ABNJ represents a historic opportunity to lay the foundation for a more integrated and cross-sectoral system of oceans governance in ABNJ and to remediate some of the gaps and deficiencies in global and regional governance of marine living resources in ABNJ. Given the growing threats and pressures on the marine environment of ABNJ and its biological diversity, the realisation of this objective is critical to ensuring the long-term survival of marine living resources in ABNJ.

CHAPTER 7

# High Seas Governance Gaps: International Accountability for Nuclear Pollution

*Günther Handl*

## Abstract

This chapter explores the existing gap in states' (and other actors') international accountability for nuclear pollution of the high seas. After assessing various pollution scenarios, the chapter reviews relevant international legal instruments, including in particular the nuclear civil liability conventions as well as UNCLOS, none of which provide for adequate or effective redress, especially as regards damage to the marine environment. It proceeds to examine the difficulties—mostly related to standing—of invoking the rules of state responsibility for the protection of the high seas against nuclear pollution as well as prospects for holding states accountable absent internationally wrongful conduct. It concludes with a call for the establishment of an international mechanism/agency with unchallengeable authority to vindicate the collective interest in the preservation and protection of the high seas.

## 1 Introduction

When it comes to nuclear pollution of the high seas,[1] indeed most other forms of marine pollution in areas beyond national jurisdiction, the 'governance gap', such as it is,[2] does not connote the absence of substantive norms for the protection or the preservation of the marine environment. After all, today,

1 High seas, as defined negatively in UNCLOS Article 86, ie all those waters other than exclusive economic zone, territorial sea, internal waters or archipelagic waters. United Nations Convention on the Law of the Sea (adopted 10 December 1982, entered into force 16 November 1994) 1833 UNTS 3 (UNCLOS) art 86.

2 Robert Beckman has framed the issue as exemplified by the absence of any 'liability or international responsibility on the part of the flag State for pollution damage to the high seas or marine biodiversity of the high seas, or any obligation to take remedial measures to combat pollution if an incident takes place, unless there is a risk of the pollution spreading to the EEZ'. Personal communication from Robert Beckman to author.

 | DOI:10.1163/9789004373303_008

as international tribunals have repeatedly confirmed,[3] states' obligations to prevent and minimise pollution of the high seas are a cornerstone of ocean governance: they are reflected in UNCLOS,[4] various regional seas conventions,[5] other relevant international legal instruments,[6] including key nuclear conventions[7] and, of course, customary international law itself.[8] Rather, it is a fragmented and still largely untested system of responsibility and liability as applied to the high seas as an international commons[9] that poses a problem.

3 See, in particular, *Responsibilities and Obligations of States Sponsoring Persons and Entities with respect to Activities in the Area* (Advisory Opinion of 1 February 2011) ITLOS Reports 2011, 10 (*Responsibilities and Obligations of States*); *Chagos Marine Protected Area Arbitration* (*Mauritius v United Kingdom*) Award (18 March 2015) PCA 2011-03, paras 320, 538; *The South China Sea Arbitration* (*The Republic of Philippines v The People's Republic of China*) Award (12 July 2016) PCA 2013–19, paras 940–48.

4 Relevant UNCLOS (n 1) provisions include Article 192 (general obligation to protect and preserve the marine environment); Article 194, para 3(a) (obligation to minimise 'the release of toxic, or noxious substances … from or through the atmosphere or by dumping'); Article 195 (duty not to transfer damage or hazards or transform one type of pollution …); Article 198 (obligation to notify 'imminent or actual'); Article 204 (obligation to monitor risks or effects pollution); Article 206 (obligation of EIA); Articles197 and 199 (obligation to cooperate generally, and on contingency planning); Articles 207–09 (obligations related to pollution from land-based sources; seabed activities on national continental shelf & in the International Seabed Area); Article 210 (obligation related to pollution by dumping); Article 212 (obligation regarding pollution from or through the atmosphere); Article 94, paras 4(c) and 5; Article 211, para 2; and Article 217 (specific flag state obligations).

5 See, e.g., the Convention for the Protection of the Marine Environment of the North East Atlantic, 1992 (adopted 22 September 1992, entered into force 25 March 1998) 2354 UNTS 67. As to nuclear pollution in the Convention area, see OSPAR Decision 98/2 on Dumping of Radioactive Waste (entered into force 9 February 1999); Radioactive Substances Strategy in the OSPAR Commission, 'The North-East Atlantic Environment Strategy. Strategy of the OSPAR Commission for the Protection of the Marine Environment of the North-East Atlantic 2010–2020' <www.cbd.int/doc/nbsap/rbsap/ospar-rbsap.pdf> accessed 23 May 2018.

6 See, e.g., Article 2 in London Protocol: 'Contracting Parties shall individually and collectively protect and preserve the marine environment from all sources of pollution and take effective measures, according to their scientific, technical and economic capabilities, to prevent, reduce and where practicable eliminate pollution caused by dumping or incineration at sea of wastes or other matter. Where appropriate, they shall harmonize their policies in this regard'. 1996 Protocol to the Convention on the Prevention of Marine Pollution by Dumping of Wastes and Other Matter, 1972 (adopted 7 November 1996, entered into force 24 March 2006) 36 ILM 1 (1997) (London Protocol) art 2.

7 See, e.g., Section 3.1 in this chapter.

8 Consider, e.g., the assertion by Australia that 'pollution of the high seas by radio-active fallout, constitute[s] infringement of the freedom of the high seas'. See *Nuclear Tests Cases* (*Australia v France*) Interim Protection (Order of 22 June 1973) ICJ Reports 1973, 99, 103, para 22 (iii).

9 This is, of course, a long-standing, yet still unresolved concern. Note in this context the mandate laid down in UNCLOS (n 1) Article 235, para 3, *infra* text at n 132. For similar exhortations to 'develop procedures regarding liability', see, e.g., London Protocol (n 6) art 15.

Thus, complex 'enabling conditions' for the effective invocation of pollution prevention or minimisation norms, on the one hand, and the unsettled state of international law in respect of transboundary harm due to lawful but hazardous activities,[10] on the other, combine to undermine polluting states' (or other actors' or polluters') accountability. Of course, 'accountability'[11] is the hallmark of any legal system,[12] including international law.[13] Indeed, it is a defining characteristic of 'good governance' itself. A complete or even partial lack thereof in respect of marine pollution, therefore, represents a potentially serious shortcoming of the high seas legal regime.

This chapter explores the dimensions of this governance gap, first, by assessing its actual, practical implications. It then examines—in limited fashion[14]—whether, and if so, in what circumstances and for what type of remedial action, international law today ensures accountability for nuclear pollution of the high seas. However, before this chapter sets out to do so, a couple of preliminary observations seem to be in order. First, 'international accountability' is being used here to denote answerability pursuant to general international law and applicable treaties. It thus refers to the answerability not only of states acting in a sovereign capacity, such as that of flag states, but also of any relevant actors under the so-called civil liability conventions.[15] Second, a state's accountability for nuclear pollution of the high seas can be the result of activities involving

The persistent lack of action in this respect has significant implications also for nuclear pollution of the oceans. See, e.g., Jon M Van Dyke, 'Ocean Transport of Radioactive Fuel and Waste' in David D Caron and Harry N Scheiber (eds), *The Oceans in the Nuclear Age* (Brill Nijhoff 2010) 161–62.

10 See Sections 3.2 and 4.

11 That is, 'the state of being accountable, liable or answerable' or responsible. See Merriam-Webster, *Webster's New Universal Unabridged Dictionary* (2nd edn, Barnes & Noble 2003).

12 See, e.g., P Reuter, *Le développement de l'ordre juridique international* (Ecrits de droit international, Economica 1995) 379: 'Dans tous les systèmes juridiques, la responsabilité apparaît au centre même des institutions'.

13 To put it in a nutshell: 'no responsibility, no (international law)'. Alain Pellet, 'The Definition of Responsibility in International Law' in James Crawford, Alain Pellet and Simon Olleson (eds), *The Law of International Responsibility* (Oxford University Press 2010) 4.

14 This chapter only documents basic impediments to accountability. Thus, it will not detail specific shortcomings of individual, potentially applicable legal instruments, such as the nuclear civil liability conventions. For such an analysis, see, e.g., Duncan EJ Currie, 'The Problems and Gaps in the Nuclear Liability Conventions and an Analysis of How an Actual Claim Would Be Brought under the Current Existing Treaty Regime in the Event of a Nuclear Accident' (2008) 35 *Denver Journal of International Law and Policy* 85.

15 These include vessel-source oil pollution as well as nuclear liability conventions, ie, international agreements that establish a system of civil liability on the part of owners or operators of a polluting or damage-causing ship or nuclear facility. For details on the nuclear liability conventions, see Section 3.1.

the state's internationally wrongful conduct, thus entailing its 'responsibility' in a technical legal sense.[16] It can also, however, occur notwithstanding the state's compliance with its prevention or minimisation obligations. In such a situation, even in the absence of wrongfulness, the state might be answerable or, technically speaking, 'liable'.[17] However, whereas state responsibility per se and the parameters of its implementation are well established,[18] 'state liability' has an uncertain status in general international law, including the law of the sea.[19] In short, the terms 'state responsibility' and 'state liability' denote two manifestations of a state's answerability. They differ in terms of the basis as well as the scope of the legal obligation to answer for nuclear pollution with, as will be discussed, significant ramifications for the operationalisation of international accountability.

## 2 Nuclear Pollution of the High Seas Scenarios

### 2.1 *Introductory Observations*

Sources of anthropogenic radioactivity in the oceans include controlled releases of low-level radioactive liquid effluents, dumping of radioactive material at

16 See Article 1 of the Articles on Responsibility of States for Internationally Wrongful Acts. United Nations General Assembly (UNGA), 'Resolution Adopted by the General Assembly. 56/83. Responsibility of States for internationally wrongful acts Articles on the Responsibility of States for Internationally Wrongful Acts' UN Doc A/RES/56/83 (28 January 2002) Annex (ILC Articles on State Responsibility) art 1.

17 See UNGA, 'Resolution Adopted by the General Assembly on 4 December 2006. 61/36. Allocation of Loss in the Case of Transboundary Harm Arising out of Hazardous Activities' UN Doc A/RES/61/36 (18 December 2006) (Allocation Principles) para 2; 'Draft Principles on the Allocation of Loss in the Case of Transboundary Harm Arising out of Hazardous Activities' with Commentaries (Commentary) in International Law Commission, *Yearbook of the International Law Commission 2006* (Vol II, Part 2, UN Doc A/CN.4/SER.A/2006/Add.l (Part 2), United Nations 2013) 60, para 8 (referencing treaty-based 'State liability'). See further G Handl, 'Liability as an Obligation Established by a Primary Rule of International Law' (1985) 16 *Netherlands Yearbook of International Law* 49. 'State liability' would also characterise a state's accountability for pollution damage/injury in circumstances precluding the international wrongfulness of the polluting state's conduct. See ILC Articles on State Responsibility (n 16) art 27 (b). Finally, 'state liability', as the ITLOS Advisory Opinion points out, confusingly might also 'refer to the state's secondary obligation, namely, the consequences of a breach of the primary obligation'. *Responsibilities and Obligations of States* (n 3) para 66.

18 See ILC Articles on State Responsibility (n 16) Part Three: The Implementation of State Responsibility. Also referred to as the '*mis-en-oeuvre* of state responsibility'. See James Crawford, *State Responsibility: The General Part* (Cambridge University Press 2013) 541.

19 See Section 4.

sea, effects from weapons tests both in the atmosphere and underwater, accidents and losses at sea, and accidental releases from coastal nuclear installations.[20] As a category, land-based sources of radioactivity are of little concern in the context of this inquiry. For, irrespective of whether they involve intentional or accidental discharges, past or present, such as from the Sellafield[21] or Fukushima[22] nuclear installations, or discharges that might occur in the future, such as from the Runit Island dome in the Marshall Islands,[23] significant radiological effects, if any, are likely to be registered only in sea areas close to land, not on the high seas.[24] By contrast, nuclear weapons testing on the high seas or in locations—underwater or in the atmosphere—potentially affecting the high seas has been a source of significant radioactive contamination of the oceans in the past.[25] Today, however, such testing is unlikely to occur[26] and, in any event, would be illegal.[27] Any harmful consequences of the resulting

---

20 See, e.g., IAEA, 'Inventory of Radioactive Material Resulting from Historical Dumping, Accidents and Losses at Sea for the Purposes of the London Convention 1972 and London Protocol 1996' IAEA-TECDOC-1776 (2015) 5. Another, albeit minor, source of anthropogenic radioactivity in the oceans is the offshore oil and gas industry.

21 Of course, discharges affect directly the Irish Sea, which is semi-enclosed and does not include any high seas area. Thus emissions from Sellafield should be of no consequence in the present context.

22 Although ocean currents dispersed Fukushima-origin radionuclides very widely, radioactivity concentrations were generally very low, except temporarily in waters east of Japan. See Arctic Monitoring and Assessment Programme, *AMAP Assessment 2015: Radioactivity in the Arctic* (AMAP 2016) 66. See also Karl Mathiesen, 'Is It Safe to Dump Fukushima Waste into the Sea?' *The Guardian* (13 April 2016) <www.theguardian.com/environment/2016/apr/13/is-it-safe-to-dump-fukushima-waste-into-the-sea> accessed 24 May 2018; Junwne Wu, 'Impacts of Fukushima Daiichi Nuclear Power Plant Accident on the Western North Pacific and the China Seas: Evaluation Based on Field Observation of $^{137}$Cs' (2018) 127 *Marine Pollution Bulletin* 45.

23 See, e.g., Coleen Jose, Kim Wall and Jan Hendrik Hinzel, 'This Dome in the Pacific Houses Tons of Radioactive Waste—And It's Leaking' *The Guardian* (3 July 2015) <www.theguardian.com/world/2015/jul/03/runit-dome-pacific-radioactive-waste> accessed 24 May 2018.

24 Moreover, at least regionally, routine discharges of radioactive substances into, for example, the Northeast Atlantic are progressively being reduced. See OSPAR Commission, 'Fourth Periodic Evaluation of Progress Towards the Objective of the OSPAR Radioactive Substances Strategy' (2016) 3.

25 For details see, e.g., Remus Prăvălie, 'Nuclear Weapons Tests and Environmental Consequences: A Global Perspective' (2014) 43 *Ambio* 729.

26 But note North Korea's recent threat to conduct a hydrogen-bomb test over the Pacific. See Reuters, 'North Korea Ramps up Threat to Test Hydrogen Bomb over Pacific' *The Guardian* (26 October 2017) <www.theguardian.com/world/2017/oct/26/north-korea-threat-test-hydrogen-bomb-pacific> accessed 24 May 2018.

27 The Partial Test Ban Treaty of 1963 specifically outlaws any nuclear testing in the atmosphere, underwater or on the high seas. See Treaty Banning Nuclear Weapon Tests in the

nuclear pollution of the high seas would therefore trigger the testing state's international legal responsibility.[28]

The dumping of nuclear wastes or other radioactive material at sea has also been a matter of long-standing international concern.[29] While such practice was common throughout the first three decades of the nuclear age,[30] the option of legally disposing radioactive materials at sea has been essentially eliminated. The 1972 Convention on the Prevention of Marine Pollution by Dumping of Wastes and Other Matter (London Convention) initially prohibited only the dumping of high-level radioactive waste or other high-level radioactive matter.[31] A ban on the dumping of all other nuclear wastes and radioactive

---

Atmosphere, in Outer Space and under Water (adopted 5 August 1963, entered into force 10 October 1963) 480 UNTS 43, art 1(a). Article 1(b), moreover, prohibits testing 'in any other environment if such explosion causes radioactive debris to be present outside the territorial limits of the State under whose jurisdiction or control such explosion is conducted'. France conducted a last atmospheric test at Mururoa atoll in 1974 in defiance of Australia and New Zealand, arguing that such testing was then lawful while China did not stop atmospheric testing until 1980. See David E Sanger and William J Broad, 'Prospect of Atmospheric Nuclear Test by North Korea Raises Specter of Danger' *The New York Times* (22 September 2017) <www.nytimes.com/2017/09/22/world/asia/north-korea-atmospheric-nuclear-test-risks.html> accessed 24 May 2018. Even though the Comprehensive Test Ban Treaty is unlikely to enter into force any time soon, it would be difficult to deny that nuclear testing in the atmosphere or on the high seas would be a violation of general international law.

28 Thus a North Korean test on the high seas would, apart from being illegal already on account of a series of North Korea-specific UN Security Council resolutions, be a breach also of its customary international law obligations originating in the 1963 Treaty. Beginning with Resolution 1718 (2006) up to Resolution 2395 (2017), the UN Security Council has adopted a total of nine unanimous resolutions condemning North Korea's nuclear weapon (and missile) tests and programs as constituting a threat to international peace and security.

29 See ED Brown, 'International Law and Marine Pollution: Radioactive Waste and Other Hazardous Substances' (1971) 11 *National Resources Journal* 221; Leslie Roberts, 'Ocean Dumping of Radioactive Waste' (1982) 32 *BioScience* 773; Dominique P Calmet, 'Ocean Disposal of Radioactive Waste: Status Report' (IAEA Bulletin 4/1989) 4; G Handl, 'Nuclear Waste Disposal: International Cooperation and Regulation' in R Bernhardt (ed), *Encyclopedia of Public International Law* (North-Holland Publishing Company 1995); Hjalmar Thiel, 'Deep Sea Impacts', in Caron and Scheiber (n 9) 19. See also Lasse Ringius, *Radioactive Waste Disposal at Sea: Public Ideas, Transnational Policy Entrepreneurs, and Environmental Regimes* (MIT Press 2000).

30 See IAEA, 'Inventory of Radioactive Waste Disposals at Sea' IAEA-TECDOC-1105 (1999) 12–20.

31 See London Convention, art IV, para 1(a) and Annex I, 6: 'High-level radio-active wastes or other high-level radio-active matter, defined on public health, biological or other grounds, by the competent international body in this field, at present the International Atomic Energy Agency, as unsuitable for dumping at sea'. Convention on the Prevention of Marine Pollution by Dumping of Wastes and Other Matter (adopted 29 December 1972, entered into force 30 August 1975) 1046 UNTS 120 (London Convention).

matter followed in 1993[32] and entered into force in 1994. However, the London Convention as amended by the 1996 London Protocol[33] has continued to permit the dumping of some material, provided it contains only *de minimis* levels of radioactivity.[34] In October 2016, contracting parties of the London Convention and London Protocol decided to keep this regime in place.[35] Thus, for all practical purposes, nuclear dumping at sea today remains internationally wrongful. A dumping state would, therefore, be directly responsible for any harmful polluting effects or, if such operations involve private actors, the flag state or the actor's national state would be internationally accountable, provided the state has failed to exercise due diligence in preventing these illegal operations.[36]

Nuclear materials dumped at sea before the entry into force of the relevant bans[37]—so-called 'legacy wastes'—pose a different governance issue. While many such operations were carried out haphazardly, with incomplete or without any site documentation, and at varying depths and distances from the coast, experts seem to agree that today the radiological and environmental impact of such dumping, especially in areas of the high seas, is negligible.[38] However, given inherent uncertainties associated with some past dumping practices,[39]

---

32 Amendments to the Annexes to the Convention on the Prevention of Marine Pollution by Dumping of Wastes and Other Matter, 1972, Concerning Disposal at Sea of Radioactive Wastes and Other Radioactive Matter, IMO Res LC.51 (16) in 'Report of the Sixteenth Consultative Meeting' IMO Doc LC16/14 (15 December 1993) Annex 5.

33 The London Protocol (n 6) reverses the approach by outlawing the dumping of any material unless specifically authorised in Annex 1.

34 See London Protocol (n 6) Annex 1, para 3. See also IMO, 'Annex 9. 2015 Guidelines for the Application of the de minimis Concept under the London Convention and Protocol' IMO Doc LC 37/16 (22 October 2015).

35 See IMO, 'Report of the 38th Consultative Meeting of Contracting Parties (London Convention 1972) and 11th Meeting of Contracting Parties (London Protocol 1996)' IMO Doc LC38/16 (18 October 2016) 35, para 10.5. This decision followed a mandatory review of the ban within 25 years after the amendment's entry into force in 1994. However, no contracting party expressed any interest in allowing a resumption of dumping at sea. Nor did the IAEA, the agency responsible under the London Convention and London Protocol for defining non-high level wastes unsuitable for sea disposal, 'encourage in any way the return to dumping in the sea under the mechanism of the LC/LP'. See IMO, 'Considerations on IAEA Safety Standards and Dumping of Radioactive Waste' IMO Doc LC38/10 (10 June 2016) 5.

36 As to relevant state obligations, see UNCLOS (n 1) arts 210 and 216, para 1 (b) (flag state) and para 1(c) (any other state in respect of loading of wastes/material within its jurisdiction or control).

37 See generally, IAEA, 'Inventory of Radioactive Waste Disposals at Sea' (n 30).

38 Surveys of, for example, North-East Atlantic dump sites (involving, however, only low-level nuclear waste) showed 'negligible radiological impact'. ibid 7.

39 See, e.g., Meeting of the Radioactive Substances Committee, 'Report on Historic Deep Sea Disposal of Radioactive Waste in Layperson's Language' OSPAR Doc RSC 14/9/1-E (Agenda

caution in assessing pollution risks emanating from these sites seems to be appropriate. First, as nuclear pollution of the Arctic Ocean demonstrates, even near-shore nuclear waste dumping, such as that by the former Soviet Union,[40] can pose an environmental risk not only for neighboring countries' maritime zones and resources, but also possibly for high seas areas.[41] Second, historic high seas dumping proper can be radiologically problematic notwithstanding the usual depth of the sites and the remoteness of their locations involved. Thus, in assessing the risk of inadvertent intrusion by humans into such ocean dump sites, the International Atomic Energy Agency (IAEA) has expressed concern in light of relevant experience with dump sites in shallow waters.[42] It noted that given 'currently feasible marine activities in deeper waters … additional measures could be necessary at the sites selected for disposal, including engineered barriers, in order to avoid or substantially reduce all possible consequences of such potential intrusions'.[43] It should be evident, therefore, that the state that is internationally responsible for the original dumping at sea, would also be obliged to adopt such measures as necessary. In this vein, the Joint Convention on the Safety of Spent Fuel Management and on the Safety of Radioactive Waste Management[44] specifically requires parties to review 'the results of past practices in order to determine whether any intervention is needed for reasons of radiation protection bearing in mind that the reduction in detriment resulting from the reduction in dose should be sufficient to justify the harm and the costs, including the social costs, of the intervention'.[45]

In sum, land-based sources, weapons tests and waste dump sites either contribute insignificantly to nuclear pollution of the high seas or, to the extent they do or might do, are subject to clear normative expectations. This would

Item 9, RSC Meeting, London, 11–13 February 2014) 24. The report referred to dumped nuclear waste in the North-East Atlantic as a 'reality … of hundreds of thousands of waste drums disposed over a wide area and over many years, made up of different types of packages and containing a wide variety of waste in terms of both physical condition and radionuclide composition'.

40 See, e.g., Alexey Yablokov, 'Radioactive Waste Disposal in Seas Adjacent to the Territory of the Russian Federation' (2001) 43 *Marine Pollution Bulletin* 8.

41 See Arctic Monitoring and Assessment Programme (n 22) 77: 'Studies indicate that specific sources (such as radioactive waste dumped in the Barents and Kara seas) have the potential to increase contamination levels at a local and possibly regional level'.

42 IMO Doc LC38/10 (n 35) 5, para 26.

43 Ibid.

44 IAEA, 'Joint Convention on the Safety of Spent Fuel Management and on the Safety of Radioactive Waste Management' IAEA Doc INFCIRC/546 (24 December 1997).

45 Ibid art 12, para (ii).

bring the polluting activities into the ambit of state responsibility rules, thereby render them relatively less problematic from a governance perspective.[46]

### 2.2 *Accidental Nuclear Pollution of the High Seas*

Sea-borne activities of potential relevance from a high seas nuclear impact perspective[47] include the operation of 'nuclear ships', i.e., vessels using nuclear power as a means of propulsion, of nuclear armed ships, of vessels carrying radioactive materials as cargo and of vessels serving as floating nuclear power installations. Over the years quite a number of nuclear-powered military ships and sea-borne nuclear weapons have ended up by accident on the bottom of the sea,[48] within either coastal states' territorial sea or exclusive economic zone. However, these losses, with a few exceptions,[49] have tended not to pose a problem for adjoining high seas areas either because of their location and their source-terms,[50] or because vessels and/ or weapons were recovered from the seabed.[51] As regards the sinking on the high seas proper of nuclear-powered or armed submarines, such as the USS *Thresher* and USS *Scorpion*,[52] as well as the *Soviet K-129* and *K-278* submarines,[53] it is leakage from the vessel's reactors or the weapons carried onboard that pose a potential threat to the marine

46 See Section 3.3.1.

47 To be sure, incidents, such as the 1966 Palomares accident or the crash of nuclear-powered satellites into the sea, prove that air-borne or space activities can result in a nuclear contamination of the oceans. However, rather than raising an ocean governance issue, these incidents pose issues for international aviation and outer space law.

48 For a partial listing, see IAEA, 'Inventory of Radioactive Material' (n 20) 18–20.

49 These might include Soviet era submarines *K-27* and *K-159*. 'The K-27 is arguably more hazardous than its other radioactive cousins in the region. A scientific expedition to the vessel in 2012 concluded that its liquid metal cooled reactor was vulnerable to an uncontrolled chain reaction and a significant radioactive release'. Charles Digges, 'Russia Hints at New Plans to Raise Sunken Nuclear Subs by 2022' *Bellona* (20 June 2017) <http://bellona.org/news/nuclear-issues/2017-06-russia-hints-at-new-plans-to-raise-sunken-nuclear-subs-by-2022> accessed 24 May 2018.

50 I.e., the types and amounts of radioactive material that could be released to the environment following an accident.

51 As was the case with the Kursk. See Sophia Kishkovsky, 'In 15 Hours, Submarine Kursk Is Raised From Sea Floor' *The New York Times* (9 October 2001) <www.nytimes.com/2001/10/09/world/in-15-hours-submarine-kursk-is-raised-from-sea-floor.html> accessed 24 May 2018.

52 See IAEA, 'Inventory of Radioactive Material' (n 20) 17–19.

53 See, e.g. ibid; Praveen Duddu, 'Peril in the Depths—The World's Worst Submarine Disasters' *Naval Technology* (6 March 2014) <www.naval-technology.com/features/featureperil-in-the-depths-the-worlds-worst-submarine-disasters-4191027> accessed 24 May 2018.

environment. Thus far, none of these accidents appear to have resulted in significant releases of radioactivity, although in the case of the *K-278* (NS *Komsomolets*), the Russian government took special steps to prevent any washout of plutonium from the submarine's missile warheads.[54]

The number of nuclear-powered civilian vessels globally is quite limited. All of them are at present Russian-flagged,[55] and most of these are icebreakers. Although other countries may plan to adopt nuclear propulsion for civilian vessels,[56] prospects for a broad launch of a civilian nuclear-powered shipping industry are at best uncertain.[57] Presently operational vessels and new generation Russian icebreakers[58] are being or will be used along the Northern Sea Route, thus in coastal waters rather than on the high seas. Any accidental releases of radioactivity associated with their operation are thus likely to be of minor interest in the present context.[59]

The same cannot necessarily be said, however, of accidents involving the sea transport of nuclear materials, including waste. For a long time, the risks ascribed to such accidents have been a matter of intense concern for coastal

54 See IAEA, 'Inventory of Radioactive Material' (n 20) 17; Central Intelligence Agency, 'The Komsomolets Disaster: Burial at Sea' <www.cia.gov/library/center-for-the-study-of-intelligence/csi-publications/csi-studies/studies/95unclass/Montgomery.html> accessed 24 May 2018. 'A Reuters wire release from Moscow on 12 July 1994 stated, 'Russia said yesterday it had sealed a sunken nuclear submarine off Norway to prevent radioactive leaks. The Komsomolets ... is now embedded in mud in international waters'.

55 OECD Nuclear Energy Agency, 'Small Modular Reactors: Nuclear Energy Market Potential for Near-Term Deployment' NEA No 7213 (2016) 56.

56 For example, Iran is reported to plan for a fleet of nuclear powered ships, fuelled by Iranian-made reactors. Mark Landler, David E Sanger and Gardiner Harris, 'Europeans Move to Save Iran Deal' *The New York Times* (27 February 2018) A1.

57 See, e.g., Tammy Thueringer and Justin Parkinson, 'The Ship That Totally Failed to Change the World' *BBC News* (25 July 2014) <www.bbc.com/news/magazine-28439159> accessed 24 May 2018.

58 Presently, Russia has three new nuclear-powered icebreakers under construction: *Arktika*, the first of the giant LK-60Ya class will enter service in 2019, while the two next, *Sibir* and *Ural* are to enter service in 2020 and 2021 respectively. See Thomas Nilsen, 'Russia Mulls Additional Two Nuclear Icebreakers' *The Independent Barents Observer* (6 December 2017) <https://thebarentsobserver.com/en/industry-and-energy/2017/12/russia-mulls-additional-two-nuclear-icebreakers> accessed 24 May 2018.

59 For a discussion see Anna Nalbandyan et al, *Nuclear Icebreaker Traffic and Transport of Radioactive Materials along the Nordic Coastline: Response Systems and Cooperation to Handle Accidents* (*NORCOP-COAST*)*: Final Report* (NKS-362, Nordic Nuclear Safety Research, March 2016). Similarly, the radiological high seas impact potential of nuclear-powered light houses along Russia's Arctic coastline being 'washed out to sea' as apparently has been the case in the past, is probably negligible. For details, see, e.g., Thomas Nilsen, 'Two Nuclear Generators Missing in Arctic' *Barents Observer* (26 August 2013) <http://barentsobserver.com/en/arctic/2013/08/two-nuclear-generators-missing-arctic-26-08> accessed 24 May 2018.

states and given rise to frequently emotional debates about safety.[60] However, the maritime transport of materials of principal concern here—spent fuel assemblies, high-level waste, plutonium and mixed oxide fuel—is subject to strict requirements[61] regarding their carriage by special-built vessels and in secure packages.[62] In consequence, various safety analyses of shipments that comply with all applicable international safety standards[63] consistently show the risks associated with collisions, fire or other accident scenarios to be negligible.[64] Given that the number of these radioactive material shipments by sea is, to begin with, low[65] and that the probability of any ship collision on the high seas is inherently small, the likelihood of an accident resulting in significant human exposure or environmental effects is extremely low. Indeed, it has been suggested that such an event approaches a 'probability of $10^{-13}$' rendering it simply 'not credible'.[66]

60 See, e.g., Ron Smith, 'The Maritime Transportation of Nuclear Materials: A View from New Zealand' (2002) 54 *Political Science* 5; Jon M Van Dyke, 'The Legal Regime Governing Sea Transport of Ultrahazardous Radioactive Materials' (2002) 33 *Ocean Development & International Law* 77; Günther Handl, 'Preventing Transboundary Nuclear Pollution: A Post-Fukushima Legal Perspective' in S Jayakumar et al (eds) *Transboundary Pollution: Evolving Issues of International Law and Policy* (Edward Elgar 2015) 219–24.

61 Such as compliance with the International Code for the Safe Carriage of Packaged Irradiated Nuclear Fuel, Plutonium and High-Level Radioactive Wastes on Board Ships (adopted by resolution MSC. 88 (71) 27 May 1999) (INF Code).

62 I.e., so-called Type B packages. 'Type B packages used for HLW, used fuel, and MOX fuel are robust and very secure casks. They range from drum-size to truck-size and maintain shielding from gamma and neutron radiation, even under extreme accident conditions.... The internal structure of transport casks (using multi-purpose canisters or not) is designed to maintain separation of fuel assemblies even in extreme accidents, and the external structure is designed to maintain safe containment in extreme accidents'. World Nuclear Association, 'Transport of Radioactive Materials' (July 2017) <www.world-nuclear.org/information-library/nuclear-fuel-cycle/transport-of-nuclear-materials/transport-of-radioactive-materials.aspx> accessed 24 May 2018.

63 See generally, IAEA, *Regulations for the Safe Transport of Radioactive Material, Specific Safety Requirements* (No SSR-6, IAEA 2012).

64 IAEA, 'Severity, Probability and Risk of Accidents during Maritime Transport of Radioactive Material: Final Report of a Co-Ordinated Research Project 1995–1999' IAEA-TECDOC-1231 (2001) 5. See also statement by the IAEA representative at OSPAR Convention for the Protection of the Marine Environment of the North-East Atlantic, Meeting of the Radioactive Substances Committee, 'RSC 2017 Summary Record' (Meeting of the Radioactive Substances Committee, Geneva, 14–16 February 2017) OSPAR Doc RSC 17/17/01, para 13.3.

65 See, e.g., IAEA, 'Communication Dated 15 April 2014 Received from the Resident Representative of Norway to the Agency Regarding the Working Group on Best Practices for Voluntary and Confidential Government-to-Government Communications on the Transport of MOX Fuel, High Level Radioactive Waste and, as Appropriate, Irradiated Nuclear Fuel by Sea' IAEA Doc INFCIRC/863 (5 May 2014) Annex 1.

66 IAEA, 'Safe and Secure Transport of Radioactive Materials: The Next Fifty Years, Summary of an International Safety and Security Conference' (Vienna, 16–21 October 2011) (2016) 53.

Finally, floating, i.e. barge-mounted, nuclear power plants (FNPPS),[67] pose a risk of accidental pollution of the high seas.[68] Such floating installations also raise a number of complex regulatory, hence accountability issues.[69] Two types of FNPP accident scenarios are readily conceivable. First, an incident on the high seas during transportation—on the way to deployment and start-up—with reactors fully fuelled but fuel assemblies not yet irradiated. An incident of this kind, however, is unlikely to result in any significant environmental effects on the high seas given that fresh (uranium) fuel assemblies[70] will be only mildly radioactive. That said, the risk potential associated with a loss of or damage to fuel assemblies might be viewed differently by a coastal state through whose maritime zones such a transportable power plant would be moving.[71] A second and potentially more threatening scenario is an incident involving a stationary, operational FNPP in a location in which the accidental release of radioactivity could affect also adjoining high seas areas.[72] Although overall the probability of a significant environmental release of radioactivity from, say a KLT-40 type reactor has been deemed to be low, such an occurrence cannot be ruled out, especially during its periodic refueling.[73] Nevertheless, a severe accident at such a FNPP anchored

---

67 For details of Russia's efforts regarding floating nuclear power plants, see OECD Nuclear Energy Agency (n 55) 56–63; Arctic Monitoring and Assessment Programme (n 22) 67–68.

68 For an acknowledgment of the special regulatory, safety and security challenges posed by transportable nuclear power plants, see 'Measures to Strengthen International Cooperation in Nuclear, Radiation, Transport and Waste Safety' IAEA Doc GC(61)/RES/8 in IAEA, 'Resolutions and Other Decisions of the General Conference. Sixty-first Regular Session 18–22 September 2017' IAEA Doc GC(61)/RES/DEC(2017) (February 2018) 22.

69 See generally, IAEA, 'Legal and Institutional Issues of Transportable Nuclear Power Plants: A Preliminary Study' IAEA Doc NG-T-3.5 (2013) 34–35. See also Catherine Redgwell and Efthymios Papastavridis, 'International Regulatory Challenges of New Developments in Offshore Nuclear Energy Technologies' in Donald Zillman et al (eds), *Innovation in Energy Law and Technology: Dynamic Solutions for Energy* (Oxford University Press 2018) 103.

70 The Russian reactors will be using slightly enriched uranium fuel. See OECD Nuclear Energy Agency, 'Current Status, Technical Feasibility and Economics of Small Nuclear Reactors' (June 2011) 34.

71 In this vein, in 2017, Norway objected to Russia's original plans to tow a fully fuelled FNPP, the *Akademik Lomonosov*, around Norway towards Pevek, Eastern Siberia, its final destination. Similarly, some Baltic states expressed concern over the FNPP's proposed voyage from St. Petersburg. See Charles Digges, 'Rosatom Agrees Not to Fuel Floating Nuclear Plant in Russia's Second Biggest City' *Bellona* (7 August 2017); Bill Chappell, 'Russia Launches Floating Nuclear Power Plant; It's Headed to the Arctic' *NPR* (30 April 2018) <www.npr.org/sections/thetwo-way/2018/04/30/607088530/russia-launches-floating-nuclear-power-plant-its-headed-to-the-arctic> accessed 24 May 2018. The *Akademik Lomonosov* houses two 35 MW KLT-40S nuclear reactors, similar to those used in Russia's nuclear-powered icebreakers.

72 See, e.g., Mark Dowdall and William John Fisher Standring, 'Floating Nuclear Power Plants and Associated Technologies in the Northern Areas' (15 Strålevern Rapport, Norwegian Radiation Protection Authority 2008) 47–48.

73 Ibid 48.

in internal waters or in the territorial sea is unlikely to have an appreciable environmental impact on the high seas as its effects would be 'mainly local'.[74]

The environmental risk profile would be different, however, if the FNPP were stationed on the high seas itself or in waters adjacent thereto. Such a situation might arise if China deployed small modular reactors[75] either on or in close proximity to structures in the sea that the country has built up into artificial islands in the South China Sea.[76] Several locations potentially designated to receive such power sources are, according to the 2016 judgment of the tribunal in the South China Sea arbitration,[77] low-tide elevations (LTEs).[78] Normally LTEs would be incapable of generating even a territorial sea. Any accident involving a nuclear power source in such a location would, by definition, be on the high seas, hence of interest in the present context. However, because of special geographic circumstances,[79] some of these potential FNPP sites qualify as 'rocks', thus are entitled to a 12 nautical mile (NM) territorial sea, or are located in the EEZ of another state.[80] In any event, the fact remains that in these locations the nuclear power reactor might be operating at only a few miles from the high seas, therefore a severe accident could pose an appreciable risk of environmental effects in international waters.

## 3 Accountability for Accidental Nuclear Pollution of the High Seas under International Law

### 3.1 *Limited Remedies under Nuclear Legal Instruments/Liability Conventions*

Among the various international treaty instruments that bear on the issue here under review, it is, of course, the nuclear civil liability conventions that are

74 Arctic Monitoring and Assessment Programme (n 22) 69.

75 Reportedly China plans to build up to 20 FNPPs with more powerful 200 MW(e) reactors. See The Economist, 'Atomic Power Stations out at Sea May Be Better Than Inland Ones' *The Economist* (10 August 2017).

76 See e.g., Shunsuke Tabeta, 'China Eyes Offshore Reactors as Next Step in Nuclear Goals' *Nikkei Asian Review* (5 January 2018); David Stanway, 'China Close to Completing First Offshore Nuclear Reactor' *Reuters* (30 October 2017).

77 See *The South China Sea Arbitration* (n 3).

78 They include Hughes Reef, Gaven Reef (South), Subi Reef, Mischief Reef and the Second Thomas Shoal. See ibid paras 358, 366, 373, 378 and 381 respectively.

79 In respect of several of these LTEs, the Tribunal concluded that they were located within 12 NM of a high-tide elevation, hence entitled to a territorial sea. This is true of Subi Reef, Hughes Reef and Gaven Reef (South). See ibid para 384.

80 As is the case with Mischief Reef and the Second Thomas Shoal, both of which are, according to the Tribunal, located in the Philippines' EEZ.

of special interest. They include the Paris Convention,[81] the Brussels Supplementary Convention (BSC)[82] and their respective 2004 Protocols (not yet in force);[83] the Vienna Conventions of 1963[84] and 1997;[85] and the 1997 Convention on Supplementary Compensation for Nuclear Damage (CSC).[86] From an accountability perspective, the conventions offer several attractive core features: strict, albeit limited liability for nuclear damage; channeling of liability to the operator; mandatory financial coverage of liability; and the vesting of exclusive jurisdiction for the judicial handling of claims. However, some of their other features offset these advantages. In that they primarily focus on damage in the territory and maritime zones of contracting parties, rather than on the high seas, they may ensure accountability for nuclear contamination of the high seas only partially, if at all. Moreover, they characteristically address only compensation of damage occasioned by a nuclear incident,[87] rather than other forms of redress, let alone the full range of reparation associated with pollution caused by internationally wrongful conduct.[88]

---

81 Convention on Third Party Liability in the Field of Nuclear Energy of 29 July 1960, as amended by the Additional Protocol of 28 January 1964 and by the Protocol of 16 November 1982 (adopted 19 July 1960, entered into force 1 April 1968) 956 UNTS 251 (Paris Convention).

82 Convention of 31 January 1963 Supplementary to the Paris Convention of 29 July 1960, as Amended by the Additional Protocol of 28 January 1964 and by the Protocol of 16 November 1982 (adopted 31 January 1963, entered into force 4 December 1974) 1041 UNTS 358 (BSC).

83 2004 Protocol to Amend the Convention on Third Party Liability in the Field of Nuclear Energy of 29 July 1960, as Amended by the Additional Protocol of 28 January 1964 and by the Protocol of 16 November 1982 (adopted 12 February 2004, not in force) (2004 Protocol to the Paris Convention); 2004 Protocol to Amend the Convention of 31 January 1963 Supplementary to the Paris Convention of 29 July 1960 on Third Party Liability in the Field of Nuclear Energy, as Amended by the Additional Protocol of 28 January 1964 and by the Protocol of 16 November 1982 (adopted 12 February 2004, not in force) (2004 BSC).

84 Vienna Convention on Civil Liability for Nuclear Damage (adopted 21 May 1963, entered into force 12 November 1977) 1063 UNTS 265, IAEA Doc INFCIRC/500 (20 March 1996) (1963 Vienna Convention).

85 Protocol to Amend the 1963 Vienna Convention on Civil Liability for Nuclear Damage (adopted 12 September 1997, entered into force 4 October 2003) 2241 UNTS 302, IAEA Doc INFCIRC/566 (22 July 1998) (1997 Vienna Convention).

86 Convention on Supplementary Compensation for Nuclear Damage (adopted 12 September 1997, entered into force 15 April 2015) 36 ILM 1473 (1997), IAEA Doc INFCIRC/567 (22 July 1998) (CSC).

87 Article I(l) of the 1963 Vienna Convention (n 84), for example, defines 'nuclear incident' as 'any occurrence or series of occurrences having the same origin which cause nuclear damage'.

88 Namely, restitution and satisfaction. See ILC Articles on State Responsibility (n 16) art 34.

The first task, therefore, will be to verify the extent to which the liability instruments are of relevance at all in a high seas governance context. Article 2 of the Paris Convention[89] is silent on the specific issue of whether and in what circumstances the Convention would apply to the high seas. However, early on, the OECD/NEA Steering Committee commenting on the Convention's territorial scope of application emphasised that states 'in adopting measures to apply the Convention, should take [into account that] the Paris Convention [was] … applicable to nuclear incidents on the high seas or to damage suffered on the high seas'.[90] On the other hand, the BSC does apply to the high seas, albeit in limited fashion, and on condition that nuclear damage is being suffered on board a ship that is registered in the territory of a contracting party, or by a national of a contracting party.[91] The 2004 Protocol to the Paris Convention unequivocally extends coverage to the high seas, yet applies only to nuclear damage suffered on board a ship registered by a contracting party; a non-contracting state that is also a party to the Vienna Convention and the 1988 Joint Protocol;[92] a non-contracting state that has no nuclear installations; or, finally, any other non-contracting state whose national nuclear liability legislation provides 'equivalent reciprocal benefits'.[93] Moreover, the Convention does not prevent a contracting installation state from expanding the scope of application of the convention under its own legislation.[94] The 2004 BSC also adjusts its geographical scope to any 'maritime areas beyond the territorial sea', thus covering the high seas, provided nuclear damage is suffered on board a ship that flies the flag of, or is registered in a contracting party.[95] It also would apply to nuclear damage suffered on or by an artificial island or installation/structure on the high seas under the jurisdiction of a contracting

89 'This Convention does not apply to nuclear incidents occurring in the territory of non-Contracting States or to damage suffered in such territory, unless otherwise provided by the legislation of the Contracting Party in whose territory the nuclear installation of the operator liable is situated …'. Paris Convention (n 81) art 2.

90 'Recommendation of the Steering Committee of 25.4.1968 [NE/M(68)1]' in OECD Nuclear Energy Agency, 'Paris Convention: Decisions, Recommendations, Interpretations' (OECD 1990) 9.

91 BSC (n 82) art 2, paras a(ii) 2 and 3 respectively. Likewise, damage to a ship is compensable only if registered in a Contracting Party.

92 Provided the Paris Convention state in whose territory the accident-struck installation is located is also a party to the Joint Protocol Relating to the Application of the Vienna Convention and the Paris Convention (adopted 21 September 1988, entered into force 27 April 1992) 1672 UNTS 302.

93 2004 Protocol to the Paris Convention (n 83) art 2(a).

94 See Ibid art 2(b).

95 BSC (n 82) art 2(a)(ii).

party, as well as the latter's nationals anywhere on the high seas, victims of a nuclear incident.[96]

The 1963 Vienna Convention does not specify the geographical scope of its application. However, it defines the 'installation state' as including a state under whose authority an installation is being operated outside 'the territory of any state'.[97] Thus, a FNPP operating at a high seas site would be covered by the Convention.[98] The 1997 Vienna Convention unambiguously declares that it applies to nuclear damage 'wherever suffered'.[99] This principle, so the IAEA commentary emphasises, 'is the opposite of the one embodied in the existing text of Article 2 of the Paris Convention'.[100] Finally, Article V of the CSC adopts a provision that is virtually identical to that of the 2004 BSC, thus conditionally covering nuclear damage suffered on the high seas.[101]

As an indication of international public policy supportive of coverage of the high seas, reference might be made here also to the 'nuclear ship' conventions, although none of them is in force or presently applicable. For example, the Brussels Convention on the Liability of Operators of Nuclear Ships, 1962, which is not in force, covers 'any ship equipped with a nuclear power plant'[102] and applies 'to nuclear damage caused by a nuclear incident occurring in any part of the world and involving the nuclear fuel of, or radioactive products or waste produced in, a nuclear ship flying the flag of a Contracting State'.[103] Other bilateral agreements covering liability for damage associated with the

96 Ibid paras 2(a)(ii)(1)–(2).

97 1963 Vienna Convention (n 84) art I(1)(d).

98 See Ibid art I, para 1(j), which excludes from among the 'nuclear installations' covered only a seaborne reactor that is being used as a means of propulsion of, or 'as a source of power' for, the vessel itself.

99 1997 Vienna Convention (n 85) art I A, para 1. Under paragraph 2, the installation state has a limited right to exclude the application of the convention, but not in respect of damage suffered on the high seas. See also Andrea Gioia, 'Maritime Zones and the New Provisions on Jurisdiction in the 1997 Vienna Protocol and in the 1997 Convention on Supplementary Compensation' (1999) 63 *Nuclear Law Bulletin* 25, 29.

100 IAEA, *The 1997 Vienna Convention on Civil Liability for Nuclear Damage and the 1997 Convention on Supplementary Compensation for Nuclear Damage: Explanatory Texts* (IAEA International Law Series No 3 (Revised) IAEA 2017) 31–32 (IAEA, *Explanatory Texts*) (footnotes omitted).

101 CSC (n 86) art V, para (b)(ii)–(iii).

102 Convention on the Liability of Operators of Nuclear Ships (adopted 25 May 1962, not in force) (1963) 57 *The American Journal of International Law* 268 (Nuclear Ship Convention) art I, para 1.

103 Ibid art XIII. The Nuclear Ship Convention has not entered into force and may never do so.

NS *Savannah*[104] and *Otto Hahn*,[105] two no longer operational nuclear ships, would have provided redress for nuclear damage sustained in foreign ports and territorial seas. However, they also would have applied to nuclear damage suffered by a port state's vessel on the high seas resulting from an incident during the nuclear ships' voyage to or from the foreign port concerned.[106] By contrast, the Convention relating to Civil Liability in the Field of Maritime Carriage of Nuclear Material, 1971[107] adds little to this inquiry. It merely reinforces the principle that the provisions of the Paris and Vienna Conventions govern liability in respect of damage caused by a nuclear incident occurring in the course of maritime carriage of nuclear material.[108]

In sum, given that the 2004 Protocols to the Paris Convention and BSC have yet to enter into force and that the nuclear ship conventions are of no relevance today, it is clear that the nuclear civil liability instruments cover only sparingly nuclear incidents on the high seas or related injurious effects experienced thereon. Another major shortcoming from an accountability perspective is the fact that these conventions primarily focus on traditional counts of damage compensable, and only secondarily, if at all, on damage to

104 See, e.g., the US-Netherlands Agreement on Public Liability Caused by N.S. Savannah (signed 6 February 1963, entered into force 22 May 1963) *United States Treaties and Other International Agreements* (Vol 14, Issue 1, US Government Printing Office 1964) TIAS 5357, 787.

105 Treaty Between the Republic of Liberia and The Federal Republic Of Germany on the Use of Liberian Waters and Ports by N.S. 'Otto Hahn' (signed 29 January 1971) <http://dipbt.bundestag.de/doc/btd/06/017/0601790.pdf> accessed 25 May 2018.

106 For example, the US-China Exchange of Notes Constituting an Agreement Relating to Liability during Private Operation of N.S. Savannah, Tai-pei, 24 March and 8 June 1967, 254 UNTS 1969, provides that '[T]he United States Government shall provide compensation by way of indemnity for any legal liability … for any damage deriving from a nuclear incident … in which the N.S. Savannah may be involved within Republic of China territorial waters, or outside of them on a voyage to or from Republic of China ports if damage is caused in the Republic of China or on ships of Republic of China registry'. By contrast, pursuant to its Article 21, the Federal Republic of Germany-Liberia Agreement (n 105) does not cover nuclear damage sustained outside Liberian territory and waters.

107 Convention Relating to Civil Liability in the Field of Maritime Carriage of Nuclear Material (adopted 17 December 1971, signed 15 July 1975) 11 ILM 277 (1972).

108 Ibid preamble and art 1. Thus the rules of the nuclear liability conventions take precedence over otherwise applicable maritime transport liability regulations: a person liable potentially liable under the latter for damage in a nuclear transport incident will be exonerated if the operator is liable for the same damage under the Paris or Vienna Conventions. The Convention does not, however, affect the liability of the operator of a 'nuclear ship' in respect of damage caused by a nuclear incident involving the nuclear fuel of or radioactive products or waste produced in such ship. ibid art 3.

the marine environment. Of course, the Paris and the 1963 Vienna Conventions predate the dawn of the international environmental era—usually considered to coincide with the 1972 Stockholm Conference on the Human Environment.

Thus, the earlier an instrument was adopted, the more likely it will cover 'damage' in a narrow sense only. A perfect illustration of this is the 1962 Nuclear Ship Convention, which defines compensable 'nuclear damage' as 'loss of life or personal injury and loss or damage to property'.[109] More importantly, both the 1960 Paris Convention and the 1963 Vienna Convention only provide the same limited coverage.[110] The BSC does not espouse a different approach to 'damage compensable' as it is governed in this respect by the provisions of the Paris Convention.[111] It was thus not until the revision of the Vienna Convention in 1997 that a multilateral nuclear liability instrument finally featured a broad definition of 'nuclear damage',which apart from traditional counts (and a new one of pure economic loss), included environment-related costs: the costs of measures of reinstatement of the impaired environment; loss of income deriving from an economic interest in the use of enjoyment of the environment, incurred due to a significant impairment of the environment; and the costs of preventive measures.[112] The CSC offers the same environment-related definition of nuclear damage.[113] The 2004 Protocol to the Paris Convention follows these examples and expands its definition of 'nuclear damage' in very similar fashion,[114] but, of course, it has yet to enter into force.[115]

Given this fact and the absence of environment-related provisions in the earlier liability instruments, today, environmental damage as such would be covered only following incidents involving installation states that are parties to the 1997 Vienna Convention or the CSC. The CSC provides compensation supplementary to any national legal system that either implements the Vienna or Paris Conventions[116] or meets the requirements laid down in the Annex to the CSC.[117] Although the CSC thus anticipates the coexistence of different

109 Nuclear Ship Convention (n 102) art I, para 7.

110 See Paris Convention (n 81) art 3, para (a)(i) and (ii); 1963 Vienna Convention (n 84) art I, para (k)(i). However, the latter includes also 'any other loss or damage' provided for by the law of the competent court. See 1963 Vienna Convention (n 84) para (k)(ii).

111 BSC (n 82) art 1.

112 1997 Vienna Convention (n 85) art I, paras (k)(iv)–(vi).

113 CSC (n 86) art I, para (f)(iv)–(vi).

114 2004 Protocol to the Paris Convention (n 83) art 1, paras (a) (4–6).

115 Art 19(b) of the 2004 Protocol to the Paris Convention (n 83) requires ratification by a minimum of five states to bring the Protocol into force. To date only two countries, Norway and Switzerland have ratified the Protocol.

116 CSC (n 86) art II, para 1(a)

117 Ibid para 1(b).

liability regimes, any state seeking to join the CSC regime, whether a party to the Paris Convention, the 1963 Vienna Convention, or not a party to either, would be required to adjust its domestic legal definition of 'nuclear damage' and embrace environmental effects as a new category of damage compensable.[118] Nevertheless, the resulting shift in international public sentiment towards wider internalisation of the costs of a nuclear incident is relatively modest: as of today, only 18 countries[119] formally support the compensability of environmental effects as an intrinsic aspect of nuclear damage.

The civil liability conventions' concern with compensating nuclear damage addresses an extremely important part, though, of course, not the complete spectrum of international accountability for nuclear pollution of the high seas. Thus, they do not explicitly establish other obligations to redress the consequences of a nuclear accident, such as in particular, to take remedial action, engage in clean-up activities, etc. However, the 1997 Vienna Convention and the CSC indirectly support a wider notion of the polluter's answerability at law, by creating incentives for remedial action. They define 'nuclear damage' as including the 'costs of preventive measures',[120] i.e., 'reasonable measures taken *by any person* after a nuclear incident ... to prevent or minimize damage'.[121] Similarly, the 1973 Protocol Relating to Intervention on the High Seas in Cases of Pollution by Substances Other Than Oil[122] authorises—but does not require—parties to take measures on the high seas to 'prevent, mitigate or eliminate grave and imminent danger to their coastline or related interests from pollution or threat of pollution' by radioactive substances such as $^{60}$Co, $^{137}$Cs, $^{226}$Ra, $^{239}$Pu and $^{235}$U.[123] Provided an intervening state's measures are proportionate to the actual or threatened damage, the costs of such intervention would have to be borne by the nuclear polluter,[124] thereby once again creating an indirect

118 See, e.g. IAEA, *Explanatory Texts* (n 100) 74: 'Some degree of harmonization in the definition of nuclear damage is also required'. See also Ben McRae, 'Convention on Supplementary Compensation for Nuclear Damage (CSC) and Harmonisation of Nuclear Liability Law within the European Union' (2015) 95 *Nuclear Law Bulletin* 74, 79.

119 The CSC has 10 states parties, and the 1997 Vienna Convention has 13 parties, with 5 states being parties to both instruments.

120 See CSC (n 86) art I, para (f)(vi); 1997 Vienna Convention (n 85) art I, para (1) (k)(vi).

121 CSC (n 86) art I, para (h) (emphasis added); 1997 Vienna Convention (n 85) art I, para (1)(n).

122 Protocol Relating to Intervention on the High Seas in Cases of Pollution by Substances other than Oil (adopted 2 November 1973, entered into force 30 March 1983) 1313 UNTS 4 (Intervention Protocol).

123 Ibid Annex, para 4.

124 By virtue of Article II of the Intervention Protocol (n 122), matters related to costs and other remedies are governed by the International Convention Relating to Intervention on

incentive for the latter to take steps on the high seas to prevent or minimise harm.

In sum, the special international nuclear legal instruments of relevance here[125] do not adequately, let alone fully, cover the high seas consequences of a nuclear accident. The civil liability conventions may provide for redress for damage on the high seas, but when they do so, they limit their scope of application *ratione personae* and/or *materiae*. The net effect of this is that convenient as the conventional regimes might be in theory from the perspective of a would-be claimant, more often than not, they will simply fail to provide a remedy, hence may leave nuclear pollution damage uncompensated. Claimants may be forced to rely on other legal instruments or customary international law to obtain redress.

### 3.2 *Redress for High Seas Pollution Pursuant to UNCLOS*

UNCLOS's direct contribution to an effective system of accountability for high seas pollution is relatively modest.[126] As noted, UNCLOS does lay down generic obligations for states to take measures to prevent the spread of pollution,[127] to notify other states of imminent or actual damage,[128] and to cooperate with others and the competent international organisations towards eliminating the effects of pollution and minimising damage.[129] However, in terms of operationalising the core principle of accountability—a state's responsibility and liability for damage incidental to marine pollution—UNCLOS offers relatively little guidance. A key provision of UNCLOS is Article 235.[130] It, relevantly, calls upon states to ensure the availability in their domestic law of recourse 'for prompt and adequate compensation or other relief in respect

the High Seas in Cases of Oil Pollution Casualties, 1969, whose provisions, however, are without prejudice to 'any remedy otherwise applicable'. See International Convention Relating to Intervention on the High Seas in Cases of Oil Pollution Casualties, 1969 (adopted 29 November 1969, entered into force 6 May 1975) 970 UNTS 211, art VII.

125 The Convention on Nuclear Safety, of course, applies only to land-based 'nuclear installations'. See Convention on Nuclear Safety (adopted 17 June 1994, entered into force 24 October 1996) 1963 UNTS 293, art 2 (ii).

126 On the other hand, UNCLOS (n 1) Article 218, para 1 provides an opening for the assertion of port state jurisdiction over high seas polluters. It is being discussed in Section 3.3.2, within the context of state responsibility-based claims for redress.

127 UNCLOS (n 1) Article 197, para 2 per se does not appear applicable to our high seas accident scenarios. However, Article 197, para 3 (a-d), would be relevant in that it requires states to take measures 'dealing with emergencies'.

128 UNCLOS (n 1) art 198.

129 Ibid art 199.

130 Other UNCLOS articles addressing responsibility and liability are Articles 139 and 304.

of damage caused by pollution of the marine environment'.[131] However, it also blandly exhorts states to develop further international law related to responsibility and liability for, and compensation of, damage to the marine environment,[132] and reiterates the principle that states are responsible for fulfilling their international obligations and liable in accordance with international law.[133]

On the other hand, in its Advisory Opinion on *Responsibilities and Obligations of States*,[134] the ITLOS Seabed Dispute Chamber discussed several aspects of UNCLOS's state responsibility-liability interface that is of interest here. In reviewing the import of key Article 139 relating to a sponsoring state's responsibility to ensure a contractor's compliance with applicable seabed mining laws and regulations and liability for damage from mining activities,[135] the Chamber, first, affirmed that a state's obligation to ensure was an obligation of conduct, hence a due diligence obligation.[136] A sponsoring state, therefore, would incur responsibility, not for the fact of a contractor's non-compliance, but only for its own failure to exercise due diligence—a contextually determined standard of care—in preventing such non-compliance.[137] Second, the Chamber also addressed the issue of whether a sponsoring state might be accountable or liable absent any wrongdoing on its part. More specifically, could the sponsoring state incur liability residually and as a matter of general international law, if the sponsored contractor had caused damage but was unable to meet its liability in full?[138] The Chamber roundly rejected such a possibility.

> [I]f the sponsoring State has not failed to meet its obligations, there is no room for its liability under article 139, paragraph 2, of the Convention even if activities of the sponsored contractor have resulted in damage. A gap in liability which might occur in such a situation cannot be closed by having recourse to liability of the sponsoring State under customary international law. The Chamber is aware of the efforts made by the International Law Commission to address the issue of damages resulting from

131 UNCLOS (n 1) art 235, para 2.
132 See Ibid art 235, para 3.
133 Ibid art 235, para 1.
134 See n 3.
135 Other UNCLOS articles of relevance are Article 159, para 4 and Annex III, Article 4, para 4.
136 *Responsibilities and Obligations of States* (n 3) paras 110–11.
137 Or, for that matter, in a situation of direct responsibility, for its own injurious and internationally wrongful conduct.
138 *Responsibilities and Obligations of States* (n 3) para 203.

> acts not prohibited under international law. However, such efforts have not yet resulted in provisions entailing State liability for lawful acts.[139]

At the same time, however, the Chamber acknowledged the non-static nature of UNCLOS's existing rules, and that further international law rules on responsibility and liability might emerge through conventional and customary international law[140] supportive of some form of state liability.

In conclusion, UNCLOS itself contains few rules that specifically assist in operationalising accountability for marine pollution. Thus, unless other special conventional rules apply, it is customary international law that determines whether and to what extent nuclear pollution of the high seas is amenable to redress. Flag states and states otherwise in a position to exercise jurisdiction or control over the polluting activity or event will not be accountable except for a failure to exercise due diligence in respect of any obligations related to the protection of the marine environment. In other words, a state will not be 'liable' in the technical sense,[141] to redress nuclear pollution of the high seas. Of course, if a state's failure to meet its obligations is causally related to the nuclear damage, the state will be accountable to make reparation in accordance with the rules of state responsibility.

### 3.3 *Invocation of State Responsibility for Environmental Damage on the High Seas*

#### 3.3.1 The Issue of Standing

There is no need here to discuss in detail the relatively straightforward relationship between a state's breach of an international obligation and legal consequences flowing therefrom. After all, the concept of state responsibility, or some version thereof, has been part of international law from its very beginnings[142] and has been recently modernised and codified by the International Law Commission (ILC) into Articles on Responsibility of States for Internationally Wrongful Acts (ILC Articles on State Responsibility).[143] If a state fails to meet its obligations bearing on the protection and preservation of the oceans, such failure entails an obligation to make reparation as 'a principle of international law, and even a general conception of law'.[144] Reparation, in

139 Ibid para 209.

140 Ibid para 211.

141 See n 17.

142 See Crawford (n 18) 3–20.

143 See ILC Articles on State Responsibility (n 16).

144 *Case Concerning The Factory at Chorzow (Germany v Poland)* (Merits) 1928 PCIJ (Series A) No 17 (Judgment 13 September 1928) para 73. As the Court explained further: 'The essential principle ... is that reparation must, as far as possible, wipe out all the consequences

turn, may 'take the form of restitution, compensation and satisfaction, either singly or in combination'.[145] Moreover, the responsible state will be obliged to cease the injurious, wrongful conduct and, if circumstances so require, offer assurances and guarantees of non-repetition.[146]

Thus, in respect of nuclear pollution of the high seas, there is no lack of clarity as to the basic mechanics of, or the scope of remedies available in, the system of responsibility. Rather it is uncertainty over who might be entitled to invoke the polluting state's responsibility that represents a key obstacle to international accountability. Of course, to the extent that such pollution causes individualised personal or property damage on or to, say, a radiation-exposed vessel at sea, standing is unlikely to be an issue. The situation, however, is quite different when it comes to damage to the environment per se given the high seas or international commons status of the waters concerned. Indeed, this was precisely the type of claim the ICJ was facing in the *Nuclear Tests* cases[147] but which the court was not, in the end, required to address.

The question of whether a community interest might be amenable to being vindicated by whom, if at all, in a system of accountability reflective of the traditional bilateralist conceptualisation of international law[148] has long been a matter of controversy.[149] Thus, international courts have given different answers at different times, not always explainable by differences of context. For example, in the *S.S. Wimbledon* case, the PCIJ accepted evidence of 'a clear interest', rather than of actual 'prejudice to any pecuniary interest', sufficient for purposes of standing.[150] Whereas in the *South West Africa* cases, the ICJ rejected the idea of an *actio popularis* in international law, as a 'right resident in any member of the international community to take legal action in vindication of a public interest',[151] in its dictum in the *Barcelona Traction* case the Court reversed itself—to a degree—by noting that

of the illegal act and re-establish the situation which would, in. all probability, have existed if that act had not been committed'. ibid para 125.

145 ILC Articles on State Responsibility (n 16) art 34.

146 Ibid art 30.

147 See n 8.

148 See, generally, Bruno Simma, 'Bilateralism and Community Interest in the Law of State Responsibility' in Yoram Dinstein (ed) *International Law at a Time of Perplexity: Essays in Honour of Shabtai Rosenne* (Martinus Nijhoff Publishers 1989) 821.

149 For a general overview, see Christine D Gray, *Judicial Remedies in International Law* (Clarendon Press 1987) 210–19; Brown Weiss, 'Invoking State Responsibility in the Twenty-First Century' (2002) 96 *American Journal of International Law* 789.

150 *Case of the S.S. 'Wimbledon' (United Kingdom, France, Italy and Japan v Germany)* 1923 PCIJ (Series A) No 1 (Judgment of 17 August 1923) para 19.

151 *South West Africa Cases (Ethiopia v South Africa; Liberia v South Africa)* (Judgment of 18 July 1966) ICJ Reports 1966, 6, 47, para 88.

> … an essential distinction should be drawn between the obligations of a State towards the international community as a whole, and those arising vis-à-vis another State in the field of diplomatic protection. By their very nature the former are the concern of all States. In view of the importance of the rights involved, all States can be held to have a legal interest in their protection; they are obligations *erga omnes*.[152]

The ILC Articles on State Responsibility follow this approach by distinguishing between the injured state (Article 42) and a state other than injured state (Article 48). The former connotes a state (1) to which the obligation breached is owed individually or (2) which, as a member of a group of states or the international community as a whole to which the obligation is owed respectively, is specially affected by the breach.[153] It includes also a state whose position regarding the further performance of the obligation, is—like that of all other members of the group or the international community at large to which the obligation breached is owed—radically affected by the breach.[154] Only the injured state will be entitled to claim the full range of remedies available vis-à-vis the responsible state. By contrast, a 'state other than an injured state' is entitled, albeit for more limited purposes, to invoke another state's responsibility, provided (1) the obligation breached is owed to a group of states, including the claimant state, and is established for the protection of a collective interest of the group; or (2) the obligation breached is owed to the international community as a whole.[155] Remedies are limited to claims for cessation of the internationally wrongful act; assurances and guarantees of non-repetition; and the performance of the obligation of reparation, in the interest of the injured state or of the beneficiaries of the obligation breached.[156]

Welcome as this adjustment of the traditional bilateralist model is, in the view of many commentators this shift simply does not go far enough.[157] At any rate, there is no denying that the text of Article 48 suffers from an inherent weakness in that it requires that the obligation concerned meet the test of being either 'established for the protection of a collective interest' of a group of states, or 'owned to the international community as a whole'. Such a determination may not be easy to make, be that in relation to the import of a

152 *Barcelona Traction, Light and Power Company, Limited* (Judgment of 5 February 1970) ICJ Reports 1970, 3, 32, para 33.
153 See ILC Articles on State Responsibility (n 16) art 42, paras (a) and (b) (i).
154 Ibid art 42, para (b) (ii).
155 Ibid art 48, para 1.
156 Ibid art 48, para 2.
157 See, e.g., Brown Weiss (n 149).

treaty provision or of a rule of customary international law. Indeed, it raises the question of who would be entitled to make that determination. It might be worth recalling, therefore, Bruno Simma's decades-old warning against new community interest-related conceptions being crafted upon the international law of state responsibility, while corresponding institution-building in support of such substantive change is lagging.[158]

This institutional issue is, of course, a key challenge that the management of the high seas and its resources generally nowadays poses. It lies at the heart of rendering operational the system of state responsibility for internationally wrongful acts resulting in pollution damage to the marine environment. It is also a key issue in ongoing negations about the governance regime for the conservation and sustainable use of marine biological diversity in areas beyond national jurisdiction.[159] Finally, it is the subject of some interesting observations in the Advisory Opinion on *Responsibilities and Obligations of States*. In reviewing who might be entitled to claim compensation for 'damage to the Area and its resources constituting the common heritage of mankind, and damage to the marine environment',[160] the Chamber listed the International Seabed Authority (ISA) as one entitled claimant. While the Chamber acknowledged that UNCLOS does not explicitly authorise ISA to make such a claim, it believes that an entitlement to this effect 'is implicit in Article 137(2) of the Convention'.[161] Other states not meeting the technical qualification of 'injured states' may also, according to the Advisory Opinion, be 'entitled to claim compensation in light of the *erga omnes* character of the obligations relating to preservation of the environment of the high seas', as supported by Article 48 of the ILC Articles on State Responsibility.

Finally, and directly related, is the question as to whether today international law might permit a third state or state other than the directly injured state, to take countermeasures against a state responsible for high seas nuclear pollution. Such measures, by definition, involve the non-performance of an international obligation on the part of the responding state.[162] The ILC Articles on State Responsibility themselves deny such a possibility only indirectly. They do this by stating that the provisions on countermeasures do not prejudice the right of a third state 'to take *lawful measures* against that State to ensure

158 Simma (n 148) 844.

159 See fn 211–12.

160 *Responsibilities and Obligations of States* (n 3) para 179.

161 Ibid para 180.

162 See ILC Articles on State Responsibility (n 16) art 49, para 2; the *Naulilaa Arbitration (Portugal v Germany)* (Arbitral Award of 31 July 1928) *UN Reports of International Arbitral Awards* (Vol 2, United Nations 1949) 1011.

cessation of the breach and reparation in the interest of the injured State or of the beneficiaries of the obligation breached'.[163] The Commission's known hesitancy in proposing this article[164] comes to the fore in its commentary, in which it characterises state practice involving third-state countermeasures as both controversial and embryonic.[165] However, given the latent uncertainty about standing to invoke a polluting state's responsibility in respect of environmental damage to the high seas, as discussed, third-party countermeasures should, as a matter of public policy, be permissible, precisely because they offer a means to put pressure on the polluter when other more formal avenues of redress might not be available.

#### 3.3.2 Port State Enforcement under Article 218(1) of UNCLOS

Finally, to round out our review of the invocability of state responsibility, mention ought to be made here of another element of the international accountability architecture that might be of relevance to some high seas nuclear pollution scenarios. It involves UNCLOS Article 218(1), which establishes port state enforcement jurisdiction in respect of polluting discharges by vessels on the high seas, a variant of universal principle jurisdiction or a form of 'community interest enforcement'.[166] Thus, provided the vessel is voluntarily in port, and the discharge concerned involves a 'violation of applicable international rules and standards established through the competent international organization or general diplomatic conference', any such event or conduct can be the subject of investigation and legal proceedings on the part of the port state.[167] Clearly, any release of nuclear materials, including the dumping of

---

163 ILC Articles on State Responsibility (n 16) art 54 (emphasis added).

164 See 'Draft Articles on Responsibility of States for Internationally Wrongful Acts' in International Law Commission, *Yearbook of the International Law Commission 2001* (Vol II, Part 2, UN Doc A/CN.4/SER.A/2001/Add.1 (Part 2), United Nations 2007) 26, 139, paras 6–7.

165 Ibid 128, para 8: 'Occasions have arisen in practice of countermeasures being taken by other States, in particular those identified in article 48, where no State is injured or else on behalf of and at the request of an injured State. Such cases are controversial and the practice is embryonic'.

166 Christian J Tams, 'Individual States as Guardians of Community Interest' in Ulrich Fastenrath et al (eds) *From Bilateralism to Community Interest: Essays in Honour of Judge Bruno Simma* (Oxford University Press 2011) 398; Henrik Ringbom, *The EU Maritime Safety Policy and International Law* (Brill Nijhoff 2008) 214. But see Alla Pozdnakova, *Criminal Jurisdiction over Perpetrators of Ship-Source Pollution: International Law, State Practice and EU Harmonisation* (Brill Nijhoff 2013) 154–55 (challenging the notion that pollution of the high seas might be a universal jurisdiction crime).

167 UNCLOS (n 1) art 218, para 1: 'When a vessel is voluntarily within a port or at an off-shore terminal of a State, that State may undertake investigations and, where the evidence so warrants, institute proceedings in respect of any discharge from that vessel outside

wastes, in violation of applicable IAEA standards or the ban under the London Convention and London Protocol, could give rise to port state measures.

Leaving aside evidentiary issues and opportunity costs[168] associated with law enforcement efforts targeting wrongful polluting activities on the high seas, few states have actually availed themselves of this UNCLOS authorisation and adopted legislation that would allow them to investigate and, where the evidence so warrants, prosecute such conduct.[169] For example, Japan, which is, of course, a party UNCLOS, has no domestic enabling legislation. By contrast, the United States, which might have a keen interest in going after polluters of the high seas, does not have legislation that would permit its law enforcement agencies to do so. The reason given is that as a non-party to UNCLOS the USA could not invoke the conventional authority; nor could such enforcement action, it has been suggested,[170] be justified on the basis of customary law, as the provision of Article 218(1) has not yet become a rule of customary international law.[171] On the other hand, in the wake of several catastrophic vessel-source pollution incidents, the European Union, with Directive 2005/35/EC,[172] launched an aggressive enforcement strategy that required member states to, inter alia, prosecute non-member state vessels' discharges of polluting substances on the high seas.[173] However, as the sequel to the *Intertanko* case[174] shows, states or, for that matter, regional organisations, such as the European Community, may

the internal waters, territorial sea or exclusive economic zone of that State in violation of applicable international rules and standards established through the competent international organization or general diplomatic conference'.

168 One disincentive for targeting non-national vessels for polluting the high seas is the costs of using scarce national investigative and judicial resources for the sake of protecting and preserving the high seas marine environment.

169 For a (by now somewhat dated) survey, see, e.g., Ho-Sam Bang, 'Port State Jurisdiction and Article 218 of the UN Convention on the Law of the Sea' (2009) 40 *Journal of Maritime Law & Commerce* 291, 302–09.

170 Personal communication from the US Coast Guard to author.

171 Commentators differ on this question. See, e.g. Alan E Boyle and Christine Chinkin, *The Making of International Law* (Oxford University Press 2007) 237 (suggesting that today no state might deny that Article 281 has become customary international law). More cautious, Ted L McDorman, 'Port State Enforcement: A Comment on Article 218 of the 1982 Law of the Sea Convention' (1997) 28 *Journal of Maritime Law and Commerce* 305, 320.

172 Directive 2005/35/EC of the European Parliament and of the Council of 7 September 2005 on Ship-Source Pollution and on the Introduction of Penalties for Infringements, OJ L 255/11 (30 September 2005).

173 See Ibid arts 3(1)(e) and 6.

174 *The Queen on the Application of: International Association of Independent Tanker Owners (Intertanko) and Others v Secretary of State for Transport* (Judgment of the Court (Grand Chamber) of 3 June 2008) ECJ Grand Chamber, Case C-308/06.

not go beyond the limits imposed by UNCLOS in using the port state authority granted in Article 218. Although the Grand Chamber of the European Court of Justice artfully avoided implicating Directive 2005/35/EC, a largely negative reaction to the judgment in the end persuaded the European Community to amend the Directive to make it compatible with MARPOL, and thereby the requirements of UNCLOS.[175]

Finally, some countries, including the USA, might employ an indirect approach by way of 'false statement' prosecutions,[176] as a sometimes-available alternative strategy for targeting unlawful polluting activities at sea, including the high seas. Thus, instead of prosecuting the wrongful activity on the high seas itself, the port state authorities will sanction any misrepresentation concerning the high seas pollution event that the crew might make in port, hence unquestionably within the port state's territorial jurisdiction. However, it is unclear to what extent such a strategy could work in respect of nuclear polluting activities.

## 4 Redress of High Seas Pollution Damage in the Absence of International Wrongfulness

The Seabed Chamber's negative finding in its *Responsibility and Obligations of States* Advisory Opinion regarding the status of state liability in customary international law[177] is unobjectionable. However, to it put into proper context, one ought to acknowledge the existence of extra-conventional state practice as well as judicial decisions supportive of the principle of states' strict liability for damage caused by hazardous activities under their jurisdiction or control.[178] Indeed, in one such instance, involving the Japanese fishing vessel Fukuryu

175 See Directive 2009/123/EC of the European Parliament and of the Council of 21 October 2009 Amending Directive 2005/35/EC on Ship-Source Pollution and on the Introduction of Penalties for Infringements, OJ L 280 (27 October 2009) art 5.

176 See, e.g. *United States v Royal Caribbean Cruises, Ltd.*, 11 F Supp 2d 1358 (Fla 1998), which, however, involved falsification of the oil record book on account of an unreported, illegal discharge, not on the high seas, but in the territorial sea of another state, the Bahamas.

177 See n 139.

178 For references to this practice, see, e.g., International Law Commission, 'Survey of Liability Regimes Relevant to the Topic of International Liability for Injurious Consequences Arising out of Acts Not Prohibited by International Law (International Liability in Case of Loss from Transboundary Harm Arising out of Hazardous Activities)' UN Doc A/CN.4/543 (24 June 2004) 130–43. See also Günther Handl, 'International Accountability for Transboundary Environmental Revisited: What Role for "State Liability"?' (2007) 37 *Environmental Policy & Law* 116, 119–20.

Maru,[179] the US government settled claims for compensation for personal injury and property damage due to nuclear pollution of the high seas.[180] While the evidence of such practice is insufficient to support an international customary law–based argument, it nevertheless is suggestive of a certain receptivity on the part of international society to the idea of a broader application of strict liability for harm due to hazardous activities. This tendency has found expression also in the 2004 European Environmental Liability Directive,[181] and 2010 UNEP Guidelines for the Development of Domestic Legislation on Liability, Response Action and Compensation for Damage Caused by Activities Dangerous to the Environment.[182] Importantly, the ILC's Principles on the Allocation of Loss in the Case of Transboundary Harm arising from Hazardous Activities (Allocation Principles) send a similar message. Although the Allocation Principles are merely to provide 'appropriate guidance' to states,[183] and although, upon their completion by the ILC in 2006, the United Nations General Assembly only commended them 'to the attention of Governments',[184] they

179 On 1 March 1954, the USA conducted its largest hydrogen bomb test ever near the Bikini Atoll in the Marshall Islands. The Fortunate Dragon (Daigo Fukuryū Maru) was about 90 miles from the blast and gravely affected. At least one crew member died due to direct exposure. The Japanese and US governments reached a political settlement in 1955 pursuant to which the USA paid USD 2 million, worth ¥720 million at that time, ex gratia to Japan for the injuries and damage caused.

180 A standard argument is, of course, that incidents involving ex gratia payments 'can only be invoked in support of a moral, as opposed to a legal obligation to pay compensation'. See R Lefeber, *Transboundary Environmental Interference and the Origin of State Liability* (Brill Nijhoff 1996) 169. However, it is disingenuous to dismiss the legal significance of such payments. Observable actual state conduct is a key element in the chain of evidence pointing to states' recognition of an underlying legal obligation. Particularly, where there is a pattern of states' claims and counterclaims—as is the case, e.g., in relation to transboundary radiation incidents following nuclear testing or accidents involving nuclear weapons—the compensating state's characterisation of its payments as 'ex gratia' is hardly persuasive. To this effect, see specifically the statement by Suchratikul at the 1686th meeting of the International Law Commission, *Yearbook of the International Law Commission 1981* (Vol 1, UN Doc A/CN.4/SER.A/1981, United Nations 1982) 224, para 24; Nathalie L Horbach, *Liability versus Responsibility under International Law: Defending Strict State Responsibility for Transboundary Damage* (NLJT Horbach 1996) 434–35.

181 Directive 2004/35/CE of the European Parliament and of the Council of 21 April on Environmental Liability with regard to the prevention and remedying of environmental damage, OJ L 143/56 (30 April 2004).

182 United Nations Environment Programme, 'Guidelines for the Development of Domestic Legislation on Liability, Response Action and Compensation for Damage Caused by Activities Dangerous to the Environment' Governing Council, Decision SS XI/5, Part B (26 February 2010).

183 Allocation Principles (n 17) 59, para 5.

184 Ibid para 2.

do enjoy an undeniable degree of authority as to how states ought to legally manage harm from hazardous activities. Indeed, the Allocation Principles' very existence is likely to frame international public policy in the long run, if not promote directly the emergence of international normative expectations.

In its general commentary, the Commission expressly excluded consideration of harm done to the environment in areas beyond national jurisdiction.[185] On the other hand, the Allocation Principles apply to 'activities conducted under the jurisdiction or control of a State, for example, on the high seas, with ... injurious impacts on ships or platforms of other States on the high seas as well'.[186] In any event, in setting out—in exemplary form—basic elements of a system of loss management applicable to hazardous activities absent internationally wrongful conduct, the Allocation Principles are clearly relevant in the present context. The Allocation Principles' key premise is, of course, the need for rules on the allocation of loss given that—as the preamble emphasises—'incidents involving hazardous activities may occur despite compliance by the relevant State with its obligations concerning the prevention of ... harm'.[187] Their fundamental purpose is twofold, namely to ensure prompt and adequate compensation to victims and to preserve and protect the environment.[188] To achieve this goal, the Allocation Principles espouse, much as the 'modern' nuclear civil liability conventions do, a wide notion of damage compensable.[189] However, the Allocation Principles also include a wide definition of 'environment'.[190] Most significantly, as the Commission pointed out, this definition includes damage to the environment per se. It 'thus not only builds upon trends that have already become prominent as part of recently concluded international liability regimes but opens up possibilities for further developments of the law for the protection of the environment per se'.[191] While

185 'The Commission ... carefully considered the desirability of examining the issues concerning the global commons ... [and] came to the conclusion that they require a separate treatment'. See Commentary (n 17) 60, para 7.

186 Ibid 70, para 26.

187 Ibid 61, preambular para 3.

188 Allocation Principles (n 17) Principle 3.

189 Ibid Principle 2(a): '"Damage" means significant damage caused to persons, property or the environment ...'.

190 Thus damage to the environment includes 'loss or damage by impairment of the environment'; the costs of reasonable measures of reinstatement of the ... environment, including natural resources; and the costs of reasonable response measures. ibid sub-paras (iii)–(v). Moreover, Principle 2, para (b) defines 'environment' as inclusive of 'natural resources, both abiotic and biotic, such as air, water, soil, fauna and flora and the interaction between the same factors; and the characteristic aspects of the landscape...'.

191 Commentary (n 17) 67, para 11 (footnote omitted).

the Commission acknowledged associated problems of standing[192] and the valuation of damage, it also pointedly referred to recent 'encouraging' trends allowing compensation for loss of 'non-use value'.[193]

The Allocation Principles recommend the imposition of strict liability on the operator,[194] and that the operator be required to provide financial security.[195] They also suggest that states, where appropriate, ought to ensure industry-wide funds at national level, and, importantly, that whenever funding through these sources is inadequate, states ensure availability of additional financial resources to provide adequate compensation.[196] Principle 5, focussing on response measures, reiterates a source state's traditional obligation to notify others of incidents and possible harmful effects, to ensure appropriate response measures, and to cooperate of other states affected or likely to be affected.[197] States, in turn, are expected to take remedial action to mitigate or eliminate the effects of damage.[198] Finally, as regards international and domestic remedies, Principle 6, inter alia, requires states to ensure—in non-discriminatory fashion[199]—the availability of prompt, adequate and effective remedies in the event of damage caused by 'hazardous activities located within their territory or otherwise under their jurisdiction or control'.[200]

Generally speaking, then, the Allocation Principles clearly recommend themselves as laying down the standard against which any existing or future international legal instrument of direct relevance to closing today's high seas nuclear governance gap might have to be measured.[201] However, they fall short in one significant respect: the source or controlling state's legal position as

---

192 Ibid para 14.

193 Ibid 69, para 18.

194 Allocation Principles (n 17) Principle 4, para 2. I.e., the person or entity 'in actual, legal or economic control of the polluting activity'. Commentary (n 17) 71–72, para 32.

195 Allocation Principles (n 17) Principle 2, para 3.

196 Ibid paras 4–5.

197 Ibid Principle 5, paras (a-c).

198 Ibid Principle 5, para (d).

199 Ibid Principle 6, para 2: 'Victims of transboundary damage should have access to remedies in the State of origin that are no less prompt, adequate and effective than those available to victims that suffer damage, from the same incident, within the territory of that State'.

200 Ibid para 1.

201 Of course, the Allocation Principles only offer model elements of an allocation of loss regime in a transboundary context. They, therefore, do not address the standing or institutional agency issues that present themselves in relation to the vindication of community interests in the protection of the high seas, as discussed above. Any international legislative initiative aiming at improving high seas governance would have to pay attention to these pressing issues.

'insurer of last resort'. As noted, Principle 4—in itself already formally a mere soft law pronouncement—only recommends that states 'take all necessary measures to ensure that prompt and adequate compensation is available for victims'.[202] Unlike some provisions of other Principles, which are couched in mandatory language,[203] all compensation provisions use hortatory language. Such a soft approach, especially as regards the financial role of the state, is, however, quite unwarranted. The state ought to be legally required to make up any shortfall in available compensation, or at least a significant portion thereof, given the nature of the damage-causing activity as a priori, recognisably 'hazardous', the implicit degree of control the state consequently exercises or must be deemed to exercise over the activity because of its role in licensing or approval, and considering the need for an equitable international distribution of the risk of harm and the benefits associated with the activity.[204] It is precisely these or very similar considerations that persuaded states to accept in the Convention on Supplementary Compensation for Nuclear Damage a subsidiary obligation for any contracting party to provide public funds to help defray the costs of a nuclear accident.[205]

In avoiding this, a more enlightened allocation of risk approach, the Allocation Principles merely follow a sentiment previously expressed by the Commission's special rapporteur on the topic of Prevention of Transboundary Harm from Hazardous Activities, who dismissed state liability as 'a case of misplaced priority',[206] maintaining that 'State liability for transboundary damage for either ultrahazardous activities or hazardous activities does not appear to have support even as a measure of progressive development of law'.[207] As discussed in detail elsewhere,[208] this assumption is both misguided and, in the end, logically untenable when tested against the import of key Prevention

202 Allocation Principles (n 17) Principle 4, para 1: 'Each State should take all necessary measures to ensure that prompt and adequate compensation is available for victims of transboundary damage caused by hazardous activities located within its territory or otherwise under its jurisdiction or control'.

203 For example, Allocation Principles (n 17) Principle 5, paras 1–2, which deal with response measures.

204 For a detailed review, see Handl (n 178) 119–22.

205 See CSC (n 86) art III, para 1(b).

206 Pemmaraju Sreenivasa Rao, 'First Report on the Legal Regime for Allocation of Loss in Case of Transboundary Harm Arising out of Hazardous Activities' UN Doc A/CN.4/531 (21 March 2003) 7.

207 Pemmaraju Sreenivasa Rao, 'Third Report on the Legal Regime for Allocation of Loss in Case of Transboundary Harm Arising out of Hazardous Activities' UN Doc A/CN.4/566 (7 May 2006) para 31.

208 See Handl (n 178).

Article 3's obligation 'to prevent significant harm or at any event to minimize the risk thereof'.[209]

## 5 Summary and Conclusions

Overall, international accountability for nuclear pollution of the high seas is limited. As seen above, accountability varies with the pollution scenario, the type of damage sustained, the claimant involved and the form of redress that is being sought. Accountability for loss of life, personal injury or property damage in conjunction with nuclear pollution on the high seas tends to be relatively strong. By contrast, accountability for damage to the environment per se is either severely limited, or altogether non-existent. It is this problematic redressability of environmental harm then that epitomises the high seas governance gap.

The nuclear liability conventions may not apply to damage on the high seas at all, providing redress only in the form of compensation, or exclude damage to the environment per se. Other international legal instruments, including UNCLOS, offer little direct support for holding the high seas polluter accountable. It is therefore customary international law and, more specifically, the technical rules of state responsibility, which provide the framework though which accountability for nuclear pollution damage to the high seas might most likely be realisable. Accordingly, a state that wrongfully fails to prevent or minimise nuclear pollution of the high seas will be required to make full reparation, which obligation includes, but is not limited to, restitution as well as compensation. To date, as a matter of customary international law, a state's causal conduct, absent international wrongfulness, does not engage its accountability even in a compensation-of-damage sense only. However, both the *Responsibilities and Obligations* Advisory Opinion and the Allocation Principles directly or indirectly testify to growing expectations that as a matter of general international law, hazardous activities ought to be subject to a regime of strict liability, that damage to the environment per se be compensable, and that the source or controlling state be held accountable financially, at least on a subsidiary basis, for compensating damage.

Unfortunately, a problem that remains unresolved in the present system of state responsibility, thus seriously undermining its effectiveness, is the

209 UNGA, 'Resolution Adopted by the General Assembly on 6 December 2007. 62/68. Consideration of Prevention of Transboundary Harm from Hazardous Activities and Allocation of Loss in the case of Such Harm' UN Doc A/RES/62/68 (8 January 2008) Annex.

continuing lack of certainty regarding who would be entitled to act on behalf of the international community at large. Evidently, there is a compelling need for an international mechanism or agency capable of vindicating fully the collective interest in preserving the high seas, including its protection against nuclear pollution. This need presents itself also in the context of any strict liability regime that covers damage to the high seas marine environment.[210] However, as the negotiations on the proposed international instrument on the conservation and sustainable use of marine biological diversity of areas beyond national jurisdiction clearly show, the pathway towards any institutional innovation can be a steep one.[211] Of course, the benefit of a breakthrough on the high seas institutional front could be huge. The presently existing high seas governance gap could significantly narrow, or might even close altogether.[212]

---

210 Because of their transboundary focus, the Allocation Principles of course do not discuss this issue.

211 See, e.g. UNGA, 'Report of the Preparatory Committee Established by General Assembly Resolution 69/292: Development of an International Legally Binding Instrument under the United Nations Convention on the Law of the Sea on the Conservation and Sustainable Use of Marine Biological Diversity of Areas Beyond National Jurisdiction' UN Doc A/AC.287/2017/PC.4/2 (31 July 2017). Institutional arrangements are listed in Section B, thereby indicating an issue on which there is divergence of views among participants.

212 Alex Oude Elferink has suggested that governance principles might include the concept of 'common concern'—not to detract from UNCLOS's freedom of the high seas or the common heritage principle—but rather to 'reinforce the fundamental notion that all States have a shared interest in *and responsibility* for the governance of the ABNJ'. Alex Oude Elferink, 'Governance Principles for Areas Beyond National Jurisdiction' (2012) 27 *International Journal of Marine and Coastal Law* 205, 244–45 (emphasis added). However, while this would be a step in the right direction, it would not be sufficient to fully address the existing governance deficit.

CHAPTER 8

# Liability and Compensation Regimes: Pollution of the High Seas

*Nicholas Gaskell*


## Abstract

The chapter considers potential legal gaps relating to the liability of private companies for pollution of the high seas arising from traditional shipping activities and the exploitation of the seabed, but also in the context of discussions about biological diversity of areas beyond national jurisdiction (BBNJ). It examines possible compensation models for the high seas, such as the IMO compensation systems (backed by liability insurance), voluntary industry schemes, administrative sanctions (e.g. through the ISA) and the various types of environmental trust funds envisaged in the BBNJ discussions (and tried, with varying degrees of success, in relation to Antarctica and climate change).


## 1 Introduction—What Are the Gaps?

The aim of this chapter is to consider some potential legal gaps relating to liability and compensation for pollution of the high seas, particularly those relating to shipping activities. There has already been much discussion about whether there are gaps, or simply failings in implementation of current instruments.[1] Gaps could be both actual and theoretical. There may be proven types of environmental harm that are not being compensated (e.g. from ocean noise and tourism),[2] or there may be limitations or restrictions in existing legal instruments or doctrine that might prevent recovery if any future (perhaps not yet identifiable) harm occurs.

Threats to the environment in or on the high seas may be broadly similar to those in coastal waters, but are exacerbated by distances from land. There are

1 Louise de La Fayette, 'A New Regime for the Conservation and Sustainable Use of Marine Biodiversity and Genetic Resources beyond the Limits of National Jurisdiction' (2009) 24 *International Journal of Marine and Coastal Law* 221, 253.

2 Ibid 257.

 | DOI:10.1163/9789004373303_009

well-recognised commercial activities—including shipping, where pollution might arise from the ship itself, its bunkers or cargo—and activities such as fishing or dumping. There is also offshore exploration and exploitation for oil and gas, with associated decommissioning. Seabed activities almost certainly involve some assistance from maritime craft. These craft may be ships, e.g. a mother ship supporting divers and submersibles,[3] or some form of fixed or floating platform. New or developing industries may also have an impact in the high seas, including carbon capture and storage,[4] mining of seamounts[5] and the extraction of so-called 'flammable ice'.[6]

All such activities may also have a direct effect on the conservation and sustainable use of marine biological diversity of areas beyond national jurisdiction (BBNJ).[7] The issue of biological diversity may raise broader issues about exploitation, e.g. where samples of living organisms in the sea column are removed for commercial development or bioprospecting, and where cultures are developed in situ (e.g. like fish farms). The BBNJ discussions themselves raise the controversial question of the best mechanism for addressing the legal gap in relation to marine biodiversity and marine genetic resources.

The legal gaps go beyond those raised in the BBNJ discussions, as there are already problems caused by dividing up the water column from the seabed Area under Part XI of the United Nations Convention on the Law of the Sea (UNCLOS). While the International Seabed Authority (ISA) can regulate

3 There are already 'gaps' in international law in respect of unmanned craft. See the Comité Maritime International (CMI) International Working Group on Unmanned Ships, 'Position Paper and Questionnaire' (29 March 2017); also IMO, 'Maritime Autonomous Surface Ships Proposal for a Regulatory Scoping Exercise' IMO Doc MSC 98/20/2 (27 February 2017); see also IMO, 'Proposal for a Regulatory Scoping Exercise and Gap Analysis with respect to Maritime Autonomous Surface Ships (MASS)' IMO Doc LEG 105/11/1 (19 January 2018).

4 See e.g., Viktor Weber and Michael N Tsimplis, 'The UK Liability Framework for the Transport of CO2 for Offshore Carbon Capture and Storage Operations' (2017) 32 *International Journal of Marine and Coastal Law* 138.

5 On the risk of plumes of dust from extraction, see David Shukman, 'Renewables' Deep-sea Mining Conundrum' *Principia Scientific International* (11 April 2017) <http://principia-scientific.org/renewables-deep-sea-mining-conundrum> accessed 24 May 2018, <www.bbc.com/news/science-environment-39347620> accessed 24 May 2018.

6 Methane hydrates. See BBC, 'China Claims Breakthrough in Mining "Flammable Ice"' (19 May 2017) <www.bbc.com/news/world-asia-china-39971667> accessed 24 May 2018.

7 For the history and development from 2014 of a binding instrument on BBNJ under UNCLOS, see Special Issue 'Conservation and Sustainable Use of Marine Biodiversity: Southern Hemisphere Issues and Approaches' (2017) 32(4) *International Journal of Marine and Coastal Law* 607; Section 6 in this chapter and Chapter 1 Robert Beckman et al 'Introduction'. This chapter will refer to 'BBNJ claims' as shorthand for compensation claims that might arise as a result of harm to marine BBNJ.

activities on the seabed, its power over the superjacent waters is limited, e.g. to activities related to mining on the seabed itself, rather than activities that take place solely in the column.[8] From an ecosystem viewpoint, such a distinction is a hindrance to a holistic approach,[9] and recourse has to be made to the general provisions of Part XII.

Further, the regulation of shipping activities on the high seas that might cause environmental harm is hindered by the traditional UNCLOS emphasis on flag state powers, and by the jurisdictional limitations of bodies such as the ISA and the International Maritime Organization (IMO). The absence of a single administrative body with a comprehensive marine environmental remit, as well as a single dedicated disputes tribunal for all high seas issues, creates difficulties in devising appropriate liability regimes.

Thus, implementing a satisfactory environmental compensation framework for the high seas is potentially messy, because there are two broadly different models available. First, the IMO has produced tort-like compensation systems (backed by liability insurance) to deal with unintentional harm, or 'accidents', caused by ships lawfully exercising rights of passage. Those systems essentially provide private law solutions, but they do not presently extend to the high seas.[10] Secondly, UNCLOS Part XI is designed for operational activities, e.g. where any harm is a natural consequence of carrying out activities such as mineral exploitation and extraction. These activities have traditionally been regulated[11] in an administrative and public law framework applied to contractors who apply for permission to operate. The planned activity may be controlled by regulation (e.g. prohibiting or curtailing the activity in question), while financial consequences of harm may be addressed through licensing conditions enforced by administrative or contract mechanisms (including the need to provide evidence of insurance cover). The public international law backstop may well be state liability.[12]

8 United Nations Convention on the Law of the Sea (adopted 10 December 1982, entered into force 16 November 1994) 1833 UNTS 3 (UNCLOS) art 135. There is also uncertainty about the status of waters above the extended continental shelf beyond 200 NM. See also Section 6.4.

9 See de la Fayette (n 1) 253–54, 258.

10 See Section 2.4.1.

11 E.g. by state authorities in the EEZ, and the International Seabed Authority (ISA) in the Area. See Section 6.4. A similar model might be suitable for regulation of activities in the water column relating to BBNJ. See Section 7.

12 For further discussion of state liability, see Günther Handl, Chapter 7 'High Seas Governance Gaps: International Accountability for Nuclear Pollution' and Erik Røsæg, Chapter 9 'Marine Pollution Preparedness, Response and Cooperation in the Arctic High Seas'.

The larger question is therefore how to integrate compensation rules covering three potential regimes for the high seas: namely those for the navigation of ships (IMO), the seabed (ISA) and the water column (e.g. BBNJ discussions).

It is not possible here to consider every potential environmental risk or liability, and an element of uncertainty is inevitable in any scoping exercise.[13] I propose, though, to consider first the development of existing maritime liability and compensation systems, before considering alternative possible systems for the high seas.

## 2 Need for Maritime Liability and Compensation Systems

### 2.1 *Introduction and UNCLOS*

Although the need for compensation for marine environmental damage was recognised before 1982, the framework is now set out in UNCLOS Article 235.

Article 235(1) places direct environmental responsibilities on states. The possibility of state liability for environmental damage is an important part of the context for any discussions concerning the high seas, but will not be dealt with in any detail in this chapter. Suffice it to note that the law on state responsibility generally is still developing, but the International Law Commission (ILC) Articles on the Responsibility of States for Internationally Wrongful Acts[14] are generally accepted as authoritative, and many of them may already represent customary international law.[15] It seems that the more prevalent view, for now, is that state liability is more likely to arise from the state's own 'wrongful act' in regulating the activities of its citizens and companies[16] than as a direct (in effect, vicarious) responsibility for the actions of private companies or individuals.

It may seem that the easiest way to ensure that there is no 'gap' in compensation remedies is to concentrate on state liability rather than that of private

---

13 The scope of future regulatory development in the 2017 ISA discussions is equally uncertain. See Section 6.4.

14 As attached to Responsibility of States for Internationally Wrongful Acts, UNGA Res A/RES/56/83 (28 January 2002). See also e.g. UNGA, 'Draft Resolution. Responsibility of States for Internationally Wrongful Acts' UN Doc A/C.6/71/L.28 (7 November 2016) for moves to consider a convention on the subject at the 74th Session of the General Assembly.

15 See e.g. Responsibility of States for Internationally Wrongful Acts (n 14) art 36.

16 See also Responsibilities and Obligations of States Sponsoring Persons and Entities with Respect to Activities in the Area (Request for Advisory Opinion submitted to the Seabed Disputes Chamber) (Advisory Opinion of 1 February 2011) ITLOS Reports 2011, 10 (ITLOS Advisory Opinion) paras 100–05.

actors.[17] This may be fine in theory, but if there were a liability, how would it be enforced if the state is developing, largely without means, or is in crisis?[18]

For that reason, this chapter will concentrate on liabilities of private actors, addressed in UNCLOS Article 235(2). There is no exclusion of the high seas in the general obligation on states to protect the marine environment (Article 192) and in the specific obligations in Article 235(3) to implement and develop international law on compensation. Article 235(3) is a reminder that it is not enough to create liability rules, but it is also necessary to consider the practicalities of ensuring payment, e.g. through 'compulsory insurance or compensation funds'.

### 2.2 *IMO Conventions: Introduction*

The history and development of maritime liability conventions at the IMO is well known and has been described elsewhere.[19] So this chapter will provide only a brief introduction in order to assess what is best for the high seas.[20]

Since the *Torrey Canyon* disaster in 1967,[21] the IMO has developed a suite of conventions that create liabilities to pay compensation for pollution damage. Those liabilities are not the liabilities of states, e.g. for the activities of

---

17 See also Julien Rochette et al, 'Seeing beyond the Horizon for Deepwater Oil and Gas: Strengthening the International Regulation of Offshore Exploration and Exploitation' (Study No 1/2014, 14 February 2014, IDDRI) Section 3, dealing with compensation for oil pollution damage resulting from offshore oil and gas exploitation.

18 This is not a fanciful scenario where such a state has been deliberately chosen by commercial parties as a place to incorporate or base a company because of that state's weakness; the flag of convenience problem is already well known.

19 I have explained the background to the IMO conventions in the following publications: Nick Gaskell, 'Compensation for Offshore Pollution: Ships and Platforms' in Malcolm Clarke (ed) *Maritime Law Evolving* (Hart 2013) 63–93; V Cogliati-Bantz et al, 'Marine Oil and Gas Pollution Spills in Australian Waters' in Aldo Chircop, Norman Letalik and Ted McDorman (eds) *The Regulation of International Shipping: International and Comparative Perspectives: Essays in Honor of Edgar Gold* (Martinus Nijhoff 2012) 371–400; Nick Gaskell and Craig Forrest 'The Wreck Removal Convention 2007' [2016] *Lloyd's Maritime and Commercial Law Quarterly* 49; Nick Gaskell 'The Bunker Pollution Convention 2001 and Limitation of Liability' (2009) 15 *Journal of International Maritime Law* 477; Nick Gaskell and Craig Forrest, 'Marine Pollution Damage in Australia: Implementing the Bunker Oil Convention 2001 and the Supplementary Fund Protocol 2003' (2008) 27 *University of Queensland Law Journal* 103.

20 For the key features of the conventions, see Section 3.

21 The supertanker sank off the coast of the UK with nearly 120,000 tons of oil, causing massive pollution in the UK and France. Owing to the legal complexities and uncertainties, the compensation claims were settled at levels well below the claimed losses. See e.g. Great Britain Home Office, *The Torrey Canyon* (Cmnd 3246, HMSO 1967); Paul Burrows, Charles Rowley and David Owen, 'Torrey Canyon: A Case Study in Accidental Pollution'

their nationals or companies (including ships flagged in their state). Indeed, although there may be cases where the concern is about state actors and entities controlled by them, it is inevitable that the commercial exploitation of the high seas (including their biodiversity) will involve privately controlled companies in the developed world. If there is a concern about commercialisation of biodiversity, then in practice it will be necessary to deal with companies that may have few assets in the state of their seat or nationality.

### 2.3 *Torrey Canyon Lessons*

The inadequacy of national or international law to deal with pollution disasters was exemplified by the *Torrey Canyon*, not only in relation to intervention powers, but also more particularly in respect of compensation for pollution clean-up.

In practice, claimant states had to resort to ordinary fault-based principles of liability in national law. The potential problems that became evident included who was to be liable when the defendant might be a foreign shipowner with few resources (e.g. if, as often occurs, it is a single-ship company whose only asset was a ship now sunk); whether there was a liability, or other insurer that might respond;[22] what type of environmental damage was recoverable; whether the shipowner could limit liability under traditional maritime law; and which courts would have jurisdiction if there was transboundary pollution.

### 2.4 *Narrow Scope of Application of the IMO Liability Conventions*

It is problems such as those referred to in Section 2.3 that the IMO has confronted with its suite of liability conventions covering matters such as pollution form oil and hazardous and noxious substances, and wreck raising.[23]

---

*Scottish Journal of Political Economy* (1974) 21(3) 237; HC Deb (11 November 1969) vol 791, cols 199–206.

22 Even if there was Protection and Indemnity (P & I) cover in existence, would the P & I Club have a defence under the 'pay to be paid' clause where the shipowner was insolvent? Any payout under a hull insurance policy might be difficult to identify and pursue.

23 *CLC 1992*. The International Convention on Civil Liability for Oil Pollution Damage 1969 ((adopted 29 November 1969, entered into force 19 June 1975) 973 UNTS 3) (CLC 1969) was amended to create the CLC 1992 by the Protocol of 1992 to Amend the International Convention On Civil Liability for Oil Pollution Damage, 1969 ((adopted 27 November 1992, entered into force 30 May 1996) 1956 UNTS 255) (CLC Protocol 1992). See Section 3.1.

*Fund Convention 1992*. The International Convention on the Establishment of an International Fund for Compensation for Oil Pollution Damage 1971 ((adopted 18 December 1971, entered into force 16 October 1978) 1110 UNTS 57) (Fund Convention 1971). It was amended to create the Fund Convention 1992 by the Protocol of 1992 to Amend the International Convention on the Establishment of an International Fund for Compensation

But before the substance of these conventions can even be considered, there is a major limitation inherent in all of them: they were designed for coastal waters, including the exclusive economic zone (EEZ), and not the high seas.

### 2.4.1 Geographical Restrictions on Damage

Thus, the CLC Article II provides that the convention shall apply exclusively to pollution damage caused in the territory, territorial sea or EEZ of a contracting state. Pollution damage on or in the high seas would not therefore be covered. However there is a limited exception: Article II(b) continues by applying the CLC to 'preventive measures *wherever taken*, to prevent or minimise *such* damage'.[24] This means that there could be liability for clean-up costs on the high seas, but only if that is designed to prevent damage in the territorial sea or EEZ. If the threatened damage only occurs in the high seas, then there is no liability. Of course, it may not be difficult to argue that clean-up operations on the high seas are needed, because of a threat that oil might drift inshore, but the shipowner could require proof that this is likely, e.g. through drift studies. Moreover, the preventive measures would not cover environmental reinstatement on or in the high seas, e.g. for oil which is deposited on the seabed of the high seas, unless again it can be shown that this threatens to cause damage in the EEZ of a contracting state.

---

for Oil Pollution Damage, 1971 ((adopted 27 November 1992, entered into force 30 May 1996) 1953 UNTS 330) (Fund Protocol 1992). In addition the Supplementary Fund 2003 was created by the Protocol of 2003 to the International Convention on the Establishment of a Supplementary Fund for Oil Pollution Damage 2003 ((adopted 16 May 2003, entered into force 3 March 2005) 2012 UKTS 48) (Fund Protocol 2003). See Section 3.1.

*HNS Convention 2010*. The International Convention on Liability and Compensation for Damage in Connection with the Carriage of Hazardous and Noxious Substances by Sea 1996 ((adopted 3 May 1996, not yet in force) 35 ILM 1415) (HNS 1996), as amended by the Protocol of 2010 to the International Convention on Liability and Compensation for Damage in Connection with the Carriage of Hazardous and Noxious Substances by Sea, 1996 ((adopted 30 April 2010, not yet in force) 2010 ATNIF 55) (HNS Protocol 2010). The HNS Convention 2010 is the HNS Convention 1996 as amended by the 2010 Protocol. Active attempts to facilitate its entry into force have been underway at the IMO (see e.g. IMO, 'Facilitation of the Entry Into Force and Harmonized Interpretation of the 2010 HNS Protocol' IMO Doc LEG 105/3 (16 February 2018)), but a concerted effort by key EU states will probably be needed. See Section 3.2.1.

*BPC 2001*. The International Convention on Civil Liability for Bunker Oil Pollution Damage 2001 (adopted 23 March 2001, entered into force 21 November 2008) (BPC 2001). See Section 3.2.2.

*WRC 2007*. The Nairobi International Convention on the Removal of Wrecks 2007 (adopted 18 May 2007, entered into force 14 April 2015) (Nairobi WRC 2007). See Section 3.2.3. For limitation of liability (relevant generally and to the BPC 2001 and WRC 2007), see n 49.

24 Emphasis added.

Similar restrictions on scope exist in respect of the Fund Convention 1992, the Supplementary Fund Protocol 2003 and the Bunker Pollution Convention (BPC) 2001.[25]

The HNS Convention 2010 Article 3 has a similar scope of application, although Article 3(c) does apply the HNS Convention to 'damage, *other than damage by contamination of the environment*, caused outside the territory, including the territorial sea, of any State'[26] if caused by a substance carried on board a ship registered in a state party. The HNS Convention is not restricted to pollution claims and can also include personal injury and property damage caused by HNS.[27] This provision thus allows death and physical property claims that arise on the high seas, but specifically *excludes* typical environmental damage. In effect, the HNS Convention would rarely apply to environmental damage on the high seas, including BBNJ claims. The only caveat to that is if an expansive interpretation is given to Article 3(d).[28] This applies the HNS Convention to preventive measures, wherever taken, to prevent or minimise damage (including by contamination of the environment) in the territory, territorial sea or EEZ of a coastal state party. Some chemicals may be more persistent pollutants than oil. In theory, it might then be easier to say that heavily polluting chemicals leaching from a ship might cause transboundary pollution in a state's territorial sea or EEZ.

The reason for such limitations in the scope of application of the IMO conventions seems both pragmatic and legal. In the 1960s the oil pollution problem was seen as one affecting coastlines and the territorial sea, and the CLC 1969 was restricted to damage in the territorial sea (then 3 nautical miles (NM)). By the time that the 1992 CLC and Fund Protocols were agreed upon, the wider scope of the coastal state rights in the EEZ had been recognised, along with the environmental obligations on states under Part XII of UNCLOS.

### 2.4.2 Gaps for the High Seas

There have been a number of examples of tanker pollution on the high seas where a potential CLC gap was demonstrated, although the effects were largely

25 The International Convention on Salvage 1989 also deliberately restricts the Article 1(d) definition of 'damage to the environment' to 'coastal or inland waters or areas adjacent thereto'. The intention was deliberately to restrict the possibility of salvors claiming special compensation under Article 14 for events occurring on the high seas (in particular). Subsequent suggestions within the CMI to expand the geographical scope have received little support, other than from the salvage industry.

26 Emphasis added.

27 See HNS Convention (n 23) art 1(6).

28 Equivalent to Article II(b) of the CLC.

unknown. The *Atlantic Empress* (with an oil cargo of 287,000 metric tonnes (mt)) sank 10 NM off Tobago in the West Indies in 1979 after a collision with another tanker. It was towed 300 NM to sea where it exploded and sank. This was 'the largest ship-source spill ever recorded. No impact studies were carried out, so it is not known what quantity of oil was burned or sank. Only very minor shore pollution was reported on nearby islands'.[29]

Similarly, in 1983 the *Castillo de Bellver* (with 252,000 mt of oil cargo) broke in two off Cape Town, and the bow section was towed away from the coast and was sunk. Oil was caught by the Benguela Current and drifted to sea. The International Tanker Owners Pollution Federation (ITOPF) reported that there was 'little requirement for clean-up (there was some dispersant spraying) and environmental effects were minimal. The only visible damage was the oiling of some 1,500 gannets'.[30] In other words, from a liability viewpoint if oil did not hit a beach, then there was no real claimant and no damage. Of course, it may be that a natural product such as oil might be 'expected to be broken up by high seas with little or no environmental impact'.[31] From a shipowner's perspective, the spill volume so far offshore may not be directly relevant to the cost of clean-up. The environmental impact could be more serious from HNS cargoes, but the liability position would be the same.

The point is that, in any high seas pollution case involving ships where there is no specialist convention, similar problems of liability and enforcement might arise to those experienced after *Torrey Canyon*. This would be so regardless of whether any environmental claim results from a traditional shipping casualty or is related to a BBNJ claim involving a ship.[32]

### 2.5 *Gaps beyond the Specialist Maritime Liability Conventions*

Gaps in environmental compensation in relation to the high seas could arise, even if the IMO convention liability regimes were extended there.[33] There are

29 ITOPF, 'Atlantic Empress, West Indies, 1979' (19 July 1979) <www.itopf.com/in-action/case-studies/case-study/atlantic-empress-west-indies-1979> accessed 24 May 2018.

30 ITOPF, 'Castillo de Bellver, South Africa, 1983' <www.itopf.com/in-action/case-studies/case-study/castillo-de-bellver-south-africa-1983> accessed 24 May 2018.

31 On the sinking of the *ABT Summer* (260,000 mt) off Angola in 1991, see ITOPF, 'ABT Summer, off Angola, 1991' <www.itopf.com/in-action/case-studies/case-study/abt-summer-off-angola-1991> accessed 24 May 2018.

32 There would also be additional problems, including the identification of the person(s) with the standing to bring a suit in respect of high seas pollution. See also Section 4.1.

33 E.g. some liability conventions are not in force for a particular state or internationally, and there may be exceptions or defences applicable. See Section 3.

also environmental compensation claims that would still fall outside of these regimes. Two examples may illustrate the problems.

#### 2.5.1 Montara Offshore Well Disaster 2009

On 21 August 2009 a blowout occurred at the Montara Wellhead Platform, approximately 250 km off Australia.[34] Oil flowed into the Timor Sea for some 74 days, and patches of oil could have affected an area as large as 90,000 km$^2$. The precise amount of oil discharged is unknown, but a conservative estimate was at least 4,736 mt.[35]

A compensation lawsuit by 13,000 Indonesian seaweed farmers was filed against the Australian operator PTTEPAA (a subsidiary of a Thai company) in the Federal Court of Australia in 2016, alleging negligence and apparently seeking more than AUD 200 million in damages. In preliminary skirmishes PTTEPAA raised a local three-year time bar and class action issues,[36] but appears to be denying that any loss of seaweed was actually caused by oil from the spill. Meanwhile, in May 2017 the Indonesian government was reported to have launched a civil suit in Jakarta against PTTEPAA claiming IDR 27.4 trillion (approximately USD 1.9 billion) for environmental damage and restoration in respect of 1,200 ha of mangrove and 700 ha of coral reef.[37] Indonesia could in theory also seek compensation from Australia, presumably on the basis of state liability for some of the regulatory failures identified in the official inquiry into the incident, although such an action may be politically unlikely.

The present relevance of the Montara incident to the high seas debates is fourfold. First, had such pollution in the EEZ occurred from an oil tanker, there would have been in force a regime of strict liability of the shipowner backed by compulsory insurance.[38] This shows the fragmentary response of the law to marine environmental incidents from operational activities.

Second, between 2010 and 2017 Indonesia failed to obtain agreement at the IMO to create a specific multilateral convention to deal with liability for

34 See generally, David Borthwick, 'Report of Montara Commission of Inquiry' (Commonwealth of Australia 2010).

35 Ibid 301. Compare the 17, 280 mt spilled from the tanker *Kirki* in 1991.

36 *Sanda v PTTEP Australasia (Ashmore Cartier) Pty Ltd* [2017] FCA 14; *Sanda v PTTEP Australasia (Ashmore Cartier) Pty Ltd* (No 3) [2017] FCA 1272. In the latter case an extension of time was granted.

37 Tempo, 'Gov't Files Lawsuit against Thai Company over Montara Oil Spill' <https://en.tempo.co/read/news/2017/05/05/056872668/Govt-Files-Lawsuit-against-Thai-Company-over-Montara-Oil-Spill> accessed 24 May 2018.

38 On the CLC 1992, see Section 3.1. Indonesia is not a party, though, to the Fund Convention 1992.

offshore transboundary pollution from installations and platforms.[39] Objections related both to the jurisdictional competency of IMO[40] and also to the substantive issue of whether such a convention was needed. Ultimately, the IMO Legal Committee decided that 'bilateral and regional arrangements were the most appropriate way to address this matter; and that there was no compelling need to develop an international convention on this subject'.[41] This demonstrates an unfortunate reluctance by, or inability of, UN institutions to take an active lead in areas at the edge of their usual competence. It also shows the strong degree of opposition to increased regulation that might come from vested commercial interests (here the offshore industry) and also states defending those interests or resisting any perception of creeping coastal state jurisdiction. There was even resistance to the development of IMO guidelines on the scope of any regional convention dealing with transboundary pollution,[42] although some rather anodyne voluntary guidelines were approved in 2017.[43]

Third, the existing Australian compensation regime for damage to its EEZ from offshore installations was somewhat inadequate. Its licensing system had required the operator to maintain insurance against 'liabilities', but such administrative requirements can be vague on the type of insurance, the level and scope of cover, and sanctions for non-compliance. After Montara, the legislation had to be tightened to make clear that although the operator itself would be strictly liable without limits, it had to maintain 'sufficient' insurance to deal with all costs including remediation, and in a form acceptable to the state.[44]

Fourth, had the Montara incident occurred on the high seas, the licensing would have been under the authority of the ISA, whose liability and insurance

39 For initial attempts to reform the system, see Gaskell, 'Compensation for Offshore Pollution' (n 19) 80.

40 The IMO Convention 1948 refers repeatedly to 'shipping' in the context of the 'purposes of the Organisation' in Article 1. Platforms are not usually considered as ships.

41 IMO, 'Report of the Legal Committee on the Work of Its Ninety-Ninth Session' IMO Doc LEG 99/14 (24 April 2012).

42 See e.g. IMO 'Report of the Legal Committee on the Work of Its One Hundred and Third session' IMO Doc LEG 103/14 (15 June 2016) para 13.14.

43 IMO, 'Any Other Business. Liability and Compensation Issues Connected with Transboundary Pollution Damage from Offshore Exploration and Exploitation Activities. Guidance for Bilateral/Regional Arrangements or Agreements on Liability and Compensation Issues Connected with Transboundary Oil Pollution Damage Resulting from Offshore Exploration and Exploitation Activities' IMO Doc LEG 104/14/2 (21 February 2017); IMO, 'Report of the Legal Committee on the Work of Its 104th Session' IMO Doc LEG 104/15 (10 May 2017).

44 It appears that no particular form of insurance (nor any specific policy level) has been specified in legislation.

rules are still under development.[45] It has to face the same sort of liability and insurance issues as Australia and other states, but perhaps with more enforcement difficulties. It is easy to create a strict liability rule, but what sort of insurance is required and obtainable, for which claims, and available for which claimants? A national regulatory regime can also impose hefty criminal fines, but this also becomes more difficult to enforce in practice where the high seas are concerned, because the basic principle of jurisdiction is that ships on the high seas are subject to the exclusive jurisdiction of the flag state.

#### 2.5.2 Shen Neng 1

When the coal carrier *Shen Neng 1* went aground in 2010 on the Great Barrier Reef (GBR),[46] the claim was not for pollution by its coal cargo or bunkers, but for physical damage caused by the ship to the GBR. Only up to 4 mt of bunkers were spilled,[47] but the GBR Marine Park Authority claimed a sum varying from AUD 70 million to AUD 194 million for the cost of removing rubble and 750–1500 kg of anti-fouling tributyltin paint that had been scraped off. The extreme location meant that work could only take place at certain times of the year, and this influenced the scale of the costs. In 2016, a settlement was reached mid-trial for AUD 35 million towards the costs of removing polluted rubble, plus AUD 4.3 million for immediate clean-up costs.[48]

This case is a good example of the sort of difficulties that might be expected for high seas claims where physical reinstatement may be difficult or even impossible, both practically and, perhaps, financially. The proportionality of the response was raised, and some evidence indicated that the shoal had recovered better than initially expected; i.e. six years was not really too long to assess the damage and its remediation. All of these factors make claims settlements long and drawn out.

Moreover, in the absence of any applicable convention to cover liability, the shipowner was entitled to limit its liability under the Convention on

---

45 See Section 6.4.

46 See Australian Transport Safety Bureau, 'Independent Investigation into the Grounding of the Chinese Registered Bulk Carrier *Shen Neng 1* on Douglas Shoal, Queensland' (ATSB Transport Safety Report No 274, Marine Occurrence Investigation MO-2010-003, Australian Transport Safety Bureau 2011); see V Cogliati-Bantz and M White, 'Marine Oil and Gas Pollution Spills in Australian Waters' in Aldo E Chircop et al (eds), *Regulation of International Shipping: International and Comparative Perspectives* (Martinus Nijhoff Publishers 2012) 393–99. From a cargo of 68,052 mt, 19,000 mt of coal were offloaded.

47 For which there would have been liability under the BPC 2001. See Section 3.2.2.

48 Minister for the Environment and Energy, 'Owner of Sheng Neng 1 to Pay Commonwealth $39 Million Following out of Court Settlement' (19 September 2016) <www.joshfrydenberg.com.au/guest/mediaReleasesDetails.aspx?id=258> accessed 24 May 2018.

Limitation of Liability for Maritime Claims (LLMC) 1996.[49] This convention (and equivalents in national law) allows a shipowner to limit its liability for a list of specified maritime claims[50] calculated according to the tonnage of the ship. The LLMC limit for the ship was in effect the sum for which settlement was made, i.e. nearly AUD 39.5 million. While limitation of maritime liability may be harder to rely on in some states, e.g. the USA, its existence is strongly defended by shipowners and has to be taken into account in any regime that seeks to make shipowners liable on the high seas.[51]

## 3 Models of Compensation Schemes for the High Seas

If there is no specific international compensation regime for the high seas (beyond general UNCLOS provisions), are there models that might be looked to for any reform? There are, first, the IMO maritime liability conventions, and second, compensation schemes produced voluntarily or as conventions by other international bodies and fora such as the United Nations Environment Programme (UNEP) or the Antarctic Treaty Consultative Meeting.

### 3.1 *IMO: CLC and Fund Model*

The IMO conventions were designed to solve the legal problems about compensation for pollution claims presented initially by the *Torrey Canyon*. What are the key common features of these conventions, and how far do they provide a model that might be extended to the high seas? I will discuss only some main features that derive from the two key conventions currently in force, namely the CLC 1992 and the Fund Convention 1992.[52]

---

49 The LLMC 1996 is the Convention on Limitation of Liability for Maritime Claims, 1976 (adopted 19 November 1976, entered into force 1 December 1986) 1456 UNTS 221) (LLMC 1976), as amended by the Protocol of 1996 to Amend the Convention on Limitation of Liability for Maritime Claims, 1976 (LLMC Protocol 1996) (adopted 2 May 1996, entered into force 13 May 2004; 2004 ATS 16). The LLMC uses the special drawing right (SDR) of the International Monetary Fund (IMF) as its unit of currency. The SDR is also used in virtually all of the modern maritime liability conventions. The IMF issues daily conversion rates for most international currencies, so examples referred to later in this chapter will generally be prefaced by 'about'.

50 Including most (non CLC) environmental damage claims, e.g. for clean-up, and economic loss claims. For more details of the LLMC and its application to bunker claims, see Gaskell, 'The Bunker Pollution Convention 2001 and Limitation of Liability' (n 19). For examples of applicable limits of liability, see Nick Gaskell, 'Limitation of Liability and Division of Loss in Operation' in Simon Gault et al (eds), *Marsden and Gault on Collisions at Sea* (14th edn, Sweet and Maxwell 2016) app 4, 973–1020.

51 See also the 'Antarctic' solution in Section 6.3.

52 See n 19, in particular Gaskell, 'Compensation for Offshore Pollution' 65.

The CLC creates liability of a shipowner for oil pollution damage from oil tankers. The liability is strict with limited defences (such as war). There is a limit of liability calculated according to the tonnage of the ship that is separate from, and higher than, the limits in the LLMC 1996.[53] Liability is channelled to the registered shipowner only; suits are not allowed against others such as charterers and salvors. The reason is the key innovation introduced by the CLC, namely, compulsory insurance of the shipowner, coupled with a right of direct action against the insurer. This combination combats the problem of the single-ship company which has no assets. In practice, the vast majority of tankers are covered by members of the International Group of Protection and Indemnity (P & I) Clubs, i.e. reputable mutual insurers with massive reinsurance cover. Compliance with insurance requirements is enforced directly by the flag state, but indirectly by port states that can demand a certificate of insurance for entry into or exit from their ports. The damages recoverable include clean-up of contamination, economic losses, compensation for impairment of the environment to cover reinstatement, and costs of preventive measures. Jurisdiction is given to the courts of a contracting state where damage occurs in its territorial sea or EEZ,[54] and there are provisions for mutual recognition and enforcement of judgments.

The CLC provides the first tier of liability. The drafters always envisaged that further protection would be needed to provide additional compensation to cover larger claims, e.g. those above the shipowner's CLC limit of liability. This second tier is provided through the Fund Convention, which reflects an acceptance that the polluter who ought to pay is not only the shipowner (who is merely a transporter and may be liable even without fault) but also the commercial interests responsible generally for the oil itself, i.e. the oil companies.

The Fund Convention does not create a liability of the owner of the oil being carried, but spreads the loss more widely over the whole industry. This is achieved by the creation of the International Oil Pollution Compensation Fund (IOPC Fund), a body with separate legal personality. The IOPC Fund operates by means of levies raised (after a particular casualty) against oil importers in all states parties, in proportion to the amount of oil received in a given year. This post-accident levy is, in effect, spread to the ultimate users of the oil, such as car users, by what are tiny increases in the cost of fuel. The levy is not against states, but against those companies importing the oil. The immediate burden of collection and payment therefore falls on those companies in the states that have caused the carriage of pollutants in the first place, particularly

53 The maximum would be about USD 126 million, but less for smaller tankers.

54 For the scope, see Section 2.4.1.

in the developed world.[55] Importantly, the system means that if a tanker on route to the developed world sinks and causes pollution in a developing state, e.g. in Africa or Asia, the CLC and IOPC Fund protection is available to that state (provided it is a state party). Moreover, developing states that are oil *exporters* do not contribute directly to the IOPC Fund, while developing states that are importers pay relatively little, especially as there is an annual import threshold for contributions of 150,000 mt per company.

The Fund has a limit of liability of around USD 285 million, while parties to the third-tier Supplementary Fund 2003 have access to an aggregate of about USD 1 billion. The Funds' liability is broadly the same as in the CLC, but in some circumstances the Funds' protection might operate down to the first dollar of any claim.[56]

In my opinion, the CLC and Fund Convention system is very successful; it is in force and it works to pay claims. Both sides of the shipping industry—shipowners and cargo owners—can contribute to claims. There is a surprisingly high degree of cooperation between the P & I Club insurers and the IOPC Funds. The IOPC Funds were set up by states to pay claims and have developed international environmental claims practice and provided helpful guidelines for claimants, including a claims manual[57] designed to flesh out which claims are likely to be admissible.

### 3.2 *Other IMO Pollution Liability Conventions*

Subsequent IMO conventions have made use of boilerplate text and principles from the CLC and Fund Convention system; thus, here are similar definitions (e.g. of pollution damage) and defences, while administrative experience in dealing with one convention has been extended to the others.

#### 3.2.1 HNS Convention 2010

The CLC and Fund Conventions apply only to oil cargoes. The HNS Convention 2010 aims to fill a gap, by providing similar protection for a list of environmentally hazardous substances other than oil. The list includes both bulk chemicals[58] and packaged cargo (e.g. containers). It follows the two-tier

55 In most years since the IOPC Fund entered into existence, Japan has been the largest contributor to the Fund, e.g. for casualties that may have happened anywhere in the world. Most recently India has become the joint highest contributor: see IOPC Funds, 'Annual Report 2017' 22, 24.

56 E.g. when the shipowner or its insurer is insolvent, or where the shipowner can rely on one of the limited defences to strict liability under the CLC Art III(2). The IOPC Funds are not liable if the pollution results from war or warships.

57 See IOPC Funds, 'Claims Manual' (October 2016).

58 E.g. the styrene aboard the chemical tanker *Ievoli Sun,* which sank off France in 2000.

liability structure (albeit in a single convention). The shipowner would have a maximum first-tier liability of about USD 140 million, backed by compulsory insurance;[59] the second tier HNS Fund would provide cover up to about USD 351 million. It concentrates only on carriage of HNS cargo[60] by sea; i.e. it contemplates a carrying voyage, rather than say a ship carrying HNS for the purposes of the operation of the ship.

#### 3.2.2 The Bunker Pollution Convention 2001

The BPC 2001 deals with another particular problem, the bunkers (fuel) carried in all ships (e.g. any cargo or passenger ship). It has a similar strict liability scheme to the CLC backed by compulsory insurance, but has no second-tier 'fund'; there is only a single-tier shipowner liability system for shipowners. To that extent, its pollution liability coverage may be severely limited by the ordinary ship's limits of liability, which moreover, might have to be shared with property claimants (e.g. cargo owners).[61] In the *Pacific Adventurer* casualty in 2009, costs of over AUD 30 million were potentially caught by the ship's LLMC limit of about AUD 18 million, although the case settled for slightly more.[62] The BPC creates some liability of non-shipowners (e.g. bareboat charterers, managers and operators), but they are not required to have insurance.

#### 3.2.3 Wreck Removal Convention 2007

The Wreck Removal Convention (WRC) 2007[63] contains a mixture of public and private law provisions. It imposes obligations on shipowners to remove wrecks in the EEZ, as well as creating liabilities. It is *not* a pollution convention as such, but recognises that removal of the hull and (more importantly) the cargo can prevent environmental harm. It creates strict liability, but only for the costs of removal of a wreck and its cargo (including hazardous substances).[64]

### 3.3 *Evaluation of IMO Liability Convention Model*

Overall the IMO suite provides a very workable compromise system that actually pays claims, and can start to do so relatively quickly without having to

---

59 Or USD 160 million where packaged cargo is being carried.

60 HNS Convention (n 23) art 1(5).

61 See Gaskell and Forrest, 'Marine Pollution Damage in Australia' (n 19); Gaskell, 'The Bunker Pollution Convention 2001 and Limitation of Liability' (n 19).

62 See Gaskell, 'The Bunker Pollution Convention 2001 and Limitation of Liability' (n 19).

63 See Gaskell and Forrest, 'The Wreck Removal Convention 2007' (n 19).

64 Even before the HNS Convention comes into force. See Gaskell and Forrest, 'The Wreck Removal Convention 2007' (n 19) 96–97.

worry about fault. It achieves a balance between environmental concerns (e.g. of states) and the cost concerns of industry. Compulsory insurance and direct action work, and the IOPC Funds provide cover for most disasters. There is a surprising degree of cooperation between insurers and the IOPC Funds in administering claims.[65] The Funds have developed key workable principles over many years.[66]

The system is not perfect, and it does not apply to the high seas. The damage definition is rather more restrictive than many environmentalists would like. Limits of liability have continually needed updating; in many ways the emphasis of the CLC/Fund Convention has moved from an environmental system to one dealing with economic losses claims (e.g. fishing and tourism) that have swamped the funds available after major casualties. The system does not always provide immediate response funds. There is a dependence on national courts for interpretation and application, which can risk a lack of uniformity in application.

In general terms a major criticism is that the IMO has, albeit understandably, produced a piecemeal series of solutions with separate conventions for oil, HNS, bunkers and wrecks. This means that there is a lack of uniformity, with different state parties and some conventions not in force (e.g. the HNS Convention). The IMO conventions only deal with ships and not pollution from oil rigs. They do not cover wider environmental damage by ships, e.g. physical damage caused by ships to reefs. And there are defences and exclusions from cover, e.g. the HNS Convention does not apply to coal cargoes.

## 4 New 'High Seas Protocols' to the Existing IMO Shipping Liability Regimes?

Should the existing IMO shipping conventions, which create liabilities for commercial parties, be extended to the high seas? Article 235(2) of UNCLOS[67] requires states to provide for legal recourse within their domestic legal systems[68] for compensation arising from damage to the marine environment (including the high seas) caused by persons under their jurisdiction. But this does not specify the type of regime required, so long as it is prompt and provides

65 Although note the tension caused by the *Nissos Amorgos* casualty. E.g. *Assuranceforeningen Gard Gjensidig v The International Oil Pollution Compensation Fund* [2014] EWHC 3369 (Comm) [2014] EWHC 1394 (Comm) [2014] 2 Lloyd's Rep 219.

66 E.g. in IOPC Funds, 'Claims Manual' (n 57).

67 See Section 2.1.

68 Compare UNCLOS (n 8) Part XII art 209.

'adequate compensation'. So it would be difficult to argue that a failure to extend the IMO conventions would itself somehow be a breach of Article 235(2).

It would not be too difficult to draft provisions extending the scope of, say, the CLC to damage outside the territorial sea and EEZ. But, even if there were a political appetite for such action, it would also be necessary to settle some general jurisdictional issues that may be raised.

### 4.1 *Jurisdictional Issues*

Although Article 235(3) encourages the development of existing compensation regimes, an extension of the IMO regimes might still be opposed. There might be objections on the basis that UNCLOS places limits on the extent to which non-flag states (e.g. coastal states) could impose liabilities on foreign-flagged ships on the high seas. For instance, during the negotiations for the WRC 2007, there was strong opposition (e.g. by the USA) to any extension of UNCLOS powers exercised by non-flag states over foreign-flagged ships outside territorial waters (even in the EEZ). If high seas protocols were agreed, states that did not become parties might challenge the application of such protocols to their ships. A particular problem might be objections that any system of compulsory insurance for ships exercising rights of navigation on the high seas would be contrary to the law of the sea (at least to the extent that there were attempts to enforce compliance on the high seas). In those states party to any high seas protocol with compulsory insurance requirements, their port state control officers can no doubt enforce the requirement to have an appropriate 'high seas protocol' insurance certificate.

It is, of course, arguable that the very extension of the CLC and Fund Convention to the EEZ is an example of Article 235(3) being used to justify measures against foreign-flagged ships outside of territorial waters. In practice, if a lack of uniformity is to be avoided, there would probably be a need to rely on flag states to become parties to any new protocols, or at least to acquiesce in their use by other states.

The jurisdictional sensitivities would be heightened if the CLC and Fund Convention model were continued and national courts were be given jurisdiction to decide environmental liabilities on the high seas. It is at least worth raising the possibility of establishing an international forum for the resolution of disputes, e.g. a chamber of the International Tribunal for the Law of the Sea (ITLOS), or the International Court of Justice (ICJ),[69] or the establishment of a specialist Arbitration panel.[70]

69 Although it is arguable whether the ICJ or ITLOS is perfectly adapted for claims about damage resulting from offshore platforms accidents.

70 See Section 6.3 on the 'Antarctic' approach.

It is curious that none of the IMO liability conventions define who can be a claimant. The restriction, if any, is in relation to the damage. In practice, claimants are governments (or those who pay for clean-up) and private companies or individuals who incur physical or economic losses. In any protocol there might be questions of standing to sue, e.g. if a state with no obvious interest claims remediation costs. If it incurs such costs, then why should it not be able to claim on the basis of preservation of the common heritage of mankind?[71] In principle, any extending protocol would need to state expressly that a remediating state can claim.[72]

### 4.2 *Evaluation of 'High Seas Protocols' Solution*

It is inevitable that, as part of any process of negotiating high seas protocols, some states and non-governmental organisations might want to 'improve' the underlying conventions. Experience at the IMO shows that this is likely to be fatal to any negotiations. For better or worse, there might need to be minimal changes beyond altering the scope of application. If shipowners are to be liable, then it may be difficult to avoid limits of liability. There is no reason why current CLC, HNS or LLMC limits cannot apply—it would be odd to have higher limits for the high seas when there are likely (initially at least) to be fewer economic loss claims. But the operation of the LLMC (e.g. with protocols to the BPC 2001 and WRC 2007) might result in relatively small amounts being available.

Moreover, there are differing parties to each of the IMO conventions. It would be difficult to get a buy-in from flag states, shipowners (and their P & I Clubs)—or from major contributing states to the IOPC Funds.[73] Many flag states might refuse to ratify if there are significantly increased risks for owners and that, in turn, might present enforcement problems.

The CLC and Fund Convention boilerplate definition of damage had been deliberately restricted, in part because of concerns by payers (shipowners and receivers). Environmentalists would prefer a more modern, expanded definition that reflects full principles of remediation as well as environmental impact assessments (EIAs). Any attempt to widen the definition of environmental damage would be controversial, given the huge potential costs

71 The ITLOS Advisory Opinion in 2011 (n 16) provides some support for a wider list of possible claimant states. In the context of deep seabed mining, this included the ISA, entities engaged in deep seabed mining, other users of the sea and coastal states. See paras 179–80.

72 See Section 6.3 on the Antarctic Protocol Annex VI solution.

73 States like Japan, a major contributor to the IOPC Fund, might be reluctant to agree to an extension (e.g. of the Supplementary Fund) where significant costs could be incurred, especially as the world's largest importers of oil, the USA and China, are not parties.

of remediation[74] on the high seas.[75] If I am right that economic loss claims have warped the CLC and Fund Convention system by relegating the (limited) amounts available away from environmental remediation, it might even be open to restrict any high seas protocols so as to exclude economic loss claims completely.[76] This would be controversial, but might perversely provide more funds for remediation or research.

Extension of the HNS Convention to the high seas may be more problematic in that chemical pollution from a ship's cargo could well be more persistent than oil, and thus more difficult and expensive to remediate. Given the difficulties in achieving entry into force of the HNS Convention (especially in ensuring a viable second tier), the cost sensitivities might result in further delay of entry into force of any HNS high seas protocol.

This 'high seas protocols' approach does not produce a holistic solution. As a matter of principle, it merely multiplies the existing difficulties among the individual liability conventions. It would not extend beyond shipping accidents to activities such as physical damage to a reef, or disturbances to ecosystems by shipping activities. But, would it not be better than no action? And, if it were linked to developments with the ISA and BBNJ discussions, would that not also add political force to encourage change at the IMO? Experience at the IMO shows that without political momentum, it would be very difficult to achieve high seas protocols at the organisation.

## 5 New Single Multilateral Maritime Environmental Liability Regime for the High Seas?

As an alternative to high seas protocols, is it worth considering something more ambitious, i.e. a new universal multilateral environmental liability convention for the high seas? Given the likely political difficulties, it may be unrealistic to consider something grander even than high seas protocols.[77] But even in theory, is such an idea feasible or worthwhile?

To some extent these compensation issues are being addressed separately in the BBNJ discussions[78] and those about the ISA Mining Code.[79] The latter is

74 Compare the costs claimed in respect of the *Shen Neng 1* in Section 2.5.2.

75 This is true even if an express proportionality principle were introduced. See WRC 2007 (n 23) art 2(2).

76 Compare the WRC 2007, which only covers costs of removal, not economic losses.

77 Compare the post-Montara problems in creating transboundary liabilities for rig spills and the potential for conflict with the developments by the ISA: see Section 2.5.1.

78 See Section 7.

79 See Section 6.4.

obviously only in the context of seabed activities and not more generally, e.g. to cover the superjacent waters. Ideally, a new single compensation convention would apply to all forms of operational and accidental pollution on the high seas. This could then cover pollution resulting from deep seabed mining as well as activities in the water column that might affect BBNJ.

Strict liability rules for private operators could be drafted fairly easily, but there may be inherent difficulties in trying to create a regime that would apply equally to, say, shipowners, rig operators and others undertaking activities in the water column. Those difficulties may relate directly to finance and insurance. The P & I Clubs would not cover platforms or seabed activities not involving ships, as these have traditionally been insured in different markets. If the CLC type system of compulsory insurance is to be kept (i.e. allowing port state control), separate insurance rules for different types of activity might be needed. There is a real risk of 'paper' insurance companies set up in less developed state parties, with the result that their insurance certificates would have to be recognised by other states.

There would also be issues about the type and level of damages claimable,[80] and whether an extensive or restrictive definition of environmental damage would be appropriate.[81] Yet CLC type definitions would not address BBNJ issues. Although itself controversial, it seems that limitation of liability is in practice going to be necessary if there is to be shipowner liability, but should there be limits for other operators?[82] The alternative is unlimited liability but with contractual policy limits on the insurance cover available. Insolvency of operators and the need for direct action against insurers are then important.

Furthermore would there be one or two tiers of liability or compensation? It is difficult to envisage an IOPC-type fund system that can identify a discrete body of contributors to cover all incidents on the high seas. The alternative might be to establish some form of quite different 'environmental trust fund', but it is difficult to see this working as part of a liability or compensation system. Rather, this might be more suited to provide discretionary payments for remediation, the undertaking of EIAs, or research.[83] It might be possible to have an environmental trust fund divided into different contribution sectors, but the complications with entry into force would be much greater than that

80 See the discussion in Section 4.2.

81 The ITLOS Advisory Opinion in 2011 (n 16) contemplated that 'other entities' could have a claim for environmental damage; this might include economic loss.

82 Compare the Antarctic solution in Section 6.3.

83 If there is a single-tier system (i.e. with only such a Trust Fund operating independently), shipowners might escape liabilities for shipping accidents. In effect, there would be cross subsidy of shipping by the exploration industries that might be obliged to contribute to the Trust Fund. See also Sections 6.2, 6.3 and 7.2 on Trust Fund issues.

presented by the HNS Convention, which is still not in force over 30 years after its original adoption.

The central question, though, is whether any 'gap' problem is viewed as being less about normal shipping transport activities and more about exploitation of the seabed and water column. On that basis, it might be better and simpler to separate the treatment of ships from operational activities on the seabed and in the water column. This would allow for more flexible responses of the kind available to an administrator of a prospecting system, like the ISA.

If that is right, any single liability convention might be restricted to cover all forms of pollution from ships (but with the usual exclusions, e.g. for nuclear material); i.e. it would bring together all the liability principles from the CLC and Fund Convention, HNS Convention and the BPC. One advantage of this approach is that some of the restrictions and overlaps between the IMO regimes could be ironed out, but it may be doubted if much would then be gained over high seas protocols to the IMO conventions. Moreover, shipowner opposition would be stronger if existing second-tier fund mechanisms are likely not to continue to operate.

Realistically, it seems unlikely that there would be much appetite at the IMO to create such an instrument for ships, at least when there is no similar unified system operating in the EEZ. If the IMO were not primarily involved, some other body would need to have carriage of any wider convention, but part of the difficulties with the BBNJ negotiations is in deciding any future governance structure.

If a new multilateral convention is highly unlikely, what other models might there be to assist the ISA and BBNJ discussions?

## 6 Other Compensation Models

The IMO liability conventions present a compensation model based on traditional maritime law. Other models have been considered internationally, with differing degrees of success.[84] Some mention will be made of voluntary industry schemes, and conventions dealing with hazardous waste and the Antarctic.

84 The nuclear conventions put liability on operators but are so specialised and, perhaps, politicised that it is not appropriate to consider them here. See Günther Handl, Chapter 7 'High Seas Governance Gaps: International Accountability for Nuclear Pollution'. For some issues concerning the loss of nuclear reprocessing cargoes, see NJJ Gaskell, 'Lessons of the Mont Louis Part Two: Prevention of Hybrid Accidents' (1986) 1 *International Journal of Estuarine and Coastal Law* 269.

### 6.1 *Industry Voluntary Schemes*

Instead of extending the existing IMO conventions, would it be possible to persuade shipowners and cargo interests to agree to new voluntary liability schemes to apply on the high seas?

Prior to the entry into force of the CLC 1969 and Fund Convention 1971, the tanker and oil industries operated two voluntary schemes, called TOVALOP and CRISTAL,[85] which provided compensation for pollution damage in a similar way to the conventions. They were designed to operate in the interim before the two conventions entered into force, although for a variety of reasons they actually lasted until 1997 in amended form.[86] The advantage for states was that such schemes could be established relatively quickly (i.e. without the need for a diplomatic conference). They provided strict liability with significant levels of compensation and were backed by the P & I Clubs (for TOVALOP) or major oil companies (for CRISTAL). From the industry perspective, the form of the schemes allowed the commercial parties a major part in their drafting and remedied some of the uncertainties they might face in a convention liability scheme. In particular, the voluntary schemes could identify the possible claimants; define the type of damages recoverable (e.g. restricting recovery to reasonable costs and excluding remote, speculative or indirect loss or damage); set out a strict (one-year) time bar; provide for unbreakable maximum limits of liability; and ensure that any disputes would be subject to a preferred governing law (in fact, English law).

Although technically the two agreements were about the assumption of liabilities *between* the respective commercial parties (shipowners or oil companies), the agreements appeared to recognise that pollution claimants would have a direct legal claim, e.g. against the participating TOVALOP owner whose tanker caused the incident. Thus, Clause VIII (D) of the revised TOVALOP deemed such an owner irrevocably to have offered to the victim to submit all disputes to International Chamber of Commerce (ICC) arbitration (presumably on the basis of the terms of the agreement). In effect, this appears to have

---

85 The Tanker Owners Voluntary Agreement concerning Liability for Oil Pollution was an agreement between tanker owners who agreed to pay for pollution from their tankers up to a first tier of liability. The Contract Regarding a Supplement to Tanker Liability for Oil Pollution was an agreement between the major oil companies that would contribute in effect to a second tier of liability where there were spills from one of their oil cargoes. See e.g. Gard, 'TOVALOP/ITOPF/CRISTAL' <www.gard.no/web/updates/content/52964/tovalopitopfcristal> accessed 24 May 2018.

86 They could even be used by states that did not become parties to the CLC 1969 and Fund Convention 1971, although one reason for their withdrawal in 1997 was to encourage states to adopt the CLC 1992 and Fund Convention 1992, which had entered into force in 1996.

been an offer for a unilateral contract (like a reward contract), offered by the industry and accepted by a claimant (state or individual) that suffered loss and made a claim.

Would it be possible to use voluntary schemes for the high seas, either (i) to create new voluntary liabilities e.g. for seabed exploitation or, (ii) more conservatively, to extend some or all of the IMO liability conventions to the high seas?

Such schemes have only come about when there is a relatively homogenous industry grouping, and perhaps in political circumstances where a threat of international instruments being produced might result in something more onerous than an industry-backed scheme. The only voluntary scheme for offshore pollution from platforms operates in northern Europe,[87] and there is no indication of this being extended geographically. It is difficult to foresee whether there could ever be a sufficient commonality of commercial interests in the mining of minerals (such as manganese, nickel or tellurium) sufficient to create a global industry organisation to come together to agree to such a voluntary shared liability scheme. It may well be, however, that offshore mineral mining becomes the province of only a few huge operators. In such an event, these operators might be in a position to agree to a voluntary scheme in the same way as the oil majors after *Torrey Canyon*. This seems unlikely at present, as the multiplicity of interests presents the same difficulties faced in the creating of the HNS Fund. Moreover, the idea of multiple schemes for each kind of mineral seems unworkable. Deep seabed activities might better be covered by licensing and insurance through the ISA.[88]

By contrast, there might be more scope for extending at least some of the IMO liability conventions to the high seas. Why not create a TOVALOP-style voluntary scheme extending the CLC, at least, to the high seas? A 'high seas TOVALOP' would be quicker to introduce than a convention, would avoid jurisdictional issues, and would build on a well-tried system. From the shipowners, and P & I Clubs', perspective, it would offer the certainty of a degree of control in drafting and enforcement. They could insist on claims being subject to the incurring of actual clean-up costs, which would avoid theoretical (and apparently indeterminate) environmental damage claims. EIAs could be performed, but limited in time and cost. From the environmental perspective, this sort of solution might seem retrograde, but would it be better than nothing (at least for the interim)? Could the oil industry be persuaded to provide a similar top-up 'high seas CRISTAL'? Again, the industry would have a degree of control

87 Offshore Pollution Liability Agreement (originally dated 4 September 1974, but currently effective as from 21 June 2017) (OPOL) <www.opol.org.uk/> accessed 24 May 2018.

88 See Section 6.4.

over costs. There might not be much liability exposure if a high seas spill is impossible to remove, or if remediation is practically impossible owing to location, or unnecessary owing to dispersal of a natural product.

If such a voluntary extension of the oil tanker regime were possible, it would not seem too great a step to provide a similar voluntary scheme for bunker pollution from non-tankers. The existing BPC 2001 scheme is a single tier of liability on shipowners alone (without any contribution from the oil industry).[89] So it would be necessary to obtain consent from a huge number of individual shipowners to join such a voluntary agreement, although there might actually be few bunker clean-up claims on the high seas (because most bunker spills are small and occur in harbour areas). Any high seas 'Bunker Pollution Agreement' could probably only be financed through the P & I Clubs or other insurers, in a similar manner to the way that they arrange to bill members for membership of ITOPF.[90] It may be more difficult, though, to persuade the industries to have a voluntary HNS type of scheme, given the existing difficulty in bringing the HNS Convention into force. Likewise, the huge potential costs would militate against a voluntary extension of the wreck removal liabilities to the high seas.

### 6.2 *Basel Convention: Liability and Compensation Protocol 1999*

The Basel Convention on Transboundary Movement of Hazardous Wastes and their Disposal 1989 (Basel Convention) has a 1999 Liability and Compensation Protocol (which is not in force). The Protocol would create strict liability for damage caused by illegal dumping of wastes or accidental spills, e.g. from a leaking container. There are CLC type defences, e.g. war, and compliance with government authorities. Depending on the stage of the movement at which damage occurs, liability is split between the 'notifier' and ultimate 'disposer'.[91] The notifier has a limit of liability of 'not less than' about USD 1.4 million per shipment (shipments up to 5 mt) or about USD 14 million (shipments up to 10,000 mt), and the disposer is liable for not less than about USD 2.8 million; there is unlimited liability if there is fault. In theory, there is to be compulsory insurance up to the limits.

The Basel Convention is perhaps of too narrow a focus to provide much assistance as a wider model for the high seas. There is an endorsement of strict

89 See Section 3.2.2.

90 ITOPF originally administered TOVALOP, but now operates as an expert advisory body to the industry on marine pollution events and damage assessment.

91 The notifier is in effect the generator or exporter of the waste, who is obliged under Article 6 of the Basel Convention to notify the exporting and importing State of the proposed transboundary movement. The notifier is liable under Art 4 of the Protocol until the disposer has taken possession of the wastes.

liability, a need to identify the person liable, and a recognition that there needs to be insurance. I can only speculate that among the reasons for the Liability and Compensation Protocol not entering into force[92] are uncertainty about the availability of insurance coverage and the somewhat vague description of the limits of liability. Presumably because of the problems of insurance, the Protocol also sets out a voluntary 'Technical Cooperation Trust Fund', which seems to be intended to operate as a second tier of liability (vaguely like the IOPC Funds) to operate as a backup. The lack of progress in creating such a fund may reflect the fact that the Basel Convention itself has reduced the underlying problem of illicit exports from the Organisation for Economic Co-operation and Development (OECD) developed states, or the underlying difficulties in financing such a fund.

### 6.3 *The Antarctic Treaty: Environmental Protection Protocol 1991*

The Protocol on Environmental Protection to the Antarctic Treaty 1959 was agreed in 1991 and entered into force in 1998. Unlike the deep seabed regime, the Protocol does not allow for mineral development, so the activities that can occur in Antarctica are fewer than in the Area under Part XI of UNCLOS, but include scientific research and tourism. The 1991 Protocol established a Committee for Environmental Protection (CEP), which meets annually.[93] It has produced a body of guidelines, e.g. about environmental impact assessments, which have been incorporated into a CEP handbook.

Annex VI to the Protocol on Environmental Protection to the Antarctic Treaty on Liability Arising from Environmental Emergencies was adopted in 2005, but will only enter into force once approved by all 29 consultative parties to the Antarctic Treaty.[94] It took 13 years to negotiate this Annex, but consensus

92 Twenty states are needed, and there are 11 parties.

93 See Committee for Environmental Protection, *25 Years of the Protocol on Environmental Protection to the Antarctic Treaty* (Secretariat of the Antarctic Treaty, Buenos Aires 2016) 14.

94 It is understood that 14 consultative parties out of 29 have now approved Annex VI, with Russia being the latest to give details of its implementing legislation at the ATCM XL-CERP XX meeting in Beijing in May 2017. States such as Australia have already implemented legislation on the statute book. See the Antarctic Treaty (Environment Protection) Amendment Act 2012 (Cth). Others (such as Finland, the Netherlands, Norway Russia and Sweden) are apparently already applying Annex VI as part of domestic law. See Antarctic Treaty Consultative Meeting, 'Final Report of the Fortieth Antarctic Treaty Consultative Meeting' (Beijing, 22 May–1 June 2017, Vol 1, Secretariat of the Antarctic Treaty 2017) 42–52.

is regarded as particularly important in the Antarctic Treaty System.[95] It seems that the scope of the liability rules was controversial, with some states proposing very wide liability rules to cover all forms of activities, including scientific research and tourism, while others wanted a more limited scope dealing only with response.[96] In the event, the Annex sets out a more restricted scope to deal with 'environmental emergencies' from the (relatively limited) activities that can occur in Antarctica.[97]

There are state obligations to take preventive measures and to require their 'operators' who organise activities in the Antarctic Treaty Area to have contingency plans and to take prompt and effective response action. Failure to do so creates strict liability[98] under Article 6 to pay the costs of the 'response action' as defined in Article 2.[99] There are limited defences, in Article 8, which have some similarities with the CLC defences, but are differently phrased.[100]

Article 9 sets out limits of liability for the operator. Antarctic activities, e.g. on the ice mass, may not necessarily involve a ship; in such cases there is a limit on response cost liability of three million special drawing rights.[101] For events involving ships there are limits based on, but legally separate from, the property limits found in Article 6(1)(b) of the LLMC 1996.[102] The implications

---

95 See Stuart Kaye, Michael Johnson and Rachel Baird, 'Law' in Marcus Howard and Tom Griffiths (eds), *Australia and the Antarctic Treaty System* (UNSW Press 2011) 101.

96 Ibid 314–15.

97 See Louise de La Fayette, 'Responding to Environmental Damage in Antarctica' in G Triggs and A Riddell (eds), *Antarctica: Legal and Environmental Challenges for the Future* (BICL 2007) 134, 145.

98 Under Article 10 of Annex VI, there is no state liability per se for the failure of private operators to take response action. Annex VI to the Protocol on Environmental Protection to the Antarctic Treaty—Liability Arising From Environmental Emergencies (adopted 14 June 2005, not in force) (Annex VI) art 10.

99 The WRC 2007 has a similar mixture of public law obligation to act, and liability for failure to do so. It also provides compensation for costs rather than wider environmental or economic losses.

100 Article 8(1)(b) includes exceptional natural disasters but less pithily that in the CLC Art III(2)(a), and has a less expansive definition of war than the CLC, as it refers to 'belligerency against the activities of the operator'. Articles 8(1)(a) and (c) introduce new defences of acts to save human life and terrorism respectively. The latter is mostly covered in the CLC defences, but if the IMO suite of conventions were being redrafted today, there seems little doubt that such an exception would be express.

101 About USD 4.2 million.

102 But the Annex VI limits do not reflect the 50 per cent increases in the LLMC 1996 limits brought into effect for the LLMC 1996 as from April 2015. Annex VI (n 98) Article 9(4) does allow for the limits to be reviewed every three years by the regular consultative party meetings, and this seem to allow for a quicker and easier revision of limits than under the LLMC.

for the Annex on the limitation of liability of shipowners were perhaps not fully appreciated in the early stages of negotiations and concerns were raised by flag states. This resulted in some last minute and rather complicated savings provisions in Article 9(2) for existing limitation conventions such as the LLMC; these require close comparisons of who is party to which instruments. It remains to be seen if these Annex VI 'stand-alone' ship limits in Article 9, deferring to existing limitation regimes,[103] are workable or satisfactory. The structure of the Article 9 limitation provisions has a number of untested differences with traditional shipowner limitation, e.g. in the definition of who can limit.[104] What seems clear is that although the Annex limits are the minimum applicable for states parties,[105] they cannot bind non-states parties[106] whose shipowners may be able to rely on other applicable limitation conventions.

There is a provision, Article 11, about insurance or other financial security. Following the pattern of national offshore operations,[107] there is a requirement on states parties to require their own operators to maintain 'adequate insurance' for response costs. The amount of insurance is to cover liability up to the limits in Article 9(1) or 9(2). This is better than leaving the amount of insurance to states parties[108] and is slightly more flexible than the IMO conventions, as it appears that the Annex VI limits can be amended more quickly.[109] By definition, this insurance requirement can only apply to operators of states

---

103 Compare the BPC 2001, where shipowner's limitation is left entirely to pre-existing regimes in national and international law, while the maximum liability of the insurer *is* linked to the limits in LLMC 1996. See Gaskell (n 19).

104 Annex VI creates liabilities of an 'operator' that organises activities, rather than the 'shipowner' under the LLMC. The Annex does not have provision for the establishment of a limitation fund.

105 Article 9(2) thus requires state parties to the Annex *and* the LLMC *1976* to apply the higher Annex limits (based as they are on the LLMC 1996 limits as they were in 2005, although not the higher limits applicable from 2015). It is arguable, though, that in a jurisdiction that applies the LLMC 1996/2015, the latter's limit will prevail over the Annex VI limit. See Antarctic Treaty Consultative Meeting (n 94) 22–23; International Group of Protection & Indemnity Clubs, 'IP 87 Liability Annex: Financial Security' (IG P & I Clubs 2017) para 13.

106 Vienna Convention on the Law of Treaties (adopted 23 May 1969, entered into force 27 January 1980) 1155 UNTS 331, art 34.

107 See Section 2.5.1; see also conventions such as the Montreal Convention for the Unification of Certain Rules for International Carriage by Air (adopted 28 May 1999, entered into force 4 November 2003) 2242 UNTS 309 (Montreal Convention 1999), art 50.

108 Compare the Montreal Convention 1999 (n 107), art 50.

109 Although revisions may in theory move through the agenda more quickly, the assumption of consensus at Consultative Meetings may mean that that shipowner states will in practice expect some linkage to IMO's LLMC revisions. In this case, the whole process may well be quite lengthy.

parties. There is no attempt to apply the requirement to non-states parties, or their ships. Unlike the IMO conventions, there is no provision for direct action against the insurer and none of the detailed provisions that go with that (dealing with possible insurer defences, such as insolvency of the assured, or wilful misconduct). As with national regulatory systems, the terms of the insurance and the reliability of the insurer are matters left to the state party. It also follows that there are no provisions on port state control, e.g. the extent to which foreign-flagged ships can be denied entry to, or exit from, ports without appropriate certificates. These are important powers in the IMO regimes that in practice can prevent ships trading internationally without insurance.

Article 7 is significant, as it allows a state party that 'has taken response action' to sue an operator that is liable. This avoids the type of dispute about standing to sue that might arise on the high seas. It also allows claims to be brought to the courts of state parties. Because operators have to be authorised under the Antarctic Treaty System, it seems easier to specify the courts that will have jurisdiction over foreign companies, e.g. those incorporated in non-state parties. In this instance they can be sued in the courts of the authorising party, although this may present problems of enforcement in practice if assets are not in that state.[110] The Antarctic Treaty 1959 has its own provisions in Article XI to deal with dispute resolution. The mechanism is consensual, allowing for the commencement of proceedings in the ICJ. The dispute settlement provision in Article 18 of the 1991 Protocol again requires consultation, and Article 19 gives parties a choice on ratification of accepting the ICJ or the 'Arbitration Tribunal' as constituted in accordance with the Schedule to the Protocol. Failure to make a choice may mean that the latter is deemed to be competent. The three-year time bar is similar to that in the CLC, but it runs from the date of the response action and with a longer 15-year longstop provision (where the identity of the operator is unknown).

The liability of the operator under Article 6 is not for some widely defined environmental damage, but merely to pay the 'costs' of preventive measures and response action as defined in Article 2. The Article 2 definitions seem more apt to cover immediate clean-up than any form of longer-scale remediation. The action has to be 'reasonable', which is defined to include actions

110 Article 7(3) does place an obligation on a party to 'ensure that there is a mechanism in place under its domestic law for the enforcement' of the liability to pay costs. Compliance might be achieved merely by enabling an action to be brought, but the real question of enforcement is whether there is a mechanism to bring claims against the insurer or provider of financial security. As already noted, there is no express provision on direct action against the insurer, although failure of an operator or insurer is a factor relevant to the Article 12 'Fund'. Annex VI (n 98) art 7(3).

that are 'appropriate, practicable, proportionate and based on the availability of objective criteria ... including technological and economic feasibility'.[111] In any event, Article 6 envisages a situation where no immediate response action is taken by anybody. In those circumstances, there is still a liability to pay the theoretical costs—ultimately into a 'Fund' to be established under Article 12.[112]

The purpose of the Annex VI Article 12 Fund is to provide a fund for the reimbursement of 'reasonable and justifiable costs' incurred by state parties, based upon 'proposals' (i.e. applications) made by state parties. These 'costs' would seem to cover liability-type post-incident expenses, but would also seem to extend to the establishment of response equipment and resources in Antarctica (i.e. independently of a specific liability incident). This Fund is therefore quite different from the IOPC Funds. It does not provide a true second *additional* tier of liability. There is no right for a state to payment and apparently no right at all for any private individual or company even to apply for funds. There are apparently no additional funds to the primary liability, except where states make voluntary contributions. There is no attempt, therefore, to spread costs among operators generally, or among wider participants. Decisions on whether to make payments are effectively discretionary, although some general criteria are mentioned. In one sense the Fund operates as a second tier, in that it can fulfil the IOPC Funds' role of stepping in as a sort of guarantor (down to the first dollar), e.g. if the operator or insurer is insolvent, or where there is an exception to liability such as terrorism.[113] In other ways, the Annex VI Article 12 Fund is more like a trust fund. At present, there are none of the detailed operational procedures for the Annex VI Fund as are set out in the IOPC Fund Conventions, or within the IOPC Funds' internal operating arrangements and practices.

It is significant, perhaps, that at the May 2017 Beijing Antarctic Treaty Consultative Meeting, there were presentations made by the International Group of P & I Clubs (IG Clubs) and the IOPC Funds' secretariat about how the IMO compensation systems operated in practice.[114] The Clubs noted that, in

111 See Annex VI (n 98) art 2.

112 Article 6(2) actually makes a distinction for non-state operators, who might pay first a state party that is itself encouraged to make an equivalent payment into the fund. Somewhat surprisingly, the Article 6(2) liability of the operator is apparently treated as a fine or penalty, with the consequence that it would not be covered by P & I Club insurers. See de La Fayette, 'Responding to Environmental Damage in Antarctica' (n 97) 146.

113 See Annex VI (n 98) art 8 and n 97 and compare the Fund Convention 1992 (n 23) art 4(1).

114 See Antarctic Treaty Consultative Meeting (n 94) 42–52; International Group of Protection & Indemnity Clubs (n 105); IOPC Funds, 'IP 88 The International Oil Pollution Compensation Funds' (IOPC Funds 2017).

principle, the IG Clubs would cover the liabilities of a commercial operator (being a shipowner) as prescribed in Article 6 of the Annex, and IG Club certificates of entry would meet the Annex VI requirements. By contrast, the definition of 'operators' within Article 2(c) was broader than that of 'shipowner', and not all operators would be shipowners covered by the Clubs.[115] Such operators would need to find other market cover.[116] The IG Club view was also that the absence of a right of direct action in Annex VI meant that insurers would be able to rely on policy defences in addition to the Article 8 exceptions, presumably including the 'pay to be paid' defence.[117]

Annex VI has only been considered in outline here, but the lesson from it for any high seas or BBNJ negotiations is perhaps that caution is required before adopting proposals on liability that are too ambitious in scope or cost. It has been noted that this Annex, relatively small in scope, had taken 'five times the effort involved in negotiating the entire Protocol and its five other Annexes'.[118] Yet the wider high-seas developments under consideration in this book go way beyond research, fishing and tourism to encompass the full spectrum of commercial exploitation. Although the Annex applies to a large area (the Antarctic Treaty Area includes all areas south of 60°S), there is a limited number of state consultative parties (29). A very limited form of cost reimbursement scheme has been set up, mostly for operators who should already have been vetted by state parties. A somewhat loosely defined 'Fund' has been envisaged, although not yet established, which appears to have no obviously identifiable source of additional financing.[119]

## 6.4 *UNCLOS Part XI: International Seabed Authority Developments 2016–17*

UNCLOS Article 139 puts a direct responsibility on states to ensure that their nationals and companies carry out activities in the Area in accordance with Part XI. UNCLOS Annex III ('Basic Conditions of Prospecting, Exploration and Exploitation') sets out the basic conditions of operating, and Annex III

115 International Group of Protection & Indemnity Clubs (n 105) para 13.

116 Russia's experience was that it was difficult to find insurance companies prepared to insure in Antarctica. See Antarctic Treaty Consultative Meeting (n 94) 49. Markets tend only to respond with products when legislation will enter into force.

117 See *Firma C-Trade SA v Newcastle Protection and Indemnity Association (The Fanti); Socony Mobil Oil Co Inc v West of England Shipowners Mutual Insurance Association (London) Limited (No 2) (The Padre Island)* [1991] 2 AC 1.

118 See Andrew Jackson and Lorne Kriwoken, 'The Protocol in Action 1991–2010' in Howard and Griffiths (n 95) 315.

119 For the use of Annex VI as a model for the high seas, see further Section 8.

Article 22 imposes liabilities on contractors for damage caused by 'wrongful acts' arising out of their operations.[120] The ISA's obligations to protect the marine environment[121] are encompassed in rules referred to as the 'Mining Code'.[122] These are gradually being developed.

In 2016 the ISA issued a number of discussion papers, including one on 'enforcement and liability issues in the Area'.[123] This paper discussed administrative sanctions for non-compliance by contractors with their obligations, and these sanctions could range from warnings to 'monetary penalties'. It noted that damages may not be the only appropriate remedy; alternatives might include declaratory relief for procedural breaches. Administrative orders could be made, perhaps backed by a 'financial guarantee', but otherwise the paper was not particularly specific about liabilities.

### 6.4.1 Draft Standard Terms 2016

In July 2016, the Legal and Technical Commission (LTC) of the ISA issued a first working draft of exploitation regulations and standard terms, including financial terms applicable to exploitation contracts. This working draft contained no provisions on liability[124] but had a draft regulation on the performance guarantees that applicants for exploitation have to deposit, and against which the ISA may make deductions. In other words, this is the typical commercial arrangement used to guarantee performance, like a letter of credit, rather than being in the form of a normal insurance policy. A draft Regulation 56 would give power for the ISA to take remedial action, with the reasonable costs and expenses incurred by the ISA to be treated as a debt due.

Among the draft contractual terms to be required would be undertakings to perform wide-ranging obligations, e.g. in relation to the marine environment. These would include requiring contractors to maintain comprehensive

---

120 Article 139(2) creates state liability for damage, but emphasises that a state is not vicariously liable for failings by a company if its administrative framework is reasonably appropriate. See Section 2.1.

121 Under UNCLOS (n 8) art 145.

122 See e.g. the Regulations on Prospecting and Exploration for Polymetallic Nodules in the Area (adopted 13 July 2000), which were later updated and adopted on 25 July 2013; the Regulations on Prospecting and Exploration for Polymetallic Sulphides in the Area (adopted 7 May 2010); and the Regulations on Prospecting and Exploration for Cobalt-Rich Crusts (adopted 27 July 2012). These Regulations also include standard contract terms.

123 See ISA, 'Ongoing Development of Regulations on Exploitation of Mineral Resources in the Area' <www.isa.org.jm/legal-instruments/ongoing-development-regulations-exploitation-mineral-resources-area> accessed 24 May 2018; Neil Craik, 'Enforcement and Liability Challenges for Environmental Regulation of Deep Seabed Mining' ISA Discussion Paper 4 (June 2016).

124 Although the terms of any exploitation contract might contain provisions about contributions to any 'trust fund'. See Section 6.4.2.

insurance and to provide the ISA with copies or access to relevant certificates and policies. This is typical of the sort of national exploitation arrangements in the EEZ, but is not quite the same as direct action as under the IMO conventions. The terms and extent of cover are not spelled out, but Section 8 of the draft standard clauses refers to 'appropriate insurance policies, with internationally recognized and financially sound carriers, on such terms and in such amounts in accordance with generally accepted international maritime practice and as is consistent with good industry practice, sufficient to cover claims made during … the duration of this contract'.[125]

### 6.4.2 Draft Environmental Regulations 2017

In January 2017 the ISA produced a discussion paper on the development and drafting of regulations for exploitation of mineral resources in the Area.[126] Describing it as a 'tentative working draft' to fulfil obligations under UNCLOS Article 145, the ISA makes clear that environmental assessment is a fundamental focus of the working draft from a procedural perspective.

A draft legal framework issued in July 2015 had proposed a restorative or rehabilitative obligation on contractors,[127] but the 2017 document is remarkably candid about the difficulties in defining how contractors are to address 'serious harm'[128] and records that response measures may not always be technically feasible or economically reasonable.[129] It may well be that this is particularly the case for restoration, although research and studies may be warranted. Like the Antarctic Treaty Environment Protocol Annex VI, the ISA 2017 draft notes a proposal that where restoration cannot (or has not) occurred, there might be equivalent compensatory payments.

Part XII of the 2017 draft concerns 'Compensatory Measures', in particular the setting up of an Environmental Liability Trust Fund to plug liability gaps (presumably when the contractual or other measures referred to above are inadequate). There are no detailed figures on costings, although a draft Regulation 69 refers to percentages of fees and penalties payable to the ISA, as well as ISA legal recoveries in any claims. It is also unclear what the relationship would be of this trust fund to others proposed in the Area, e.g. the 'Seabed Mining Sustainability Fund', proposed in the draft legal framework issued by

125 In respect of environmental liability insurance the cover might extend for, say, five years following the expiry of the contract.

126 See 'Developing a Regulatory Framework for Mineral Exploitation in the Area' (ISA, 2017). See also International Seabed Authority, 'Briefing Note to the Council on the Submissions to the Draft Regulations on Exploitation of Mineral Resources in the Area' ISBA Doc ISBA/24/C/CRP.1 (21 February 2018).

127 'Developing a Regulatory Framework' (n 126), 11 (para 7.12 and its n 20).

128 Ibid, 10 (para 7.5).

129 Ibid, 11 (para 7.11).

the LTC in March 2015. It is clear from the purposes of the trust fund, outlined in draft Article 68, that this is quite a different type of fund to the IOPC Funds. The latter have actual liabilities to claimants, while the ISA *trust* funds are conceived of as being more discretionary in nature (e.g. covering future research), and their distributions will be limited by the funds available, rather than being defined by strict rights.

Overall, what is being envisaged seems (at least for private contractors) to be a sort of two-tier scheme. Initial liabilities are defined essentially by contract (i.e. consensually by an individual company licensed to operate) and backed by commercial insurance for sums yet to be decided. A second tier would consist of a loosely defined trust fund that would be contributed to (in the main) by all contractors generally out of license fees. Ultimately, disputes (including those with contractors) would be referred to the Seabed Disputes Chamber of ITLOS under Article 187 of UNCLOS.

In many ways this type of response is very similar to that which operates at a national level. But it is concerned with those operators who choose to submit themselves to the authority of the ISA. The regime being developed does not cover those using the water column, nor ships transiting the high seas.

### 6.5 *Warsaw International Mechanism*

A recent model that addresses environmental loss and damage is the Warsaw International Mechanism for Loss and Damage associated with Climate Change Impacts (Warsaw International Mechanism) established at the 19th Session of the Conference of Parties (COP) to the United Nations Framework Convention on Climate Change (UNFCCC) 1992, as amended.[130] The Warsaw Mechanism was designed by the COP to address loss and damage associated with the impacts of climate change in developing countries that are particularly vulnerable to the adverse effects of climate change. Its importance was recognised by the Paris Agreement 2015,[131] whose parties agreed generally to support this mechanism with respect to loss and damage associated with the adverse effects of climate change.[132]

In effect, the Warsaw International Mechanism consists of an executive body with a general role of 'promoting the implementation of approaches to address such loss or damage'. Among the functions the body could undertake

130 See UN Framework Convention on Climate Change, 'Report of the Conference of the Parties on its Nineteenth Session, Held in Warsaw from 11 to 23 November 2013. Addendum Part Two: Action Taken by the Conference of the Parties at its Nineteenth Session' UN Doc FCCC/CP/2013/10/Add.1 (31 January 2014).

131 Paris Agreement (adopted 12 December 2015, entered into force 4 November 2016) UN Doc FCCC/CP/2015/10/Add.1 Decision 1/CP.21, art 8(1).

132 Ibid art 8(3).

are enhancing knowledge, providing overviews of best practice, 'enhancing action and support, including finance, technology and capacity building'. The aims seem largely aspirational and centre around education and information provision. The Warsaw International Mechanism, as yet, does not address the sort of detailed liability and compensation rules addressed elsewhere in this chapter.

By contrast, the UNFCCC Article 11 sets out a financial mechanism for providing financial resources, e.g. for technology transfer, effectively provided by developed states. One of the achievements of the UNFCCC is the establishment in 2011[133] of a Green Climate Fund, which by 2017 had approved a total of USD 2.2 billion to support developing countries through funding proposals (i.e. competitive bids), e.g. USD 80 million for a sustainable energy facility in the Eastern Caribbean.[134] The size of the Green Climate Fund is impressive, and the range of activities it can sponsor is wide, although not apparently related to clean-up and remediation, for instance. Its financial resources and aims have not been without controversy in the developed world but, even so, it can be viewed as a large-scale international environmental project that has been successfully funded to deal with global problems. A question is whether its funding model will ever be replicated in relation to the high seas, especially in the BBNJ context.

## 7 BBNJ Negotiations and Compensation

### 7.1 *Background*

It is not possible here to rehearse the full history and development of the BBNJ negotiations,[135] but at its 66th Session in December 2011 the UN General Assembly (UNGA) reaffirmed its role in the conservation and sustainable use of marine biological diversity beyond areas of national jurisdiction. It approved the preliminary study by an Ad Hoc Open-ended Informal Working

133 UN Framework Convention on Climate Change, 'Report of the Conference of the Parties on its Seventeenth Session, Held in Durban from 28 November to 11 December 2011. Addendum Part Two: Action Taken by the Conference of the Parties at its Seventeenth Session' UN Doc FCCC/CP/2011/9/Add.1 (15 March 2012) 55 (Decision 3/CP.17. Launching the Green Climate Fund).

134 See UN Framework Convention on Climate Change, 'Report of the Green Climate Fund to the Conference of the Parties' UN Doc FCCC/CP/2017/5 (25 August 2017).

135 See generally, the Preparatory Committee material at UN DOALOS, 'Preparatory Committee established by General Assembly Resolution 69/292' <www.un.org/depts/los/biodiversity/prepcom.htm> accessed 24 May 2018; Intergovernmental Conference on Marine Biodiversity of Areas beyond National Jurisdiction <www.un.org/bbnj> accessed 24 May 2018. See also Beckman et al, Chapter 1 'Introduction'.

Group, which proposed work on a package of four issues.[136] These were broadly (i) the conservation and sustainable use of marine biological diversity of areas beyond national jurisdiction; (ii) marine genetic resources, including questions on the sharing of benefits; (iii) measures such as area-based management tools, including marine protected areas, environmental impact assessments; and (iv) capacity-building and the transfer of marine technology. The Working Group reported in February 2015 and proposed that there be an international legally binding instrument (ILBI) and that a preparatory committee work on substantive recommendations on it.[137]

At this stage compensation issues were not specifically raised, as the focus was naturally on the larger environmental and political issues, although the general need to identify legal gaps was recognised.[138] At its 69th Session the UNGA decided on 19 June 2015[139] that an ILBI should be developed. It approved the creation of a slightly more formal Preparatory Committee, which met for four sessions. For the Third Session in March 2017, there was a Chair's non-paper[140] on elements of a draft text of the ILBI.[141] A much shorter, streamlined version of the Chair's non-paper was presented to the Fourth Session in

136 UNGA, 'Resolution Adopted by the General Assembly on 24 December 2011' UN Doc A/RES/66/231 (5 April 2012) 29.

137 UNGA, 'Letter dated 13 February 2015 from the Co-Chairs of the Ad Hoc Open-ended Informal Working Group to the President of the General Assembly' UN Doc A/69/780 (13 February 2015).

138 Ibid para 17.

139 Development of an International Legally Binding Instrument under UNCLOS on the Conservation and Sustainable Use of Marine Biological Diversity of Areas Beyond National Jurisdiction, UN Res A/RES/69/292 (6 July 2015). See also the Ocean Conference (New York, 5–9 June 2017) <www.un.org/en/conf/ocean> accessed 24 May 2018; the general problem of financing at UNGA, 'Preparatory Process of the United Nations Conference to Support the Implementation of Sustainable Development Goal 14: Conserve and Sustainably Use the Oceans, Seas and Marine Resources for Sustainable Development' UN Doc A/71/733 (27 February 2017) 16.

140 Chair of Preparatory Committee, 'Chair's Non-Paper on Elements of a Draft Text of an International Legally-Binding Instrument under the United Nations Convention on the Law of the Sea on the Conservation and Sustainable Use of Marine Biological Diversity of Areas Beyond National Jurisdiction.' (28 February 2017) (Chair's Non-Paper) 5–6 (Part II A) <www.un.org/depts/los/biodiversity/prepcom_files/Chair_non_paper.pdf> accessed 24 May 2018. The idea of such a non-paper had been used in the Informal Working Group when its Co-Chairs presented a non-paper on 17 December 2014. It is a useful way to start discussions when there is no formal secretariat and no particular state wants to be seen as taking a lead.

141 The Chair's Non-Paper (n 140) was later supplemented by a Chair's Overview of the session (unnumbered and undated). This Overview contains oral reports of informal working groups, one of which touched on liability issues (Overview, Appendix 5, 28).

July 2017.[142] At this session, the Preparatory Committee produced a report[143] of the work of its four sessions for the 72nd Session of the UNGA in December 2017. The UNGA decided on 24 December 2017[144] to convene an intergovernmental conference to consider the elements, and to elaborate the text, of an ILBI under UNCLOS on the conservation and sustainable use of marine biological diversity of areas beyond national jurisdiction, with a view to developing the ILBI as soon as possible. It is envisaged that there will be four sessions of this intergovernmental conference, with the first starting in September 2018, two further sessions in 2018 and one in 2020.[145] A first, organisational, meeting was scheduled for April 2018.

### 7.2 *Elements of the Draft ILBI in 2017*

The Preparatory Committee recommended that 'elements' in Sections A and B of its Report be considered as part of the development of the ILBI. It recorded that these sections do not reflect consensus, although Section A includes 'non-exclusive elements that generated convergence among most delegations', while Section B 'highlights some of the main issues on which there is divergence of views'.[146] Section A contains the broad structure of an instrument; it has headings (and sub-headings) under which are some comments in bullets about the nature of the text that would be included. In the two of its Parts relevant to this chapter, it is noticeable that Section A merely has headings and no real indication of the detail of what the text of the ILBI might cover. Thus, under Part VI 'Financial resources and issues', there is merely a statement that the 'text *would* address financial issues relating to the operation of the instrument', whereas under Part IX 'Responsibility and liability', the statement is only that the 'text

142 Chair of Preparatory Committee, 'Chair's Streamlined Non-Paper on Elements of a Draft Text of an International Legally-Binding Instrument under the United Nations Convention on the Law of the Sea on the Conservation and Sustainable Use of Marine Biological Diversity of Areas beyond National Jurisdiction' (2017) (Chair's Streamlined Non-Paper) <www.un.org/depts/los/biodiversity/prepcom_files/Chairs_streamlined_non-paper_to_delegations.pdf> accessed 24 May 2018. See UNGA, 'Report of the Preparatory Committee Established by General Assembly Resolution 69/292: Development of an International Legally Binding Instrument under the United Nations Convention on the Law of the Sea on the Conservation and Sustainable Use of Marine Biological Diversity of Areas beyond National Jurisdiction' UN Doc A/AC.287/2017/PC.4/2 (31 July 2017).

143 UN Doc A/AC.287/2017/PC.4/2 (n 142).

144 UNGA, 'International Legally Binding Instrument under the United Nations Convention on the Law of the Sea on the Conservation and Sustainable Use of Marine Biological Diversity of Areas beyond National Jurisdiction' UN Doc A/RES/72/249 (19 January 2018).

145 See Intergovernmental Conference on Marine Biodiversity of Areas beyond National Jurisdiction (n 135).

146 UN Doc A/AC.287/2017/PC.4/2 (n 142) para 38.

*could* address issues relating to responsibility and liability'.[147] Moreover, Section B lists among the subjects on which there are already divergent views, both 'funding'[148] and 'responsibility and liability'.[149]

It seems clear that there is still a great deal of uncertainty about the scope and content of any financial provisions. There is an inevitability, but also some danger,[150] in leaving the issues of finance until later in negotiations. I will attempt, though, to identify some key elements of such an instrument that might be relevant to compensation. While the 'elements' from Section A in the Report of the Preparatory Committee are too sparse to allow for much discussion, Section A itself follows almost identically the numbered structure provided in the Chair's non-papers. Both of those non-papers did indicate possible matters that might be included. To this extent, the Chair's non-papers can therefore help in identifying some of the issues that might be covered in the future ILCBI, as well as those that have obviously caused controversy. At this stage, though, care is needed to avoid placing too much reliance on these non-papers, especially as substantive text is developed into an ILBI during the sessions of the intergovernmental conference.

Even the Chair's non-papers contain relatively little discussion of the sort of liability and compensation issues addressed in this chapter, but that is no surprise as the 2011 package itself does not really highlight the subject. The bulk of the discussion in the non-papers, therefore, is naturally concerned with conservation and sustainable use of marine biodiversity. The issue of funding is raised in a number of places, both generally[151] and in a more specific context, e.g. in relation to capacity building and transfer of marine technology.[152]

In that context, the concept of a BBNJ trust fund, or funds, is touched upon in both non-papers.[153] Thus, Part VIA of the Chair's initial non-paper lists possible sources of funding for implementing the agreement generally, e.g. a voluntary BBNJ trust fund, or funds, financed by states or others (including

147 Emphasis added. I assume that the words 'would' and 'could' here are deliberate, reflecting greater uncertainty in respect of the content of Part IX 'Responsibility and Liability'.

148 'Further discussions are required on the scope of the financial resources required and whether a financial mechanism should be established'. See UN Doc A/AC.287/2017/PC.4/2 (n 142)17.

149 'Further discussions are also required'. See UN Doc A/AC.287/2017/PC.4/2 (n 142) 18.

150 See e.g. the late realisation about problems with shipowners' limitation of liability in the Antarctic liability negotiations in Section 6.3.

151 Chair's Non-Paper (n 140) 106–07 (Part VI); Chair's Streamlined Non-Paper (n 142) 51–52.

152 Chair's Non-Paper (n 140) Part IIIF(6); Chair's Streamlined Non-Paper (n 142) 43.

153 See the section on 'Financial Resources and Mechanism' in Chair's Non-Paper (n 140) Part VI and Chair's Streamlined Non-Paper (n 142) Part VI.

natural and juridical persons).[154] The Chair's 2017 streamlined non-paper also refers to a 'global trust fund' as well as an 'endowment fund' to promote scientific research.[155] The idea of these Part VIA funds appears to be to perform general environmental functions under the proposed ILBI, including education, consultation and the financing of activities of developing states (e.g. in capacity building and technology transfer). The relationship between all these funds (and, indeed, their purposes) is still to be explored, even before their viability can be assessed.

Moreover, Part VIB of both non-papers is headed 'Rehabilitation/Contingency Fund' and refers to funding mechanisms to deal with rehabilitation according to the polluter pays principle.[156] The Chair's initial non-paper refers to the establishment of some form of rehabilitation fund and suggests that private entities wishing to engage in the exploration and exploitation of BBNJ would be required to contribute to it.[157] The Chair's 2017 streamlined non-paper also considers the possibility of a rehabilitation fund.[158] They both also refer somewhat hopefully to the need for a 'mechanism for a financial arrangement' to address contingencies of marine pollution and disaster,[159] or a 'mechanism to deal with loss, damage and contingencies'.[160] The Chair's 2017 streamlined non-paper is a little more specific about possible 'mechanisms', pointing to experience from the Warsaw International Mechanism[161] and other 'similar regimes'.

The Chair's 2017 streamlined non-paper also suggests that such a Part VIB mechanism would have a 'residual nature, i.e. to enter into action only when the primary entities liable or responsible could not completely deal with the damage or rehabilitation'.[162] It adds that 'clear criteria would be established for the funding, which could come from an enterprise's up-front payment, sponsoring States' deposit of a bond, voluntary contributions, mandatory contributions, or a mix of all of these'.[163] The Part VIB 'mechanism' suggestions at least focus on funding from operational exploitation activities, and separate them

154 References to other 'sustainable' or 'special funds' appear elsewhere, e.g. in Part IIIF(6) on capacity building: see Chair's Non-Paper (n 140) 90; Chair's Streamlined Non-Paper (n 142) 43.

155 Chair's Streamlined Non-Paper (n 142) paras 276–77.

156 Chair's Non-Paper (n 140) 107; Chair's Streamlined Non-Paper (n 142) 51.

157 Chair's Non-Paper (n 140) 107.

158 Chair's Streamlined Non-Paper (n 142) para 281.

159 Chair's Non-Paper (n 140) 107.

160 Chair's Streamlined Non-Paper (n 142) para 281.

161 See Section 6.5.

162 Chair's Streamlined Non-Paper (n 142) para 281.

163 Ibid.

from rehabilitation arising from environmental disasters for which traditional liability mechanisms could operate.

The suggestions of how to deal with contingencies (e.g. disasters) and rehabilitation in Part VIB sit somewhat uneasily after the mention of the Part VIA trust fund or funds. Are the Part VIB rehabilitation or contingency funds intended as something separate from a BBNJ Trust Fund? How they might work is not specified, nor is the standing of states or others to claim against them. So, the exact relationship of the various funds, whether they are alternatives or complementary and how they might relate to other high seas liabilities (including from passing ships) are all issues still very much up in the air.

In fact, the non-papers and the elements in the 2017 Report of the Preparatory Committee all have a separate Part IX dealing with 'Responsibility and Liability'. This part is very brief[164] and is mainly concerned with possible state liability.[165] One suggestion in the Chair's non-paper, though, is to allow states parties to establish liability and seek compensation also from private entities for environmentally harmful activities involving BBNJ.[166] The Chair's 2017 streamlined non-paper[167] does make a passing reference to possible guidance from 'conventional regimes addressing liability', which presumably could include the IMO liability conventions.[168] Among the many general principles mentioned in the Chair's non papers, it is hardly surprising to find the polluter pays principle,[169] but there is still a debate on what exactly it means, at least as regards shipping.[170] There is obviously still a lot of work to integrate the varied concepts being mooted to deal with environmental harm and payment for remediation. Can trust funds work, will there need to be a separate

164 A few bullet points in the Chair's Non-Paper (n 140), and half a page in the Chair's Streamlined Non-Paper (n 142) paras 303–06. Para 304 of the latter does refer for possible guidance to the ILC Articles on Transboundary Harm from Hazardous Activities.

165 This chapter does not address state liability (see Section 2.1), but the Chair's Streamlined Non-Paper ((n 142) paras 303–06) suggests three possible options for state liability: Option 1 is a provision like Article 35 of the UN Fish Stocks Agreement (i.e. 'States Parties are liable in accordance with international law for damage or loss attributable to them in regard to this Agreement'.); Option 2 is a sui generis provision (e.g. for states to ensure that 'activities within their jurisdiction or control do not cause damage to the environment of other states or of areas beyond the limits of national jurisdiction') (para 303); Option 3 would be to have no specific provision, given the authoritative ILC articles (see n 14).

166 Chair's Non-Paper (n 140) 111 (Part X). This also highlighted the need for a dispute resolution mechanism but made no suggestion as to what it might be.

167 Chair's Streamlined Non-Paper (n 142) para 304.

168 See Sections 2.1, 3.1.

169 Chair's Non-Paper (n 140) 18 (Part III A); Chair's Streamlined Non-Paper (n 142) paras 24, 304.

170 See the discussion in Section 3.1 about the shared responsibility of shipowner and cargo owner.

liability system for private operators (both exploiters and ships exercising ordinary rights of navigation), and how will these concepts be integrated with each other and separate provisions on state liability? Is it possible to produce a solution that is not so utopian that it will deter states from becoming parties?

The BBNJ discussions are naturally focussing on those who are seeking to exploit biodiversity, but there is a need to ensure that any BBNJ solution fits logically into other regimes that might apply to the high seas, including those of the ISA (seabed exploitation) and IMO (commercial shipping). It is clear that the wider the scope of any ILBI, the wider the potential liabilities. Should there be an activity-oriented or impact-oriented approach? If marine protected areas or marine reserves are given special protection, then environmental remediation may be more expensive and time consuming. There could be damage to biodiversity generally, e.g. to an ecosystem, or biological or marine genetic resources. In addition to traditional physical damage to plant or machinery (including ships), there might be economic loss if processes are interrupted or prevented.

The draft ILBI will also have to address broader structural issues. Effective governance[171] may require a secretariat[172] and other bodies, but a key issue may be the relationship with the ISA, including whether the ISA should have an expanded role. The scope would also affect relevant relationships to other instruments[173] and international organisations, such as the IMO. Dispute settlement[174] would not be focussed on compensation claims, but might simply refer to existing mechanisms such as UNCLOS Part XV. There is the possibility of using ITLOS, i.e. with an additional mandate or as a model for a new tribunal.

At this stage, many of the liability and compensation proposals appear to be no more than a wish list. Experience elsewhere shows that on these issues alone the BBNJ discussions have a long way to go.

## 8 Conclusions

Until the ILBI or any wider regulatory system is settled, it is difficult to devise a liability or compensation system for pollution on the high seas. The Indonesian experience at the IMO with Montara shows how easy it is for states to make

171 Chair's Non-Paper (n 140) 93 (Part IV Institutional Arrangements); Chair's Streamlined Non-Paper (n 142) 45; UN Doc A/AC.287/2017/PC.4/2 (n 142)15.

172 Possibly DOALOS: Chair's Non-Paper (n 140) 94, 102; Chair's Streamlined Non-Paper (n 142) para 265; UN Doc A/AC.287/2017/PC.4/2 (n 142)15.

173 Chair's Non-Paper (n 140) Part II (D) Scope.

174 Ibid Part VIII.

objections to new liability schemes. Extension of the IMO conventions is theoretically possible, e.g. with high seas protocols, but agreement on these might be very difficult unless there is a strong commitment from the contributing states to the IOPC and HNS Funds. However China and the USA are not party to these agreements. It is debateable whether a proven need can be shown for such protocols for oil pollution, although the risks from HNS are surely greater. Any idea that a shipowner would be liable to make contributions to a fund based on theoretical costs of remediation would be strongly opposed.[175] It may well be that a more pressing need is to expand (in some way) the WRC 2007 to the high seas—given the controversy over its application in the EEZ.[176] The reality is that in the deepest oceans it may be practically impossible for ships and cargoes to be removed, even before considering environmental remediation.

The BBNJ discussions have a UNGA impetus, given the 2015 UNGA resolution to develop an ILBI, but compensation issues are, understandably, relatively undeveloped. There is a clear overlap with the parallel discussions taking place in the ISA about Part XI, and there appears to be some merit in considering whether the ISA's environmental remit could be expanded to cover the water column,[177] and in appropriate cases to engage in proportionate wreck and cargo raising activities.[178]

In any event, it is quite appropriate that the ISA is focussing on contractual and administrative arrangements to deal with offshore contractors. It means that it can have more flexible remedies. But it still needs to address the level of insurance cover and the security of the insurance to other guarantees. The ISA and BBNJ discussions have both raised the idea of creating environmental trust funds of some kind, and the Green Climate Fund[179] shows that, with a political will, funding can be arranged (although it may be that climate change was a special case). In the absence of a strong political will, care needs to be taken that there is no proliferation of overlapping funds, e.g. in the BBNJ or ISA contexts. It would be too easy for these to be established in hope rather than

175 Compare the Antarctic Protocol Annex VI solution in Section 6.3.

176 The sinking in the English Channel of the chemical tanker *Ievoli Sun* in 2000 with a cargo of some 4,000 mt of styrene shows the practical difficulties in identifying the precise toxicity of particular HNS, and then in removing it from sunken ships. Costly disasters such as that to the *Costa Concordia* will mean great resistance to any extension of the WRC 2007.

177 Although there would be objections from states that regard commercial activities, e.g. bioprospecting for marine genetic resources, as a right within UNCLOS Part VII. See also de la Fayette (n 1) 259–60, 279.

178 This would almost certainly need approval of flag states to such a system.

179 See Section 6.5.

in realistic expectation of sound financing. Whether the idea of a single super fund, initially financed from deep seabed mining, is politically practical is another matter. Developing states might object that they would lose possible ISA income to establish such a fund, but that may be a worthwhile price. The pragmatic solution is to source funding from where it can be found: in particular, profits from seabed mining and, if there is agreement, those from exploitation of marine genetic resources. Albeit within a relatively closed 'club', the Antarctic Protocol Annex VI structure provides a model of a minimum system that might work for the high seas, especially perhaps for the ISA within the Area where it can license operators.[180] In common with the Area, though, Annex VI does not prejudice rights under international law to use the high seas.[181] For those operations on the high seas or in the water column, the Annex VI liability provisions might seem too narrowly drawn to satisfy some of the BBNJ concerns.

Strict liability with limited defences is the norm in any liability and compensation system, but who is to be required to have insurance and for what? The Antarctic Protocol Annex VI avoids some of the idealism inherent in the BBNJ discussions, by only dealing with 'response costs' following an environmental emergency, rather than specifying extensive remediation. Moreover, the reference to issues such as 'technological and economic feasibility' will give some comfort to potential defendants. If insurance is required, some form of 'limit' to the insurance cover seems almost unavoidable, while it is debateable if operators should themselves have limited liability. The issue of direct action against insurers ought to be covered expressly, if only as a requirement for a state party in its national legislation to ensure that the insurer is amenable to action.

The hardest issue would be to define the extent of any liability. If an expansive definition is given to the expression 'response costs', insurers may be likely to see the idea of fund contributions as a form of ill-defined tax, and the

180 The Annex VI concept of solving 'standing' issues by giving express rights of suit to a state responder is a good one, especially in the absence of agreement on a supra-national body (e.g. the ISA) to take action.

181 The Antarctic Treaty (adopted 1 December 1959, entered into force 23 June 1961) 402 UNTS 71, art VI. The extent to which the liability rules can apply to non-state party vessels exercising rights of passage on the high seas, e.g. fishing vessels and tourist vessels, may depend on more general principles of international law. Article 1 of Annex VI declares that it applies to all activities in the treaty area that require advance notice under Article VII(5) of the Antarctic Treaty, but in relation to vessels only specifically mentions tourist vessels. The Annex's application to environmental emergencies relating to '*other* vessels' (emphasis added) falls to be decided under Article 13 (a provision allowing for amendment or medication of the Annex).

market may not develop. Otherwise, the Antarctic Protocol Annex VI Fund is what may be called a 'soft' fund, in that it is quite unclear what its resources will be, or how they will be expended. The BBNJ discussions also provide the scope for widening any BBNJ trust fund(s) to cover matters other than seabed mineral extraction, e.g. rehabilitation from wider activities to exploit BBNJ.[182] In any event, any ISA or BBNJ trust fund is quite different in scope and purpose to the successful IOPC Funds; they would be more an administrative measure than a compensatory one.

At present, it seems that discussion of high seas environmental compensation issues is split three ways: shipping, seabed and BBNJ. It would be unfortunate if the seabed and BBNJ compensation solutions were kept separate. The Antarctic Protocol Annex VI model—although unproven—might offer a way to create shipowner liability, though perhaps with inducements to shipowners[183] including unbreakable limits of liability within the new ILBI (but linked to the LLMC).[184] A properly financed environmental trust fund would provide some augmentation. Perhaps the threat of developing unlimited state liability might be the spur to find pragmatic solutions via trust funds. Finally, the international community should not entirely dismiss the idea of promoting voluntary pollution liability schemes that extend to the high seas, at least for shipping activities,[185] and even if only on an interim basis before something wider can be agreed upon.

182 But if the fund was also used in respect of shipwrecks and other transportation disasters, how would contributions be raised from the shipping and cargo industries? The WRC 2007 model provides for shipowner-only liability for removal, but the HNS Convention model might require a contribution from the HNS Fund to wider environmental remediation. As noted, there would be shipowner objections if in some way they were made liable to contribute to a new super fund.

183 E.g. requirements as to proportionality (as in the WRC 2007) and perhaps an exclusion of economic loss claims.

184 As with the Antarctic Protocol Annex VI, it would be necessary to avoid a treaty conflict with the LLMC, for it would be difficult to agree a new LLMC Protocol at the IMO.

185 See Section 6.1.

CHAPTER 9

# Marine Pollution Preparedness, Response and Cooperation in the Arctic High Seas

*Erik Røsæg*


## Abstract

When pollution prevention fails, preparedness and response is of paramount importance. Coastal states will organize this in their own interest in their adjacent waters, and states often sign cooperation agreements with other states applicable there. However, who is responsible for the high seas, and is there a right to intervene there? Is there a difference in this respect between the high seas and the EEZs? The chapter discuses different bases for intervention, including the intervention conventions, the salvage conventions and the doctrine of necessity, and, in addition, whether a basis for intervention at all is necessary. The financial position of a state requesting assistance is discussed in relation to assisting states and the polluter.


## 1 Introduction

The themes for this chapter are the rules of public international law governing clean-up. This topic includes preparedness, financial consequences of the clean-up operation (but not the financial consequences of the pollution itself), and information management. There are basically three kinds of pollution rules: prevention and safety rules, preparedness and response rules, and liability rules.

The theme here is related to the second category, preparedness and response, which we will discuss in relation to all kinds of marine pollution. However, marine pollution by drifting objects, wrecks and dumping will not be discussed. Neither will licensing of oil and mineral exploitation be discussed, although this certainly has a pollution prevention and safety aspect.

The general area of focus is the Arctic, specifically the central Arctic Ocean beyond the exclusive economic zone (EEZ) of any state, which will be used to illustrate the general issues as well as a few special issues.

Sections 2 and 3 are introductory sections. While Section 2 deals with the practical aspects of cleaning up (for example, is it possible to clean up oil

 | DOI:10.1163/9789004373303_010

spills?), Section 3 is an overview of the requirements in respect of pollution preparedness. Section 4 deals with rights and duties in respect of cleaning up, for example whether states have a duty under international law to instigate clean-up operations and whether a state can instigate a clean-up operation in the EEZ of another state. Section 5 addresses the financial settlement between states after clean-up assistance, and Section 6 briefly discusses information management after pollution incidents.

### 1.1 *Principal Instruments*

There is a general duty of cooperation in respect of marine pollution preparedness and response.[1] The principal agreement in respect of the Arctic is the Agreement on Co-operation on Marine Oil Pollution Preparedness and Response in the Arctic, 2013.[2] This agreement, also known as MOSPA, has been ratified by all Arctic states.[3] MOSPA is a regional variant of the International Convention on Oil Pollution Preparedness, Response and Cooperation, 1990 (the OPRC Convention).[4] This global instrument negotiated by the International Maritime Organization (IMO) has been ratified by 111 states,[5] and the IMO regulation in this instrument also extends to petroleum activity.[6]

Although the OPRC Convention only addresses preparedness, response and cooperation in relation to oil pollution, its scope has been extended by the Protocol on Preparedness, Response and Co-Operation to Pollution Incidents by Hazardous and Noxious Substances, 2000 (the OPRC-HNS Protocol).[7]

1 United Nations Convention on the Law of the Sea (adopted 10 December 1982, entered into force 16 November 1994) 1833 UNTS 3 (UNCLOS) art 197.

2 International Convention on Oil Pollution Preparedness, Response and Co-operation (adopted 15 May 2013, entered into force 25 March 2016) (MOSPA) <https://oaarchive.arctic-council.org> accessed 23 May 2018.

3 Artic Council, 'Ratification Completed for Agreement on Oil Pollution Preparedness and Response' <https://tinyurl.com/yd2r93z7> accessed 23 May 2018.

4 International Convention on Oil Pollution Preparedness, Response and Co-operation (adopted 30 November 1990, entered into force 13 May 1995) 1891 UNTS 77 (OPRC Convention).

5 International Maritime Organization (IMO), 'Status of Conventions' <www.imo.org/en/About/Conventions/StatusOfConventions/Pages/Default.aspx> accessed 23 May 2018.

6 Recently, it has been argued that IMO does not have a mandate that extends beyond ships. See IMO, '(i) Analysis of liability and compensation issues connected with transboundary pollution damage from offshore exploration and exploitation activities, including a re-examination of the proposed revision of Strategic Direction 7.2' IMO Doc LEG 99/13 (5 January 2012); IMO, 'Report of the Legal Committee on the Work of its One Hundred and Third Session' IMO Doc LEG 103/14 (15 June 2016) para 13.11.

7 Protocol on Preparedness, Response and Co-Operation to Pollution Incidents by Hazardous and Noxious Substances (adopted 15 March 2000, entered into force 14 June 2007) (the OPRC-HNS Protocol) <https://tinyurl.com/y9bc5cjb> accessed 23 May 2018. This protocol is

While many protocols and conventions only contain amendments to the parent conventions, the OPRC-HNS Protocol largely sets out the same clauses as the OPRC Convention. In this way, the protocol has more of the format of a separate convention than that of a protocol. However, the protocol has only 37 ratifications and has not been nearly as successful as its parent convention.[8]

In respect of nuclear pollution, there is a special Convention on Assistance in the Case of a Nuclear Accident or Radiological Emergency, 1986 (Convention on Nuclear Assistance).[9] Since this is the oldest convention and since it relates to a very specialised field, the other conventions have left it more or less untouched. It has been ratified by 115 states[10] and is in this sense successful.

The cooperation agreements on pollution preparedness and response are supported by general principles, such as the Rio Declaration on Environment and Development.[11] For other specific geographic areas, there are a number of regional agreements akin to MOSPA.[12] These regional agreements are often not very different from the OPRC Convention. In the European Union, cooperation has been taken to a higher level by the European Coast Guard and by the potential for cooperation among established institutions.[13]

sometimes thought to be a protocol the International Convention on Liability and Compensation for Damage in Connection with the Carriage of Hazardous and Noxious Substances by Sea (adopted 3 May 1996, not in force) (HNS Convention), which is a liability convention unrelated to the OPRC-HNS Protocol. Although the OPRC-HNS Protocol refers to oil in its acronym, it does not in fact deal with oil.

8 IMO, 'Status of Conventions' (n 5).

9 Convention on Assistance in the Case of a Nuclear Accident or Radiological Emergency (adopted 26 September 1986, entered into force 26 February 1986) 1457 UNTS 133 (Convention on Nuclear Assistance).

10 United Nations Treaty Collection, 'Convention on Assistance in the Case of a Nuclear Accident or Radiological Emergency' <https://tinyurl.com/yb45sdpt> accessed 23 May 2018.

11 UN General Assembly, 'Report of the United Nations Conference on Environment and Development' UN Doc A/CONF.151/26 (Vol II) (13 August 1992) para 17.33 <https://tinyurl.com/y8vgy4wj> accessed 23 May 2018. See also UNCLOS (n 1) art 197.

12 See the overview at IMO, 'Oil Spill Organizations and Resource Providers' <https://tinyurl.com/kanxgqg> accessed 23 May 2018. There is also a useful overview in Oleg Khalimonov, 'Framework for Co-operation and Recent IMO Activities in Preventing and Managing Marine Pollution' (1999) 3 *Singapore Journal of International & Comparative Law* 370, 377 and Gabino Gonzalez et al 'Conventions Relating to Pollution Incident Preparedness, Response, and Cooperation' in David Attard et al (eds), *The IMLI Manual on International Maritime Law, Volume III: Marine Environmental Law and International Maritime Security Law* (Oxford University Press 2016) 195.

13 See in particular European Maritime Safety Agency, 'Pollution Response Services' http://emsa.europa.eu/operations/pollution-response-services.html accessed 23 May 2018.

There are also a number of bilateral treaties.[14] For Norway and the Arctic, the Norwegian-Russian Cooperation Agreement is important.[15] This agreement precedes and overlaps the OPRC Convention and MOSPA. However, there are few normalised rules on how these texts should be reconciled, except for clauses stating that the newer conventions are not intended to alter existing agreements.[16]

In this chapter, the OPRC Convention and MOSPA will be emphasised.

### 1.2 *Geographic Scope*

The OPRC Convention has no explicit geographic scope and presumably applies to preventive measures and clean-up operations anywhere. Similarly, the Convention on Nuclear Assistance has no strict geographic limits:

> If a State Party needs assistance in the event of a nuclear accident or radiological emergency, whether or not such accident or emergency originates within its territory, jurisdiction, or control, it may call for such assistance from any other State Party…[17]

MOSPA applies north of a defined limit, depending on the geography of each state (in Canada, for example, the limit is at 60 degrees) and

> with respect to oil pollution incidents that occur in or may pose a threat to any marine area over which a State whose government is a Party to this Agreement exercises sovereignty, sovereign rights or jurisdiction, including its internal waters, territorial sea, exclusive economic zone, and continental shelf.[18]

Thus, MOSPA does not apply in the Central Arctic Basin. However, some of the provisions apply without regard to this limitation:

---

14 The treaties relevant to the Arctic are listed in Appendix IV, Table I in MOSPA (n 2) in the Operational Guidelines.

15 Agreement between the Government of the Russian Federation and Government of the Kingdom of Norway concerning Cooperation on the Combatment of Oil Pollution in the Barents Sea (adopted 28 April 1994, entered into force 30 January 1996) (Norwegian-Russian Agreement) <https://tinyurl.com/y74pbygv> accessed 23 May 2018 (login required). This agreement is discussed in Ragnhild Steigberg, 'Rettslig regulering av beredskapen ved sjøulykker i nordområdene. Med særlig fokus på forholdet til Russland' (2010) 388 *MarIus* 108.

16 OPRC Convention (n 4) art 11; MOSPA (n 2) art 16.

17 Convention on Nuclear Assistance (n 9) art 2.

18 MOSPA (n 2) art 3(1).

> Each Party shall also apply Articles 6, 7, 8, 10, and 15 and other provisions of this Agreement as appropriate to areas beyond the jurisdiction of any State, above the southern limit set forth in paragraph 1 of this Article, to the extent consistent with international law.[19]

The provisions referred to, which are applicable on the high seas and in the Area,[20] can also be found in the OPRC Convention (perhaps with the exception of Article 7 on monitoring). The reference to them in MOSPA is therefore of little importance.

Unlike in respect of the search and rescue of persons,[21] in respect of pollution clean-up the rather undefined geographical scope of the conventions is not supplemented by sectors of organisational responsibility for each state.[22] Thus, neither the geographic scope of the convention nor an agreement on areas of responsibilities points to one responsible state in respect of a particular incident. Therefore, to some extent the rules fill gaps in the law that is applicable if and when a clean-up operation is instigated. Nevertheless, an analysis of rights and duties will be attempted in several contexts below.

## 2 Cleaning Up—Practical Aspects

### 2.1 *The Insufficiency of the Remedies*

Before this chapter enters into a discussion on the legal framework for marine pollution preparedness, response and cooperation, it is important to stress that, regardless of the legal framework, the remedies available are inefficient. Even the best of legal frameworks would not make much difference.

To begin with, there are no methods for removing oil slicks from rough seas. Only a fraction of the oil can be removed, and the methods for removing other kinds of pollution are even less sophisticated. Oil that hits the shore must be removed manually by scrubbing or by machines that are not significantly more efficient.

In the Arctic, these problems are even greater. The clean-up operations must be carried out in an environment that may be colder, darker and more remote from the response facilities than that further south. The ice represents

19 Ibid art 3(2).

20 UNCLOS (n 1) pt XI.

21 International Convention on Maritime Search and Rescue (adopted 27 April 1979, entered into force 22 June 1985) 1405 UNTS 118 (SAR) as amended.

22 Agreement on cooperation on aeronautical and maritime search and rescue in the Arctic (adopted 12 May 2011, entered into force 19 January 2013) (SAR Arctic) Annex.

a particular problem: it hampers access to the area, pollution may freeze into the ice even on the high seas, and oil slicks may slide under the ice, where they will remain for a long time and be unavailable for clean-up. Clean-up operations in Arctic conditions can be so potentially dangerous for the crew involved that clean-up cannot be carried out.

### 2.2 *The Politics of the Remedies*

One can usually take for granted that at least all states involved would attempt to avoid and remedy pollution. However, there are exceptions, and choices sometimes have to be made. A clean-up operation is then a matter not just of resources but also of who is in charge.

One example is the decision of whether to intervene actively. It should be considered whether clean-up operations are worthwhile, given the inefficacy of clean-up operations, the damage caused by the chemicals etc. used in the clean-up operations, and the ability of nature to clean the environment for example through storms. Good preparedness is of no use if it is decided that no action should be taken. The rules discussed here do not address such issues.

Another example of the politics of clean-up is when ocean currents or other circumstances make it possible to predict where the pollution will cause damage, which may not be in the decision-maker's territory. Spills north of Norway, for example, are likely to be brought to Russian coasts by ocean currents, and spills west of Norway are likely to be brought towards Great Britain. One cannot exclude the possibility that this will influence the decisions of the governments of these countries to engage in clean-up operations and invoke the rules discussed here.

A final example of the politics of pollution prevention and response concerns ships in distress seeking an emergency shelter to avoid pollution. There are several examples of coastal states rejecting such ships in the hope of reducing the local risk, even though the overall risk posed by the pollution will be increased. The rules discussed here do not prevent bad decisions in such situations.

In summary, the politics of pollution preparedness, response and cooperation are not straightforward. The conventions only provide a framework within which politicians can act.[23]

### 2.3 *The Benefits of Cross-Border Cooperation*

The benefits of cross-border cooperation have not escaped the attention of the drafters of the conventions discussed here.

---

23 This corresponds well with the final remarks of Section 1 above.

The plan for the organisation and deployment of resources provided for under Article 6 of the OPRC Convention, discussed in Section 3 below, may include foreign elements. There is, for example, no reason why two neighbouring states should each have a large tugboat on standby if they can share one.

In respect of nuclear preparedness, resources can be pre-registered to facilitate overview.[24] Unfortunately, few states have taken advantage of this possibility,[25] perhaps because registration would expose inadequacies in preparedness.

Article 7 of the OPRC Convention deals with cooperation in actual operations. Article 7(2) even invites a state to request the IMO to assist in identifying foreign sources for provisional financing of the costs of clean-up operations.

Much can be achieved by sharing knowledge internationally, as stated in Articles 8 and 9 of the OPRC Convention.[26] The articles include a provision on the standardisation of equipment (Article 8(4), for example, the couplings of hoses) and the transfer of technology (Article 9(2)).

There is also a general clause on promoting cooperation via international agreements (Article 10).

Such provisions are at best 'soft law'. However, they express a political will and serve as a reminder for responsible officials. For my part, I think the most important effect of such clauses would be to trigger joint exercises, so that officials can learn from each other and from each other's organisations and can get used to communicating directly rather than via diplomatic channels. Most likely, the exchange of phone directories and establishment of personal contact are more important than treaty clauses in this context.[27]

## 3 Preparedness

Good preparedness is a prerequisite for a successful clean-up operation. In the following section, rules on private and public preparedness will be discussed. Obviously, if a state effectively legislates for extensive private preparedness, less will be required of the state itself in terms of preparedness.

24 Günther Handl, 'Nuclear Off-site Emergency reparedness and Response: Some International Legal Aspects' in Jonathan L Black-Branch (ed), *Legal Aspects of the Use of Nuclear Energy for Peaceful Purposes* (T.M.C. Asser Press 2016) 347.

25 Ibid.

26 The IMO plays an important role here. See Gurpreet S Singhota, 'The IMO's Role in Promoting Safety of Navigation and Control of Marine Pollution in the Straits of Malacca and Singapore' (1998) 2 *Singapore Journal of International and Comparative Law* 290, 298.

27 Are Sydnes and Maria Sydnes, 'Norwegian–Russian cooperation on oil-spill response in the Barents Sea' (2013) 39 *Marine Policy* 257, 262.

### 3.1 *Private Response*

Many states require that enterprises that are likely to pollute prepare for pollution clean-up operations. This is a natural extension of the 'polluter pays' principle, so that the polluter pays not only for the clean-up but also for the cost of maintaining preparedness.

There is a distinct difference between offshore activities and shipping in this respect. While a documented system of preparedness is quite a common condition for obtaining a licence for oil exploitation,[28] ships are allowed to trade with only very limited preparations for combatting pollution.[29] The exception is the United States, which requires that visiting ships enter into a contract with a provider of such services.[30]

The reason for these differences may be traditional (this is how things have always been) or practical (there are limits to how much ships can and should carry). However, there may also be jurisdictional considerations that cause this difference. States have always been reluctant to impose local laws on visiting ships, even in port, and can clearly not impose local pollution preparedness laws on ships engaged in innocent passage.[31] The Norwegian government also maintains that the polluter pays principle—and presumably, therefore the duty to prepare—does not apply outside the territory.[32]

The conventions require pollution emergency plans for both ships and offshore units.[33] This is a modest requirement, as the plan can be brief and can rely on external resources that have not yet been contracted for. However, a vessel's plan is subject to port state inspection and may cause the ship to be detained if deemed unsatisfactory. Regarding offshore units, the coastal state will ensure the adequacy of the plans in its own interest.

It would be important to require that vessels carry a minimum amount of equipment on board, such as oil barriers. Organising cooperation in advance would also be an advantage. Finally, certain resources, such as class assistance in calculating the stability of a damaged vessel, should also be part of the plan and should perhaps be required to be contracted for in advance.

---

28 For example, Petroleum Act 1996 (Norway) s 9-2 <www.npd.no/en/Regulations/Acts/Petroleum-activities-act> accessed 23 May 2018.

29 For example, Protocol of 1978 relating to the International Convention for the Prevention of Pollution from Ships, 1973 (adopted 17 February 1978, entered into force 2 October 1983) 1340 UNTS 61 (MARPOL), does not include extensive lists of equipment on board.

30 Code of Federal Regulations, 'Qualified Individual and Alternate Qualified Individual' 33 CFR S155.1026.

31 UNCLOS (n 1) art 21.

32 Nature Diversity Act 2009 (Norway) ss 2(3) and 11 <www.regjeringen.no/en/dokumenter/nature-diversity-act/id570549> accessed 23 May 2018.

33 OPRC Convention (n 4) art 3.

Such preparedness is particularly important in the Arctic, where the distances are great and the oil may cause damage to the ice on the high seas.

### 3.2 *State Response*

States are required to set up a pollution response unit in advance of any accidents. The minimum requirements are very modest, such as in Article 6 of the OPRC Convention:

(1) Each Party shall establish a national system for responding promptly and effectively to oil pollution incidents. This system shall include as a minimum:
  (a) the designation of:
    (i) the competent national authority or authorities with responsibility for oil pollution preparedness and response;
    (ii) the national operational contact point or points, which shall be responsible for the receipt and transmission of oil pollution reports as referred to in article 4; and
    (iii) an authority which is entitled to act on behalf of the State to request assistance or to decide to render the assistance requested;
  (b) a national contingency plan for preparedness and response which includes the organizational relationship of the various bodies involved, whether public or private, taking into account guidelines developed by the Organization.

The Organization is defined as the IMO (Article 2(6)), and the administrative aspects of oil pollution are dealt with in a handbook.[34]

Even a good organisation is of little help if there is no equipment available. However, the duty to establish depots, designate trawlers for emergency duty,[35] etc. is rather limited, as can be seen in the remainder of Article 6 of the OPRC Convention:

(2) In addition, each Party, within its capabilities either individually or through bilateral or multilateral co-operation and, as appropriate, in co-operation with the oil and shipping industries, port authorities and other relevant entities, shall establish:

---

34 Conference Resolution 1 at the Diplomatic Conference adopting the OPRC Convention (n 4) explains that the reference to guidelines is a reference to the 'Manual on Oil Pollution, Section II—Contingency Planning, developed by the Marine Environment Protection Committee of the Organization'. There are several further manuals listed at the website IMO, 'List of IMO OPRC-HNS Related Guidance and Manuals'<https://tinyurl.com/y9cprcll> accessed 23 May 2018.

35 Trawlers may be useful to set out oil barriers, to store removed oil, etc. On the use of fishing vessels, see Maaike Knol and Peter Arbo, 'Oil Spill Response in the Arctic: Norwegian experiences and Future Perspectives' (2014) 50 *Marine Policy* 171, 174.

(a) a minimum level of pre-positioned oil spill combating equipment, commensurate with the risk involved, and programmes for its use;
(b) a programme of exercises for oil pollution response organizations and training of relevant personnel;
(c) detailed plans and communication capabilities for responding to an oil pollution incident. Such capabilities should be continuously available; and
(d) a mechanism or arrangement to co-ordinate the response to an oil pollution incident with, if appropriate, the capabilities to mobilize the necessary resources.

Because of the Arctic's vast distances, pre-distributed depots would likely be useful. However, such depots would be difficult to maintain. The provisions for training the national response organisations are also important, particularly in the Arctic high seas, where the conditions are difficult and more than one state is likely to be involved.

### 3.3 *Summary*

The requirements for preparedness in relation to pollution incidents in the Arctic are rather limited.

## 4 Cleaning Up—Rights and Duties

### 4.1 *No Duty to Instigate Clean-Up*

Is there a duty for states to instigate clean-up operations, even if they see their interest better served by prioritising other tasks?

While there is a general duty for states to preserve the environment in all maritime zones,[36] UNCLOS stipulates no general duty to instigate clean-up operations.[37] The issue is addressed, but the duty is very limited:

> States shall take all measures necessary to ensure that … pollution arising from incidents or activities under their jurisdiction or control does not

36 UNCLOS (n 1) arts 192 and 194.

37 Philippe Sands et al, *Principles of International Environmental Law* (Cambridge University Press 2012) 394. Detlef Czybulka, 'Article 192. General Obligation' in Alexander Proelss (ed) *UNCLOS: A Commentary* (C.H. Beck, Hart, Nomos 2017) notes that the word 'restore' is missing from Article 192. Thor Falkanger, *Maritime Casualties and Intervention. Coastal State Measures When Casualties Pose the Threat of Pollution* (Fagbokforlaget 2011) 229, seems to assume an unspecified duty for (all?) states to intervene (always?), sanctioned by liability.

> spread beyond the areas where they exercise sovereign rights in accordance with this Convention.[38]

Apparently, this would mean that a coastal state would be responsible for containing pollution,[39] even if it stems from a ship in innocent passage or in its EEZ, although the coastal state in those cases would have had no means or jurisdiction to prevent the accident.[40] In this provision, there is no explicit duty to prevent damage to nature within the coastal zones, only to prevent its spreading beyond those zones. Seemingly, the high seas and the atmosphere as well as the coastal zones of other states are protected.

This responsibility could entail liability for a state that does not prevent the spread of pollution. However, liability requires an injured party, and on the high seas or in the atmosphere it is likely that there would be either no damage caused to particular interests or no injured state. However, in the coastal zones of other states, liability issues could easily arise.

Such a wide liability for an innocent coastal state would be quite inconsistent with the lack of a general duty to instigate clean-up measures. The cited provision should, therefore, be understood as a duty to use reasonable measures to contain the pollution.[41] In addition, the coastal state could be liable for clean-up costs if the pollution were caused by its lack of due control concerning activities in its coastal zones or on its continental shelf.[42]

---

38 UNCLOS (n 1) art 194(2).

39 Detlef Czybulka, 'Article 194. Measures to Prevent, Reduce and Control Pollution of the Marine Environment' in Proelss (n 37) notes that the duty goes beyond damage prevention, so the duty to contain the pollution is independent of the possible damage caused by it.

40 International Law Commission (ILC), 'Draft Articles on Prevention of Transboundary Harm from Hazardous Activities, with Commentaries' in *Yearbook of the International Law Commission* (Vol II Part II, United Nations 2001) Part Two does not apply to a coastal state in such situations; see 150–51. The draft articles could apply to oil spills and petroleum exploitation, but do not focus on spill response.

41 Or a 'due diligence obligation', to use the terminology of the ITLOS in Responsibilities and Obligations of States with Respect to Activities in the Area (Request for Advisory Opinion Submitted to the Seabed Disputes Chamber) (Advisory Opinion of 1 February 2011) ITLOS Reports 2011, 10 para 110.

42 Ibid; ILC, 'Draft Principles on the Allocation of Loss in the Case of Transboundary Harm Arising out of Hazardous Activities' in *Yearbook of the International Law Commission* (Vol II Part II, United Nations 2006). See also the UN General Assembly, 'Report of the United Nations Conference on Environment and Development' UN Doc A/CONF.151/26 (Vol I) (12 August 1992) Principle 2 <https://tinyurl.com/y8qs57ne> accessed 23 May 2018. UNCLOS (n 1) art 208(2), however, primarily aims at legal measures (as opposed to physical intervention), just as in the preceding and following sections of this UNCLOS Article.

This is very much in line with state practice. To my knowledge, there are no reported examples of actions against states for the failure to instigate clean-up measures caused by incidents that they are not responsible or liable for, and certainly not on the high seas. For example, Norwegian law generally does not allow liability actions against licensees for damage outside the Norwegian continental shelf, even if it is caused by Norwegian petroleum activities.[43] It is unlikely that the state responsibility and liability under public international law should fill in for licensees' lack of liability.

The International Seabed Authority has no explicit duties in respect of pollution preparedness and response beyond issuing regulations,[44] and has no pollution preparedness of its own.

Article 1(1) of the OPRC Convention seems to imply that states have a duty to instigate clean-up operations, contrary to what has been submitted here:

> Parties undertake, individually or jointly, to take all appropriate measures in accordance with the provisions of this Convention and the Annex thereto to prepare for and respond to an oil pollution incident.

The remainder of this convention does not limit response actions. In light of the undefined geographical scope of this convention, it is unlikely that this is intended as a hard law obligation for any state to clean up any oil spill wherever it occurs.

In sum, if pollution originates from an activity of the coastal state or an activity licensed by it, the no-harm principle applies, as does the duty to clean up after a pollution incident. However, there are no sources supporting the idea that a state is generally responsible for a spill just because it is located nearby. Still, there is a due diligence obligation to prevent the spread of pollution to other jurisdictions.

### 4.2 *The Right to Instigate Preventive and Response Actions on the High Seas*

A state may wish to instigate preventive and response actions related to pollution on the high seas even if not legally obliged to do so, either because of a potential threat to that state or for more altruistic reasons. For example, it may be better to take action at the source or before the pollution has spread too

---

See Frank Wacht, 'Article 208. Pollution from Seabed Activities Subject to National Jurisdiction' in Proelss (n 37).

43 Petroleum Act (Norway) (n 28) ch 7.

44 UNCLOS (n 1) art 209.

much. However, is any state at liberty to instigate such operations on the high seas (below in Section 4.2.1), even against foreign vessels (below in Sections 4.2.2 and 4.2.3)?

### 4.2.1 General Rules

No state can claim sovereignty over the high seas.[45] However, cleaning up pollution is not claiming sovereignty or jurisdiction nor is it dependent on it. The converse of the principle applies: no state can prevent another from instigating clean-up.

Clean-up is not listed among the freedoms of the high seas,[46] but the list is not exhaustive. The freedom to instigate clean-up is supported by the general duty to prevent pollution[47] and the special commitment in the conventions on pollution preparedness, response and cooperation.[48]

Ordinarily, a state would not object to such activities. However, organising a response to pollution may be interpreted as expressing too much of an interest in an area, as would be the case if a non-Arctic state were to instigate a clean-up operation in the Central Arctic Ocean when none of its vessels were involved in the spill. Clean-up may also involve actions on board a damaged vessel, which the freedom of the seas does not allow, or providing insights into the cause of the accident, embarrassing the flag state of the vessel that caused the accident.

If necessary, the state instigating clean-up can rely on the doctrine of necessity in such cases.[49] Emergency powers are also applicable to protect the environment on the high seas.[50]

The doctrine of necessity is more acceptable if invoked by a state that is clearly acting in the interest of all states than by one that perhaps seeks to further its own jurisdictional interests.[51] Clean-up on the high seas would typically be in the interest of all states, but even if a state takes action because its coastline may be threatened, intervention cannot easily be seen as creeping jurisdiction.

---

45 Ibid art 89.
46 Ibid art 87.
47 Ibid art 194.
48 See above in Section 4.1 on art 1(1) of the OPRC Convention (n 4).
49 ILC, 'Draft Articles on the Responsibility of States for Internationally Wrongful Acts' in *Yearbook of the International Law Commission* (Vol. II, Part Two, United Nations 2001) art 25.
50 Ibid art 25 Commentary.
51 International Law Commission, 'Addendum to the Eighth Report on State Responsibility, by Mr Roberto Ago' UN Doc A/CN.4/318/ADD.5-7 (29 February, 10 and 19 June 1980) 28.

Although the right to perform preventive measures and clean-up operations is arguably fairly wide, such activities subject to general rules. To the extent that the use of dispersants is prohibited, this applies also to clean-up operations on the high seas.[52]

#### 4.2.2 Intervention

In respect of intervention against a polluting or potentially polluting ship on the high seas, UNCLOS Article 221 assumes that there is such a right, in line with the general rules discussed above:

1. Nothing in this Part shall prejudice the right of States, pursuant to international law, both customary and conventional, to take and enforce measures beyond the territorial sea proportionate to the actual or threatened damage to protect their coastline or related interests, including fishing, from pollution or threat of pollution following upon a maritime casualty or acts relating to such a casualty, which may reasonably be expected to result in major harmful consequences.
2. For the purposes of this article, "maritime casualty" means a collision of vessels, stranding or other incident of navigation, or other occurrence on board a vessel or external to it resulting in material damage or imminent threat of material damage to a vessel or cargo.

The rights were specified in two international conventions prior to UNCLOS,[53] which were agreed in the aftermath of the Torrey Canyon oil spill. The conventions do not limit the scope of the doctrine of necessity.[54]

The conventions do not apply to incidents involving an 'installation or device engaged in the exploration and exploitation of the resources of the sea-bed and the ocean floor and the subsoil thereof'.[55] For such installations and devices, one has to apply the general rules.[56]

---

52 See the overview on dispersants at IMO, 'Dispersant Use' <https://tinyurl.com/yamew8x9> accessed 23 May 2018, and below in Section 4.3.1 on the relevance of continental shelf jurisdiction in this respect.

53 International Convention Relating to Intervention on the High Seas in Cases of Oil Pollution Casualties (adopted 29 November 1969, entered into force 6 May 1975) 970 UNTS 211 (Intervention Convention). Protocol Relating to Intervention on the High Seas in Cases of Marine Pollution by Substances Other than Oil (adopted 2 November 1973, entered into force 30 March 1983) (Intervention Protocol) <https://tinyurl.com/y8fkjrqc> accessed 23 May 2018.

54 Falkanger (n 37) 105–10.

55 Intervention Convention (n 53) art II 2(b).

56 In addition to a flag state, the continental shelf state also having jurisdiction over the exploitation of resources may have a say in this respect.

### 4.2.3 Salvage

The right of salvage is mostly associated with the right to a salvage award in the case of a successful salvage operation. In this context, however, the important feature of salvage is that a salvage attempt can be made and remuneration claimed without an agreement with the shipowner. Therefore, anyone, including a state, can instigate salvage operations and thereby legitimise interventions against a polluting ship.

There are two major salvage conventions[57] that together have widespread acceptance. The idea of salvage is generally accepted in the law of the sea in relation to ships. Therefore, for our purposes here, it does not matter if the flag state of the ship in danger has ratified one of the salvage conventions or the same convention as the flag state of the salvor.[58] In addition, if salvage is a part of the general law of the sea, its geographical scope is unlimited.

The right of salvage applies to ships in danger but perhaps not to offshore units, etc.[59] At least, if the right of salvage is based on the general law of the sea, it is probably limited to salvaging ships.

Although salvage is not dependent on agreement with or approval by the shipowner, the master can prohibit a salvor from handling her ship if 'reasonable'.[60] If the alternative is an environmental catastrophe, prohibition is certainly not reasonable.[61] Today, environmental considerations form an integral part of the law of salvage.[62]

Falkanger submits that prohibition of salvage is always reasonable if there is no prospect of real salvage, because the vessel is already a total loss.[63] However, it is difficult to see why bad prospects of success should prevent salvage attempts.

Falkanger submits that the flag state can revoke the right of salvage, perhaps arbitrarily.[64] It is difficult to see the basis for this, particularly if the flag

57 International Convention for the Unification of Certain Rules of Law Relating to Assistance and Salvage at Sea (adopted 23 September 1910, entered into force 1 March 1913) (Salvage Convention 1910) <https://tinyurl.com/yd9p7jqs> accessed 23 May 2018; International Convention on Salvage (adopted 28 April 1989, entered into force 14 July 1996) 1953 UNTS 165 (Salvage Convention 1989).

58 To the contrary, Falkanger (n 37) 224.

59 Salvage Convention 1989 (n 57) art 3 expressly excludes offshore units 'when such platforms or units are on location engaged in the exploration, exploitation or production of sea-bed mineral resources'. However, the starting point according to Article 1(a) is that the convention shall apply to the salvage of 'a vessel or any other property'.

60 Salvage Convention 1910 (n 57) art 3; Salvage Convention 1989 (n 57) art 19.

61 Falkanger (n 37) 224.

62 Salvage Convention 1989 (n 57) para 2 of the Preamble.

63 Falkanger (n 37) 224.

64 Ibid 225.

state is bound by a salvage convention. The right of salvage will in any event provide a basis for intervention until the flag state has attempted to prohibit such actions.

Salvage is limited to salvaging the vessel and its cargo.[65] However, salvaging the ship and the cargo may often have positive environmental effects in that spills are reduced. That is, removing oil or other pollutants already on the water is not salvage. This does not matter very much in this context, as there is no need to invoke the salvage rules in order to remove the oil. It does, however, matter in the sense that the salvor must often find a basis for remuneration for this part of the salvage operation elsewhere than in the law of salvage.[66]

Article 5 of the Salvage Convention 1989 includes a special provision on salvage operations controlled by public authorities:

1. This Convention shall not affect any provisions of national law or any international convention relating to salvage operations by or under the control of public authorities.
2. Nevertheless, salvors carrying out such salvage operations shall be entitled to avail themselves of the rights and remedies provided for in this Convention in respect of salvage operations.
3. The extent to which a public authority under a duty to perform salvage operations may avail itself of the rights and remedies provided for in this Convention shall be determined by the law of the State where such authority is situated.

This article does not exclude the view that states can operate as salvors but preserves the right for states to intervene on other bases.[67]

#### 4.2.4 Summary

A state that wishes to instigate preventive measures or a clean-up on the high seas has the opportunity to do so. Even in relation to foreign ships, the doctrine of necessity, the intervention conventions and the salvage conventions provide the necessary legal bases to intervene.

### 4.3 *The Right to Extend Clean-Up to the Source*

Pollution emergencies threatening the high seas may also extend to areas under the partial jurisdiction of states other than the intervening state, for example the EEZ. In particular, if the source of the pollution, for example the

65 Ibid.

66 There is, however, a possibility for special remuneration pursuant to the Salvage Convention 1989 (n 57) art 14.

67 Falkanger (n 37) 224.

wreck of an oil tanker, is located there, the intervening state may wish to extend its operation to such areas. The question to be discussed here is if it can do so under the protest of the coastal state.

### 4.3.1 UNCLOS Jurisdiction

It is clear that a state cannot normally instigate clean-up operations inside the territorial sea borders of another state: the coastal state has full sovereignty. The innocent passage regime is an exception to the coastal state's sovereignty,[68] but pollution prevention and response actions are clearly not passage. The doctrine of necessity[69] may, however, allow exceptions even in territorial seas. Theoretically, there can be situations in which the doctrine of necessity can be invoked to stop pollution in internal waters or even in the land territory of another state, but that would only be conceivable in extreme situations.

In the EEZ, the coastal state has limited jurisdiction, meaning that it does not have a monopoly over clean-up operations.[70] There are examples of widening of the competence that strictly follows from UNCLOS,[71] but there are no indications and no sound reasons to extend the coastal state jurisdiction so that the clean-up operations of another state can be prohibited, even subject to the doctrine of necessity. However, the limited EEZ jurisdiction of the coastal state extends to prohibiting the use of dispersants, etc.

Even when the coastal state jurisdiction does not pose a problem, the flag state of the source of the pollution may object to intervention. The doctrine of necessity may apply also in this respect and provide a basis for intervention. In addition, the Intervention Convention and Protocol[72] could form a basis for intervention. These are worded so that they only apply on the high seas. However, as they were created before the concept of the EEZ was established by UNCLOS, they could also arguably apply in the EEZ of another state.[73] Originally, they applied in the areas that are now EEZs.[74]

---

68 UNCLOS (n 1) art 17.

69 See Section 4.2.

70 See also the unrestricted wording of UNCLOS (n 1) art 221.

71 See in particular *The M/V 'Virginia G' Case* (*Panama v Guinea-Bissau*) Award, ICGJ 452, ITLOS Case No 19, 14 April 2014 (the Virginia G Case) para 217.

72 See Section 4.2.2.

73 Probably so, as argued by Colin M De La Rue and Charles B Anderson, *Shipping and the Environment. Law and Practice* (2009) 813.

74 UNCLOS (n 1) also converted some former high seas areas to territorial seas by extending the maximum territorial sea limit to 12 nautical miles. I would still hesitate to construe the Intervention Convention and Protocol so that they now apply to areas that were formerly high seas but now are territorial seas. While the EEZ only is a change of terminology as

As far as the EEZ covers the continental shelf, continental shelf jurisdiction does not call for any particular remarks in this respect. However, the continental shelf may extend beyond the EEZ, or the EEZ and continental shelf in one particular area may be under the jurisdiction of different states. In these cases, the continental shelf jurisdiction may be important. However, clean-up operations cannot be prohibited by continental shelf jurisdiction more than by EEZ jurisdiction. Still, the continental shelf state can probably prohibit the use of dispersants that will cause the pollutants (oil) to sink to the bottom of the sea, even if the superjacent waters are high seas. The continental shelf state has jurisdiction over sedentary species[75] and therefore presumably also the right to protect them from such pollution.

### 4.3.2 Conventions on Pollution Preparedness, Response and Cooperation

The conventions on pollution preparedness, response and cooperation include some provisions that could be relevant even to clean-up operations relating to the high seas.

Relating to territory, the OPRC Convention has a special rule of access for a foreign state instigating preventive or response actions relating to pollution:[76]

(3) In accordance with applicable international agreements, each Party shall take necessary legal or administrative measures to facilitate:
   (a) the arrival and utilization in and departure from its territory of ships, aircraft, and other modes of transport engaged in responding to an oil pollution incident or transporting personnel, cargoes, materials, and equipment required to deal with such an incident; and
   (b) the expeditious movement into, through, and out of its territory of personnel, cargoes, materials, and equipment referred to in subparagraph (a).

This article does not give a right of access to the territory of the coastal state, but it does legitimise a request to get access.

In relation to the rescue of people, the corresponding rule is more inviting.[77] In relation to nuclear assistance, however, the corresponding rule is somewhat more reserved.[78]

far as the EEZ jurisdiction does not apply, the extension of the territorial sea is an overall change in jurisdictional regime.

75 Ibid art 77.

76 OPRC Convention (n 4), art 7(3).

77 SAR (n 21) Ch 3.

78 Convention on Nuclear Assistance (n 9) arts 8 and 9.

In summary, the conventions on pollution preparedness, response and co-operation are of little importance in this context.

#### 4.3.3 A Right to Refuse Assistance

When there is an emergency, it is not uncommon for one state to offer assistance to another.[79] In respect of nuclear incidents, the literature indicates that a state cannot refuse an offer of clean-up assistance unless it can and will instigate such operations by other means.[80] This idea can easily be extended to other (perhaps less serious) kinds of pollution and to incidents at sea (most nuclear reactors are land based).

It is not clear what the consequence of an unjustifiable refusal of assistance would be, and unjustifiable refusal could perhaps trigger the doctrine of necessity discussed above,[81] so that intervention will not trigger the responsibility of the intervening state. However, this doctrine could also be applicable even without an unjustifiable refusal of an offer of assistance.

The lack of right to refuse assistance is important when the intervening state has to ask permission in the first place. At sea, outside territorial waters, the starting point is that any state may instigate clean-up.[82] The proposed doctrine is therefore not so important in the context here. In particular, however, if a state does not take action after an incident in its territorial seas, the proposed doctrine would be a sensible and appropriate rule.

## 5 Financial Consequences

### 5.1 *Reimbursement for Requested Assistance*

All the conventions discussed here allow a state to seek assistance from another state. In these cases, the requesting state is obliged to reimburse the assisting state for their services.[83]

The assisting state can claim compensation even if the pollution stems from its actions, omissions or coastal zones. If Russia asks for assistance from Norway after a blowout on the Norwegian continental shelf, Norway can still

79 See IMO, 'Guidelines on International Offers of Assistance in Response to a Marine Oil Pollution Incident' IMO Doc PPR 2/WP.6 (23 January 2015) Annex 1 <https://tinyurl.com/y85s8u9n> accessed 23 May 2018.

80 Handl (n 24) 350.

81 See Section 4.2.1.

82 See Section 4.3.1.

83 OPRC Convention (n 4) art 7 and Annex; Convention on Nuclear Assistance (n 9) art 7; MOSPA (n 2) art 10.

claim compensation from Russia. This has been criticised.[84] However, if the pollution is attributable to the assisting state, the requesting state can set off its liability for the assistance against its claim for compensation for the pollution damage.[85]

The liability of the requesting state means that the assisting state would not have to claim reimbursement from the polluter. The requesting state bears the risk of such recovery not being possible.

The rules on such reimbursements in other conventions are similar but not identical.[86] Here, I will comment on the provisions in MOSPA. Article 10, which deals with this matter, applies also on the high seas and in the Area.[87] The Article reads as follows:

Article 10—Reimbursement of Costs of Assistance

1. Unless an agreement concerning the financial arrangements governing actions of the Parties to deal with oil pollution incidents has been concluded on a bilateral or multilateral basis prior to an oil pollution incident, the Parties shall bear the costs of their respective actions in dealing with pollution in accordance with subparagraph (a) or subparagraph (b). The principles laid down in this paragraph apply unless the Parties concerned otherwise agree in any individual case.
    a. If the action was taken by one Party at the express request of another Party, the requesting Party shall reimburse to the assisting Party the cost of its action. The requesting Party may cancel its request at any time, but in that case it shall bear the costs already incurred or committed by the assisting Party.
    b. If the action was taken by a Party on its own initiative, this Party shall bear the costs of its action.
2. Unless otherwise agreed, the costs of action taken by a Party at the request of another Party shall be fairly calculated according to the law and current practice of the assisting Party concerning the reimbursement of such costs.
3. The assisting Party shall be prepared to provide upon request documentation and information to the requesting Party on the assisting Party's

84 Sands et al (n 37) 543.

85 On the procedural problems of set-off in public international law, see *Case Concerning the Factory at Chorzow (Germany v Poland)* (Merits) PCIJ Rep Series A No 17 1928 (13 September 1928) para 173.

86 A similar regulation is also found in OECD, 'Recommendation of the Council concerning Certain Financial Aspects of Actions by Public Authorities to Prevent and Control Oil Spills' OECD Doc C(81)32/FINAL (28 April 1981).

87 See Section 1.2.

estimated costs for the assistance and on the assisting Party's actual costs following the provision of any assistance. The Party requesting assistance and the assisting Party shall, where appropriate, cooperate in concluding any action in response to a compensation claim.

4. The provisions of this Agreement shall not be interpreted as in any way prejudicing the rights of Parties to recover from third parties the costs of actions to deal with pollution or the threat of pollution under other applicable rules of national and international law. Special attention shall be paid to international instruments and national law on liability and compensation for oil pollution damage.

### 5.1.1 When Does the Provision Apply?

The provision applies only when the requesting state has made an express request and not when the assisting state has acted on its own initiative (Paragraph 1). If the requesting state has asked for the assistance of another state but has left the decision of exactly what to do to the assisting state, it is not clear whether all assistance shall be considered as expressly requested. In my view, it suffices that the presence of the assisting state has been expressly requested. The specific actions carried out in such a situation cannot be said to be made on the assisting state's own initiative as long as they do not violate instructions given.

The provision only applies directly to actions of the assisting state and not to its subcontractors. However, it seems reasonable to reimburse the assisting state even if it has used a subcontractor and not only its (permanent) employees.

### 5.1.2 Law of the Assisting State

The cost shall be 'calculated according to the law and current practice of the assisting Party' (Paragraph 2). This is surprising: usually, one would use the law of the place where the damage has occurred in such cases.[88] Here, one considers the assisting state more as a contractor operating according to its own terms and conditions than as a claimant for pollution damage.[89]

The advantage of this way of calculating the damage is that the assisting state can follow its own procedures both in relation to which operations shall be carried out and how they shall be documented. The calculation shall be

88 See, for example, the choice of law rules in Regulation on the Law Applicable to Non-Contractual Obligations (Rome II) (EC) No 864/2007 arts 4(1), 7 and 11(3).

89 See, for example, the choice of law rules in Regulation on the Law Applicable to Contractual Obligations (Rome I) (EC) No 593/2008 art 4(1)(b).

fairly done (Paragraph 2). Presumably, not only the calculations but also the actions reimbursed must be reasonable. In all events, one would require very clear words to say that unreasonable clean-up operations were made upon express request. However, the reasonableness standard should be that of the assisting state and not that of the requesting state. Such standards may differ, for example in relation to cleaning oil from seabirds.

The greatest difference among the Arctic states in relation to oil spill compensation is perhaps found in the rules for compensation for environmental damage and punitive damages. Only costs for actions actually undertaken are reimbursable under this provision. Luckily, these differences are not relevant here.

Other differences among the tort laws of the Arctic states may be found in relation to compensation for the use of military and other government employees, compensation for value-added or goods and services tax, and the addition of overhead costs. Again, Paragraph 2 indicates that the law of the assisting state is decisive.

The bilateral agreement between Russia and Norway provides that the assisting party can choose the currency and use its own exchange rates for services.[90] The OPRC Convention and MOSPA have not changed this.[91]

#### 5.1.3 Direct Claims and Recourse

Paragraph 4 expressly allows for claims against third parties. Except for claims against the polluter, there will often be claims against the International Oil Pollution Compensation Funds.[92] However, these conventions[93] only apply for preventive measures in relation to damage within the EEZs of states parties to the relevant conventions. The United States is not a party to these conventions but has other national arrangements.[94]

If the assisting state makes a claim against the requesting state, the requesting state may have a recourse claim under Paragraph 4. The choice of law rules applicable to the recourse claim may differ from those outlined above in Section 5.1.2. The result may be that the requesting state cannot claim everything in the recourse action that it has paid to the assisting state.

90 Norwegian–Russian Agreement (n 15) art XIV.

91 See Section 1.1.

92 See for example Protocol of 1992 to Amend the International Convention on the Establishment of an International Fund for Compensation for Oil Pollution Damage (adopted 27 November 1992, entered into force 30 May 1996) 1110 UNTS 57 (Fund Convention); International Oil Pollution Compensation Funds <www.iopcfunds.org> accessed 23 May 2018.

93 Ibid art 4.

94 United States Coast Guard, 'National Pollution Funds Center' <www.uscg.mil/Mariners/National-Pollution-Funds-Center> accessed 23 May 2018.

Similarly, the recourse claim could be subject to maritime rules of limitation of liability. The requesting state is still liable to pay the assisting state in full: no rules of limitation of liability apply between them. This is a major risk for the requesting state and is inherent in the conventions discussed here. However, in many cases, the requesting state is the state where the damage occurs, and its own limitation rules can be invoked by the polluter. The requesting state can therefore adjust these rules.

In a recent Norwegian Supreme Court case,[95] it was held that some national limitation rules[96] for clean-up costs, based on an international convention,[97] did not apply if the cleaning up was left to the shipowner. This decision is of no avail here, as the assumption is that the cleaning up has been carried out by an assisting state and not by the shipowner. When Article 10 of MOSPA comes into play, this case becomes irrelevant, and limitation rules may apply.

#### 5.1.4 Payment Risk

The assisting state always runs the risk of the requesting state's not honouring its payment obligation. It is impractical for the requesting state to be sued at an international court or arbitration tribunal in such cases, and states do not accept being sued by foreign states in their own courts. The better approach for the assisting state would be to sue the polluter or a third party. However, even that may be difficult if the party must be sued in the courts of the requesting state. The conventions simply lack provisions on security and conflict resolution.

#### 5.1.5 Summary

In summary, the interests of the assisting state are well taken care of with the exception of the risk of the requesting state not paying. The requesting state, however, runs a considerable risk with these rules, as it may not have effective recourse against the polluter. These rules may make it easier to get assistance, and the requesting state can reduce its risk by providing for a full recourse claim in its own laws.

### 5.2 *Reimbursement for Unrequested Assistance*

Sometimes, the assisting state A has a basis for a claim for compensation against state B not based on request. For example Norway causes pollution by

95 *The 'Server',* Award (9 February 2017) Supreme Court of Norway HR-2017-331-A (the 'Server').

96 Maritime Code 1994 (Norway) ch 9 http://folk.uio.no/erikro/WWW/NMC.pdf accessed 23 May 2018.

97 Convention on Limitation of Liability for Maritime Claims (adopted 19 November 1976, entered into force 1 December 1986) 1456 UNTS 221 (LLMC 1976).

its negligence, violating its obligations under national law, or by having failed to establish such preparedness as required under the conventions discussed here.[98] Russia then takes action to protect its own coastline. Does MOSPA then prevent Russia from claiming compensation from Norway for the cleaning up it has performed?

The wording[99] of Articles 10(1)(b) (on unrequested assistance) and (4) (savings only for claims against third parties) of MOSPA indicates that there should be no claims in such cases. However, this seems to be very harsh, and there is no indication that this was intended. Presumably, therefore, MOSPA and similar conventions do not exclude liability for unrequested assistance if there is a basis for liability on other grounds.

### 5.3 *Claims in Relation to Accidents While Cleaning Up*

Liability issues can also arise the other way around. For example, the requesting state may seek to claim compensation from the assisting state for errors committed in the course of the assistance if the errors have caused additional pollution. Most of the conventions discussed here do not address this, and presumably, general principles of international law apply.[100]

The requesting state will be in a very strong negotiating position, as it can offset its claims against the claims for reimbursement from the assisting state.[101] Still, the lack of express provisions in this regard has been criticised.[102]

A claim against the assisting state or its agents is not necessarily straightforward. Rules of responder immunity may apply under the relevant national law or, in some cases, under international instruments.[103] In addition, some states—surprisingly, even Norway—seem to maintain that the polluter pays principle does not apply outside the territorial border or on the continental shelf.[104]

---

98 Possible bases for such claims under international law will not be discussed.

99 The wording is reproduced in Section 5.1.

100 Sometimes, the requesting parties even undertake the risk of damage caused by the assisting party towards a third party. See for example Agreement between Denmark, Finland, Iceland, Norway and Sweden about Cooperation Concerning Pollution Control of the Sea after Contamination by Oil or other Harmful Substances (adopted 29 March 1993, entered into force 16 January 1998) (Copenhagen agreement) art 11 <www.ust.is/library/Skrar/COPA/engelsk.pdf> accessed 23 May 2018.

101 On set-off, see Sands et al (n 37) 543.

102 Ibid.

103 Protocol to Amend the International Convention on Civil Liability for Oil Pollution Damage, which created the International Convention on Civil Liability for Oil Pollution Damage (adopted 27 November 1992, entered into force 30 May 1996) 973 UNTS 3 (CLC) art III (4)(e).

104 Nature Diversity Act 2009 (Norway) (n 32) ss 2(3) and 11.

The Convention on Nuclear Assistance includes a wider immunity rule for the assisting state and its agents in Article 10:

1. Unless otherwise agreed, a requesting State shall in respect of death or of injury to persons, damage to or loss of property, or damage to the environment caused within its territory or other area under its jurisdiction or control in the course of providing the assistance requested:
   (a) not bring any legal proceedings against the assisting party or persons or other legal entities acting on its behalf;
   (b) assume responsibility for dealing with legal proceedings and claims brought by third parties against the assisting party or against persons or other legal entities acting on its behalf;
   (c) hold the assisting party or persons or other legal entities acting on its behalf harmless in respect of legal proceedings and claims referred to in sub-paragraph (b), and
   (d) compensate the assisting party or persons or other legal entities acting on its behalf for:
      (i) death of or injury to personnel of the assisting party or persons acting on its behalf;
      (ii) loss of or damage to non-consumable equipment or materials related to the assistance
      (iii) except in cases of wilful misconduct by the individuals who caused the death, injury, loss or damage.
2. This article shall not prevent compensation or indemnity available under any applicable international agreement or national law of any State.

This article is supplemented by optional rules on immunity for foreign personnel on the territory where action is taken.[105]

Such provisions will perhaps make it easier to get assistance in emergencies.

## 6 Information Management

An increasingly important part of accident handling is information management. This includes information to the decision makers and threatened states as well as release of information to the press and the general public. These matters could be sensitive, both in respect of the right to obtain information and the control of its release.

The OPRC Convention includes a duty for the polluter to report a 'discharge or probable discharge', for observers of oil slicks to report them, and for the

105 Convention on Nuclear Assistance (n 9) arts 8(2) and (9).

state to warn other states.[106] This also applies to pollution on the high seas.[107] Regarding nuclear accidents, there is a special convention regarding notification of incidents.[108]

Article 6 of the Convention on Nuclear Assistance seeks to leave it to the requesting state to handle information for the press and the general public. The other conventions have no similar provisions.

None of the conventions discussed here include the right of the general public to be informed about accidents. Such a right exists in other conventions.[109] MOSPA, however, includes a provision on release to the public of general information.[110] None of the conventions include provisions for post-event fact-finding to evaluate the extent of damage and the effect of remedial actions. If such information is collected by an assisting state, the conventions do not apply.

## 7 Summary

The law of pollution preparedness and response is perhaps less developed than the law of pollution prevention and pollution liability. Still, there is a system in place for prepared and internationally coordinated efforts to combat pollution. The system includes rules for such matters as notification of spills and reimbursement of costs when one state assists another, but includes only fragmentary rules on information to the public.

States will often make private parties responsible for pollution clean-up or for paying for the clean-up. Occasionally states will be responsible for the

106 OPRC Convention (n 4) arts 4 and 5. UNCLOS (n 1) art 198, MARPOL (n 29) art 8 and Protocol 1 on Provisions Concerning the Reporting of Incidents Involving Harmful Substances include a similar duty.

107 Donald R Rothwell and Tim Stephens, *The International Law of the Sea* (Hart 2016) 390.

108 Convention on Early Notification of a Nuclear Accident (adopted 26 September 1986, entered into force 27 October 1986) 1439 UNTS 275 (Nuclear notification) <https://tinyurl.com/ya82p3on> accessed 23 May 2018.

109 Convention on Access to Information, Public Participation in Decision-Making and Access to Justice in Environmental Matters (adopted 25 June 1998, entered into force 30 October 2001) 2161 UNTS 447 (Aarhus Convention) (not ratified by the Arctic rim states of Russia, Canada and the USA); Convention on Civil Liability for Damage resulting from Activities Dangerous to the Environment (adopted 21 June 1993, not in force) (Lugano Convention) Ch III <https://rm.coe.int/168007c079> accessed 23 May 2018; Convention on the Transboundary Effects of Industrial Accidents (adopted 17 March 1992, entered into force 19 April 2000) 2105 UNTS 457 (Helsinki Convention) (not ratified by the Arctic rim states of Canada and the US) art 9.

110 MOSPA (n 2) art 12 with Appendix V.

pollution, and therefore for cleaning up. Beyond this, there is no strict duty for states to instigate clean-up operations on the high seas (and not even in their own coastal zones). The popular idea that there are islands of plastic refuse floating around on the oceans without any state feeling responsible for them is symptomatic. However, states do have a general obligation to fight pollution at sea, but not everywhere and not by all means necessary to achieve the desired result.

When states feel threatened by pollution, they are likely to organise clean-up in their own interest. In the Arctic High Seas, this may not be the case. However, a state that volunteers cleaning up services has wide possibilities to do so, even in the EEZ of other states.

CHAPTER 10

# Conclusions

*Robert Beckman, Millicent McCreath, J. Ashley Roach and Zhen Sun*

The chapters in this book have highlighted some of the gaps and loopholes in the governance of the high seas, especially with respect to the preservation and protection of the marine environment, including rare or fragile ecosystems as well as the habitat of depleted, threatened or endangered species and other forms of marine life.

## 1 The Gaps in the High Seas Governance Regime

One of the underlying themes of this book is that there are gaps in the international regime governing activities on the high seas, including activities that may cause significant pollution of the marine environment, especially rare or fragile ecosystems as well as the habitat of depleted, threatened or endangered species and other forms of marine life.

There are several reasons for the perceived gaps. First, for many years the oceans were viewed as highways for international shipping, and the major activity of states on the high seas was that they sailed ships under the principle of freedom of the seas. The principle developed that ships on the high seas are subject to the exclusive jurisdiction of the flag state. Therefore, the emphasis was to establish universally accepted rules to ensure the safety of international shipping and to prevent pollution from the operation of ships.

Second, when spills from tankers polluted the marine and coastal environment of coastal states, liability and compensation schemes were developed to provide compensation to coastal states for damage caused by the pollution as well as the costs of combating the pollution and cleaning it up. After new maritime zones were created to give coastal states rights and jurisdiction to explore and exploit the natural resources in their exclusive economic zone (EEZ),[1] the liability and compensation conventions were extended to include damage to these maritime zones. However, they were not extended to include damage to the high seas.

1 United Nations Convention on the Law of the Sea (adopted 10 December 1982, entered into force 16 November 1994) 1833 UNTS 3 (UNCLOS) Part V.

 | DOI:10.1163/9789004373303_011

Similarly, when states began to exploit oil and gas resources on the continental shelf adjacent to their coasts, they enacted laws and regulations to govern the activities of companies engaged in activities in their maritime zones. The issue of pollution of the marine environment from such activities was addressed by giving coastal states jurisdiction and control over artificial islands, installations and structures on their continental shelf or in their EEZ. No rules were developed to address the problem of the pollution spreading to the high seas from such artificial islands, installations and structures.

## 2 Fundamental Principles Governing the High Seas

It may be helpful when analysing this issue to go back to the basic principles of international law governing the high seas. The fundamental principles governing the high seas were 'codified' in the 1958 Convention on the High Seas (High Seas Convention).[2] Almost all of the provisions in the High Seas Convention were included in the 1982 United Nations Convention on the Law of the Sea (UNCLOS) without any changes. The fundamental principles governing the high seas are that no state may validly purport to subject any part of the high seas to its sovereignty, and that all states have 'freedom of the high seas'. All states have freedoms including the freedoms of navigation and overflight as well as the freedoms to lay submarine cables and pipelines and the freedom to conduct marine scientific research.[3] The High Seas Convention provides that there are two limitations on these freedoms. First, they are to be exercised under the conditions laid down in that Convention and by other rules of international law. Second, the freedoms are to be exercised with 'reasonable regard' to the interests of other states in the exercise of their freedoms.[4] This language remains essentially unchanged in UNCLOS, except that Article 87 provides that the freedoms shall be exercised with 'due regard' for the interests of other states in the exercise of their freedoms.[5]

2 Convention on the High Seas (adopted 29 April 1958, entered into force 30 September 1962) 450 UNTS 11 (High Seas Convention).

3 UNCLOS (n 1) arts 87, 89.

4 High Seas Convention (n 2) art 2.

5 Regarding whether there is any difference between 'reasonable regard' and 'due regard', see Bernard Oxman 'The Principle of Due Regard' in International Tribunal for the Law of the Sea, *The Contribution of the International Tribunal for the Law of the Sea to the Rule of Law: 1996–2016* (Brill Nijhoff, 2018), 108, 109, who states that '[t]his textual change was not substantive'.

These freedoms are held by states, and not by individuals. Therefore it seems clear that individuals, companies and other entities do not have rights or freedoms to conduct any of these activities on the high seas. The underlying assumption is that individuals and private entities may carry out activities on the high seas only under the authorisation of a state. As has been discussed throughout this book, the way we use the oceans and particularly the high seas has changed since the conclusion of both the High Seas Convention and UNCLOS, and non-state entities are increasingly seeking to undertake activities on the high seas in a private capacity.

Since we usually analyse law of the sea by examining the rights of flag states, coastal states and port states, we often assume that it is the flag state that should exercise jurisdiction and control over activities on the high seas such as the laying of submarine cables or pipelines. After all, ships on the high seas are subject to the exclusive jurisdiction of the flag state. Flag states will agree that they have an obligation to ensure that ships flying their flag comply with the generally accepted international rules, regulations and practices on the safety of navigation and ship-source pollution developed under the auspices of the International Maritime Organization (IMO). The flag state of a cable ship (a specialised ship that lays or repairs fibre optic telecommunications cables) would require that the ship comply with IMO regulations on areas to be avoided, marine protected areas, etc., especially if they are indicated on navigational charts. However, the flag state of the cable ship would not approve the delineation of the course for the laying of the cable, would not require the cable ship to have due regard to existing cables or pipelines, or have authority over the cable or pipelines once they have been deployed on the seabed. In addition, the flag state of the cable ship would not have any regulations requiring the cable company to exercise due diligence to ensure that the cable ship not lay or repair a cable in such a manner that it would threaten or harm rare or fragile ecosystems and the habitat of depleted, threatened or endangered species or other forms of marine life.

## 3 Filling the Gaps

As Nilüfer Oral suggests in Chapter 2, the preferable solution would be that the state with 'jurisdiction or control' over the companies or individuals engaged in the activities would have the responsibility to regulate their activities when they take place outside the territorial jurisdiction of any state. The national regulations of the state with jurisdiction and control should ensure that the companies or individuals fulfil their obligation to give due regard to the interests of other states in the exercise of their high seas freedoms. This can be described as 'activity state jurisdiction'.

Although the activities of non-state entities on the high seas have been considered in this book as a 'gap', perhaps it is to some extent a gap in implementation rather than in regulation. Two key obligations in UNCLOS apply to all states on the high seas, and are not limited to flag states. First, pursuant to Article 87, the high seas freedoms must be exercised *by all states* with due regard for the interests of other states. Second, states have a due diligence obligation to protect and preserve the marine environment, and to take all measures consistent with UNCLOS to prevent, reduce and control marine pollution from any source. These obligations are not limited to the exercise of jurisdiction and control over ships on the high seas, but should be considered to extend to any activity taking place on the high seas. Therefore states properly have an obligation to regulate and take administrative and enforcement measures to ensure that activities being undertaken by their nationals or entities under their jurisdiction and control on the high seas are conducted with due regard for other users and do not harm the marine environment. These obligations potentially have a very wide reach. However, what is lacking is some more specified content or details on the required standard of regulation or conduct. To that end, a resolution from the United Nations General Assembly or an agreed set of guidelines for the regulation of the particular activity on the high seas may be all that is required in most cases.

Taking this interpretation to particular contexts discussed in the book, some of the gaps in the liability and compensation schemes that have been pointed out could be addressed by implementing the provisions in UNCLOS. Article 193 provides that states have the sovereign right to exploit their natural resources pursuant to their environmental policies *and in accordance with their duty to protect and preserve the marine environment*. Article 194(2) provides that states shall take all measures necessary to ensure that pollution arising from incidents or activities under their jurisdiction and control does not spread beyond the areas where they exercise sovereign rights and jurisdiction. This would include the high seas. The state with jurisdiction and control over the companies or individuals who are intending to engage in activities on the high seas should enact laws and regulations to ensure that they fulfil their obligations under Articles 192 and 194 of UNCLOS to protect and preserve the marine environment in all areas, including the high seas. Courts and tribunals have declared that the obligation of states under these articles is one of due diligence. There is also an obligation to monitor and assess the impacts of such activities under Articles 204–206.

The state that should have an obligation to exercise jurisdiction and control over the laying of a cable is the state whose companies or nationals own the cable company (or the state where the cable is registered if the cable is owned by a consortium of companies). Cable companies should seek authorisation from the state with jurisdiction and control, and that state should register the cable and be responsible for ensuring that the company complies with the obligations of the state under UNCLOS when laying and repairing the cable. Cable companies are presently not regulated by any state when laying and repairing cables. If they were registered by a state, they would have to comply with its laws and regulations. However, they would also receive benefits. If any other state were to interfere with the laying or repair of the cable in a manner that is contrary to UNCLOS, the state of registration could exercise the right of diplomatic protection on behalf of the cable company.

This form of 'activity state jurisdiction' would also fill to a large extent the governance gap exemplified by the planned activity of The Ocean Cleanup discussed in Chapter 1 of this book. The Ocean Cleanup, a non-governmental organisation, is planning to deploy large floating arrays on the high seas of the Pacific Ocean in 2018 to collect marine debris.[6] The Ocean Cleanup is based in the Netherlands, and therefore should seek authorisation from the Netherlands, which if granted would trigger the responsibility of the Netherlands to ensure that the UNCLOS obligations are complied with. The deployment of the floating arrays on the high seas to collect marine debris would be an exercise of the Netherlands' high seas freedoms. Any interference with the arrays would be in violation of the Dutch freedoms and therefore the Netherlands would be able to exercise diplomatic protection on behalf of The Ocean Cleanup. Should arrays be deployed by other companies based in other countries, those countries would have a duty to regulate the activity.

The content of these obligations could be filled by developing guidelines or best practices on the obligations of states to regulate the activities of persons under its jurisdiction or control who carry out particular activities in areas beyond national jurisdiction (ABNJ). Private companies who intend to carry out such activities would have to register in a state and be subject to their laws and regulations. At the same time, the state of registration would be expected to exercise the right of diplomatic protection if another state interfered with the activity of the company in a manner inconsistent with UNCLOS. For example if a company registered in Australia were authorised by Australia to lay a cable across the Pacific, and the cable was damaged by fishing activities of another state in areas under the high seas, Australia could exercise diplomatic

---

6 The Ocean Cleanup, <www.theoceancleanup.com> accessed 28 May 2018.

protection in respect of the cable company and bring an action against the flag state of the fishing vessel provided the criteria are met.

Many of the current gaps in the legal regime could be filled if states were responsible for taking all measures necessary to ensure that the activities of persons under their jurisdiction and control in ABNJ are carried out in accordance with their due regard obligation as well as their obligation of due diligence to protect and preserve the marine environment. If states fail to take measures to fulfil these obligations, a dispute could arise between states parties to UNCLOS that would be subject to the compulsory binding procedures in Section 2 of Part XV. In addition, interested states could work together to request advisory opinions from International Tribunal for the Law of the Sea (ITLOS) to clarify the due diligence obligations of states over activities on the high seas by persons or entities subject to their jurisdiction and control.

Given the increased knowledge about the marine environment and recognition of the need to protect sensitive marine areas in the high seas, there is widespread agreement in the international community that marine protected areas are a valuable tool and should be established to protect such sensitive areas. Although we may agree that marine protected areas on the high seas can be effective and should be established, difficult procedural issues must be addressed, namely who has the authority to request the creation of such areas, and who has the authority to establish the areas? In addition, how can a system be established to ensure that states exercise due diligence to ensure that persons and entities under their jurisdiction and control comply with the regulations applicable within the areas? It is expected that some of these issues will be addressed in the current negotiations at the United Nations on an international legally binding instrument (ILBI) under UNCLOS on the conservation and sustainable use of marine biological diversity of areas beyond national jurisdiction (BBNJ negotiations).

## 4 Potential Role of the BBNJ ILBI

The writing of this book has taken place at the same time as the BBNJ negotiations. It was not the intention of this book to cover in any detail the BBNJ process or discussions. However it is worth mentioning briefly during these concluding remarks the extent to which any future ILBI may fill some of the gaps identified in the book. At this stage of the negotiations, with a diplomatic conference to be convened in September 2018, it is difficult to say with any certainty the precise subject matter that will be covered by the new ILBI. However, the report of the Preparatory Committee of 31 July 2017 gives a general overview of the topics that may be included.[7] There are four themes to the BBNJ negotiations, namely marine genetic resources, area-based management tools, environmental impact assessments, and capacity-building and transfer of marine technology. With such wide coverage it can be expected the new ILBI will be relevant to at least some of the gaps identified in this book. With marine protected areas being a major focus, we can expect that there will be some guidance on the identified issues of who has the authority to establish or request the establishment of marine protected areas on the high seas. There may be agreement on the criteria to be used to identify sensitive marine areas requiring the protection of either a marine protected area or some other form of area-based management tool. The Preparatory Committee report currently lists 15 criteria for the identification of sensitive areas, including many of those discussed by Youna Lyons in Chapter 4.[8]

The report of the Preparatory Committee includes in its recommendations a list of general principles and approaches to guide the conservation and sustainable use of marine biological diversity in ABNJ. The guidelines as outlined in the report are numerous and broad, and could be used to inform the content of the key UNCLOS obligations discussed above, the obligation to have due regard to the exercise of freedoms by other states and the obligation to protect and preserve the marine environment. The provisions on environmental impact assessments are also likely to have a general application to many of the gaps identified in the book, particularly activities such as geoengineering. It is likely that there will be a requirement to conduct an environmental impact assessment for such activities, and the ILBI may specify the required procedural steps, the content of the report, and the thresholds and criteria for identifying which activities require assessment. The planned activity of The Ocean Cleanup discussed in Chapter 1 would also most likely require an environmental impact assessment. The Ocean Cleanup despite its positive agenda has been subject to criticism from legal and scientific experts, some of whom consider that the environmental impact assessment has not been sufficiently

7 United Nations General Assembly (UNGA), 'Report of the Preparatory Committee Established by General Assembly Resolution 69/292: Development of an International Legally Binding Instrument under the United Nations Convention on the Law of the Sea on the Conservation and Sustainable Use of Marine Biological Diversity of Areas beyond National Jurisdiction' UN Doc A/AC.287/2017/PC.4/2 (31 July 2017). See Nicholas Gaskell, Chapter 8 'Liability and Compensation Regimes: Pollution of the High Seas'.

8 See Youna Lyons, Chapter 4 'Identifying Sensitive Marine Areas in the High Seas: A Review of the Scientific Criteria adopted under International Law'.

thorough to identify and manage the many potential risks of deploying large arrays on the high seas.[9]

Regarding the responsibility and liability issues highlighted in the book, it is possible that the future ILBI will be of relevance. This is a 'cross-cutting issue', meaning that it does not fall only within one of the four themes mentioned above. A proposal was made during the negotiations for a more developed responsibility and liability provision than that in Article 235 of UNCLOS. However, it is not clear whether this proposal will be followed at the diplomatic conference. The report of the Preparatory Committee on the BBNJ merely states that '[t]he text could address issues relating to responsibility and liability', and therefore it is not possible to say to what extent the responsibility and liability issues identified in the book may be resolved by the ILBI, except to say that it may be relevant. UNCLOS anticipates the development of further rules on state responsibility and liability, with Article 304 stating that its provisions are without prejudice to such future rules.[10]

It should be noted that the international law on state responsibility was set out by the International Law Commission in the 2001 Articles on the Responsibility of States for Internationally Wrongful Acts (ILC Articles on State Responsibility).[11] Therefore, the provisions in UNCLOS and in the 1994 and 1995 Implementation Agreements were drafted before the ILC Articles on State Responsibility were finalised. It is hoped that the provisions on responsibility and liability in any ILBI arising from the BBNJ negotiations will take into account and be consistent with the ILC Articles on State Responsibility. At a minimum, the ILBI should provide that states shall be responsible for ensuring that activities in ABNJ, whether undertaken on their behalf or by persons or entities under their jurisdiction and control, are conducted in accordance with their obligations to give due regard to the exercise of freedoms by other states and their obligations to protect and preserve the marine environment.

9 Nick Kilvert, 'Great Pacific Garbage Patch Plastic Removal System could become "World's Biggest Piece of Marine Debris", Critics Say', www.abc.net.au/news/science/2018-05-08/great-pacific-garbage-patch-plastic-ocean-cleanup-boyan-slat/9714246 accessed 22 May 2018.

10 UNCLOS (n 1) art 304.

11 UNGA, 'Resolution Adopted by the General Assembly. 56/83. Responsibility of States for internationally wrongful acts Articles on the Responsibility of States for Internationally Wrongful Acts' UN Doc A/RES/56/83 (28 January 2002) Annex.

# Index

www.ingramcontent.com/pod-product-compliance
Lightning Source LLC
LaVergne TN
LVHW021544021026
849161LV00005BA/999
* 9 7 8 9 0 0 4 3 7 3 3 1 0 *